BASIC ELECTRICAL ENGINEERING

Edited by

Draženko Macanović

Toronto Academic Press

BASIC ELECTRICAL ENGINEERING

Editor: **Draženko Macanović**

Toronto Academic Press
4164 Lakeshore Road
Burlington ON L7L 1A4
Canada
www.tap-books.com
Email: orders@arclereducation.com

ISBN: 978-1-77956-307-1

ABOUT THE EDITOR

Draženko Macanović is an accomplished Automation Engineer with extensive experience in power systems and networks. Born on July 18, 1992, in Banja Luka, Bosnia and Herzegovina, Draženko pursued his education at the University of Banja Luka, where he earned his degree in Electrical Engineering, focusing on power engineering. He later completed his master's degree at the University of Belgrade, specializing in the analysis of weakly meshed LV distribution networks with power electronic devices. In his professional career, Draženko has held significant roles, beginning as an associate in engineering and construction, where he oversaw the reconstruction of medium and low voltage distribution grids. He later advanced to an automation engineer role, where he worked on various critical projects, including the integration of Smart City concepts and the implementation of SCADA/PLC & HMI systems. Fluent in English with basic knowledge of the German language, Draženko is known for his strong communication skills, teamwork, and technical expertise. He is proficient in a variety of software tools and programming languages, including MS Office, Matlab, AutoCAD, Ladder programming, and C, making him a versatile engineer in the field of automation and electrical engineering. Outside of work, he enjoys maintaining a healthy lifestyle, often spending his free time at the gym or running in nature.

Table of Contents

7 TRANSFORMER 221

8 DC MACHINES 255

9 THREE PHASE INDUCTION MOTORS 279

List of Figures

List of Tables

PREFACE

The narrative of electrical engineering is a captivating journey that melds ancient insights with groundbreaking innovations. It begins with a nod to early civilizations and their grasp of electricity, exemplified by pioneers like Edison, Tesla, and Faraday.

Centuries ago, well before the dawn of modern electrical engineering, ancient societies displayed a rudimentary understanding of electricity and magnetism. The ancient Egyptians, for instance, crafted a mysterious artifact known as the Baghdad Battery. This device, consisting of a clay jar enclosing a copper cylinder, an iron rod, and an acidic solution, demonstrated their primitive knowledge of electricity. When the copper and iron were connected, it generated an electric current used to gild silver objects—a remarkable feat of ancient electrical application.

Similarly, the ancient Greeks coined the term "Elektron" from the Greek word for amber, highlighting their observation that rubbing amber could generate static electricity, attracting lightweight objects like feathers and leaves—a rudimentary understanding of electrostatic forces.

The 19th century marked a pivotal era for electrical engineering, driven by visionaries such as Thomas Edison, Nikola Tesla, and Michael Faraday, who elevated the discipline to new heights despite its early origins. Thomas Edison, renowned for his groundbreaking inventions including the practical electric light bulb, phonograph, and motion picture projector, founded the Edison Electric Light Company, which pioneered the widespread adoption of electric power in homes and businesses, revolutionizing daily life and work. Nikola Tesla, celebrated for his advancements in alternating current (AC) electricity, significantly improved upon the limitations of direct current (DC). His revolutionary AC system enabled efficient long-distance power transmission, transforming the global energy landscape. Michael Faraday, a British physicist, made foundational discoveries in electricity and magnetism. His work on electromagnetic induction laid the groundwork for electric motors and generators, while his elucidation of electrolysis principles facilitated the production of crucial materials like aluminum. The legacies of Edison, Tesla, Faraday, and their contemporaries established the bedrock of modern electrical engineering. Today, electrical engineers play pivotal roles in designing systems that power our homes, fuel technological advancements, and support our daily needs.

From ancient experiments to 19th-century breakthroughs, the historical trajectory of electrical engineering continues to shape our modern world. Whether through electric vehicles, smart grids, robotics, or consumer electronics, electrical engineering remains at the forefront of technological innovation, driving societal progress in myriad ways.

For today's aspiring engineers, understanding the future landscape of electrical engineering is crucial. Emerging sectors such as electric vehicles, smart grids, and advanced consumer electronics promise substantial growth and career opportunities. By exploring cutting-edge technologies and career paths within electrical engineering, students can align their studies and ambitions with the evolving demands of the field, ensuring a meaningful and impactful career in engineering.

Content/Structure of the Book

This textbook is organized into ten comprehensive chapters, each designed to cover essential concepts and principles in the field of electrical engineering.

Chapter 1 lays the foundation for understanding electric circuits. It covers the basic principles of electricity, including voltage, current, resistance, and power. The chapter also introduces Ohm's Law and Kirchhoff's Laws, which are fundamental to circuit analysis.

Chapter 2 explores various methods for analyzing complex resistor networks, such as series and parallel combinations, the use of Thevenin's and Norton's Theorems, and the application of mesh and nodal analysis techniques.

Chapter 3 shifts focus to alternating current (AC) circuits, discussing their behavior and characteristics. It covers the analysis of AC circuits using phasor diagrams, impedance, and admittance. Key topics include resonance, frequency response, and the use of complex numbers in AC circuit analysis.

In Chapter 4, the book examines the generation and distribution of AC power. It covers the principles of power generation in power plants, the types of power generation methods, and the transmission and distribution of electrical power to consumers. The chapter also discusses the components of the power grid and the challenges of power distribution.

Chapter 5 focuses on the instruments and methods used for measuring electrical quantities. It introduces various types of electrical measuring instruments, such as voltmeters, ammeters, and oscilloscopes, and explains how to use them effectively. The chapter also covers measurement techniques and error analysis.

In Chapter 6, the book explores the principles of magnetism and magnetic circuits. It explains magnetic fields, magnetic flux, and the laws governing magnetic circuits. The chapter also discusses electromechanical energy conversion, focusing on how electrical energy is converted into mechanical energy and vice versa.

Chapter 7 provides an in-depth look at transformers, essential components in electrical power systems. It covers the construction, operation, and types of transformers. Key topics include the principle of electromagnetic induction, transformer equivalent circuits, efficiency, and voltage regulation.

The book then moves into direct current (DC) machines, discussing their construction, operation, and applications. It covers different types of DC machines, including DC motors and generators, their characteristics, and the methods used to control their speed and torque.

Chapter 9 focuses on three-phase induction motors, widely used in industrial applications. It explains the construction and working principles of these motors, their performance characteristics, and methods of starting and controlling them. The chapter also covers the analysis of their equivalent circuits and efficiency.

The final chapter examines alternators and synchronous motors, key components in power generation and industrial applications. It covers the construction, operation, and types of alternators, as well as their synchronization with the power grid. The chapter also discusses the principles and applications of synchronous motors, including their operation under various load conditions.

Each chapter in this book is thoughtfully crafted to provide a detailed and coherent understanding of electrical engineering principles, ensuring that readers can grasp both fundamental concepts and their practical applications in the field.

Audience

This book is tailored for students and professionals entering or advancing in the field of electrical engineering. It serves as an essential primer for undergraduate students pursuing degrees in electrical engineering, providing a solid foundation in core concepts and fundamental principles. The content is structured to accommodate learners at various stages, from introductory courses to more advanced topics, ensuring clarity and depth in understanding key theories and applications. Beyond academia, the book also caters to technicians and practitioners seeking to enhance their practical knowledge and proficiency in electrical systems. It offers practical insights into electrical circuits, devices, and systems, bridging theoretical learning with real-world applications. Additionally, the book appeals to individuals transitioning into electrical engineering careers from related fields or seeking to refresh their understanding of foundational concepts. It covers a wide spectrum of topics including circuit analysis, electromagnetism, power systems, and electronics, making it a comprehensive resource for professional development and lifelong learning in the electrical engineering discipline. By combining theoretical rigor with practical relevance, this book aims to empower its audience with the knowledge and skills necessary to excel in both academic pursuits and professional endeavors within the dynamic and evolving field of electrical engineering.

Learning Features

- Start-of-chapter learning objectives
- Key terms in each chapter
- Summary Points at the end of each chapter
- Review Questions at the end of each chapter
- Case Studies and Activities in each chapter for solving real problems
- References at the end of each chapter

CHAPTER 1

Fundamentals of Electric Circuits

LEARNING OBJECTIVES

After studying this chapter, you will be able to:

- Give an overview of electric circuit
- Explain circuit elements and sources
- Illustrate circuit theorems

KEY TERMS FROM THIS CHAPTER

Ammeter	Amperes	Capacitor
Circuit	Closed circuit	Current
Ideal current	Inductor	Load
Mesh analysis	Open circuit	Resistance
Switch	Voltage	Voltage polarity

INTRODUCTION

The fundamentals of electric circuits form the backbone of modern electrical engineering and technology. At the core of electric circuits is the flow of electric current through a pathway composed of various electronic components. These pathways, or circuits, are driven by a potential difference known as voltage, which propels the movement of electrons, creating an electric current measured in amperes. The concept of resistance, measured in ohms, plays a crucial role in determining the extent of this current flow by opposing it. This relationship between current, voltage, and resistance is elegantly described by Ohm's Law, which states that the current passing through a conductor between two points is directly proportional to the voltage across those points and inversely proportional to the resistance, formulated as V=IR.

Key to understanding complex circuits are Kirchhoff's Laws. Kirchhoff's Current Law (KCL) asserts that the total current entering a junction in a circuit must equal the total current leaving it, ensuring the conservation of charge. Kirchhoff's Voltage Law (KVL) maintains that the sum of all electrical potential differences around a closed loop is zero, which helps in analyzing the energy conservation within the circuit. Circuit components like resistors, which impede current flow and drop voltage; capacitors, which store and release electrical energy; and inductors, which store energy in a magnetic field when current flows through them, are fundamental elements in the design and functioning of electric circuits. Power sources such as batteries or power supplies provide the necessary voltage and current to drive these circuits. Mastery of these basics enables the design, analysis, and troubleshooting of more complex electronic systems, forming the basis for advancements in electronics, communications, and a myriad of technological applications that power our modern world. Understanding these principles is essential for anyone involved in the study or application of electrical engineering.

1.1. ELECTRIC CIRCUIT

An electric circuit is a model of an actual circuit which links various devices together. Electric circuits consist of different components, and understanding these components is essential in comprehending the flow of current within the circuit.

In 1800, Alessandro Volta became the first person to conceive of an electric circuit. Through his experiments, Volta found that by connecting bowls of salt solution with metal strips, he could generate a constant stream of electricity. He later constructed a voltaic pile using layers of copper, zinc, and soaked cardboard discs, creating an early version of a battery. By linking a wire from top to bottom, Volta successfully directed the flow of electric current through the circuit.

An electric circuit is a closed loop or path through which current flows. Various components are added to electric circuits to perform specific tasks. Essentially, an electric circuit is a collection of elements arranged to guide electric current in a specific direction. The source is where the current begins, and the return points are where it ends.

The diagram of the electric circuit is show in Figure 1.1, let's learn about the electric circuit through the diagram.

Figure 1.1. The diagram of the electric circuit.

Source: https://media.geeksforgeeks.org/wp-content/uploads/20230227144415/Circuit-on-off-gif.gif.

The above electric circuit has;

- Electric bulb
- Wire (Conductor)
- A battery

When the positive end of the battery is connected to the positive terminal of the bulb, the bulb does not light up. However, when the other end of the battery is connected to the bulb, completing the circuit, the bulb lights up. This shows that current only flows in an electric circuit when it is closed, and circuits that allow current to flow are called complete circuits.

Failure to secure all connections in a circuit could potentially lead to a fire that cannot be put out using conventional methods. It is important to note that using water to extinguish an electrical fire is dangerous and can result in electric shock.

1.1.1. Types of Electric Circuits

Many types of electrical circuits are used in our daily life some of the most common electric circuits are;

- Open Circuit
- Closed Circuit
- Short Circuit
- Series Circuit
- Parallel Circuit
- **Open Circuit:** An open circuit is a circuit where the switch is open and no current is flowing through

it. Figure 1.2 illustrates an open circuit.

Figure 1.2. *An open circuit.*

Source: https://media.geeksforgeeks.org/wp-content/uploads/20230227144444/Electric-Circuit-4.jpg.

- **Closed Circuit:** An open circuit is a circuit in which the switch is closed and allows current to flow through it. Figure 1.3 illustrates a closed circuit.

Figure 1.3. *A closed circuit.*

Source: https://media.geeksforgeeks.org/wp-content/uploads/20230227144514/Electric-Circuit-3.jpg.

- **Short Circuit:** A short circuit occurs when the current in the circuit takes a shorter path than its intended route. Figure 1.4 illustrates a short circuit.

Figure 1.4. *A short circuit.*

Source: https://media.geeksforgeeks.org/wp-content/uploads/20230402112108/Electric-Circuit-5.jpg.

The result of a short circuit is the potential ignition of the electric board or system. The primary reason behind electric fires is often attributed to short circuits. The picture below depicts a fire ignited by a short circuit.

- **Series Circuit:** A series circuit is formed when all resistors are connected in a series combination, as shown in the Figure 1.5.

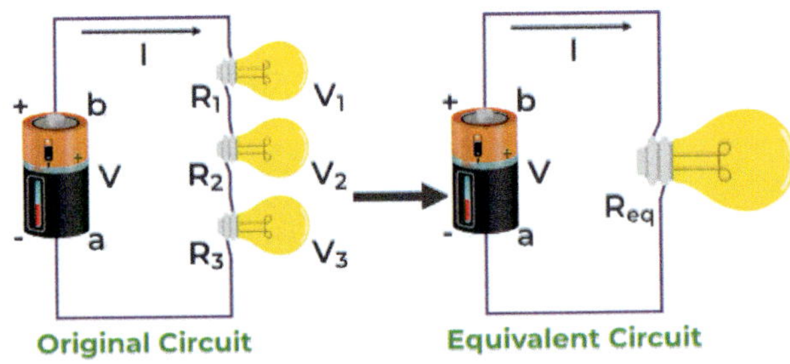

Figure 1.5. *A series circuit.*

Source: https://media.geeksforgeeks.org/wp-content/uploads/20230227151617/Comb-1-(2).png.

- **Parallel Circuit:** A parallel circuit is formed when all the resistors are connected in parallel. Figure 1.6 displays a parallel circuit.

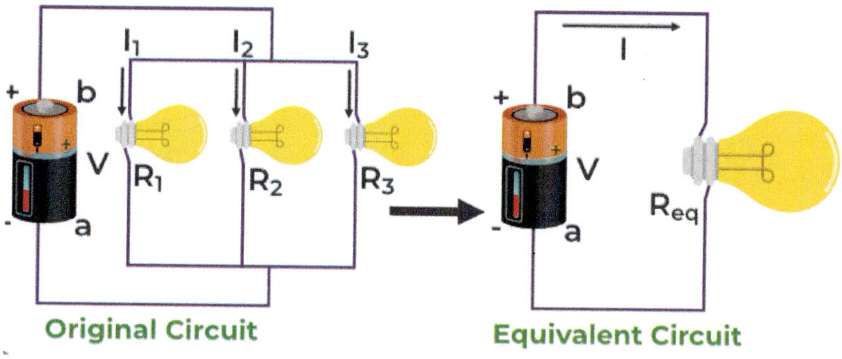

Figure 1.6. *A parallel circuit.*

Source: https://media.geeksforgeeks.org/wp-content/uploads/20230227152358/Comb-2-(2).png.

1.1.2. Domestic Electric Circuit

The electrical circuit utilized in our household is known as a Domestic circuit, designed specifically for residential use. Several components included in the domestic circuit are:

- **Mains Supply:** Electric power we receive in our houses is called the main supply.
- **Cables:** Electric power is supplied by cables which are either open or underground.

There are three types of wires used in domestic circuits which are:

- Earth Wire;
- Live Wire; and
- Neutral Wire.
- **Potential Difference (or Voltage):** In our country, the current is supplied at 220 V in domestic circuits.

Our house receives electric current at 220 volts through a circuit protection element known as a fuse. If the current supplied exceeds the allowed amount, the fuse wire will break/melt, halting the flow of current. The current then goes through the meter board.

From the meter board generally, two types of current are supplied to the circuit:

- **15 Amp Current:** This current is used by appliances with higher power ratings, like geysers, ACs, refrigerators, electric motors, and others; and
- **5 Amp Current:** This current is used by appliances with lower power ratings, like TV, fans, bulbs, computers, smartphones, and others.

1.1.3. Electric Circuit Symbols

In the drawing of electric circuits, a variety of symbols are utilized. Among the key symbols used in physics for this purpose are (Figure 1.7):

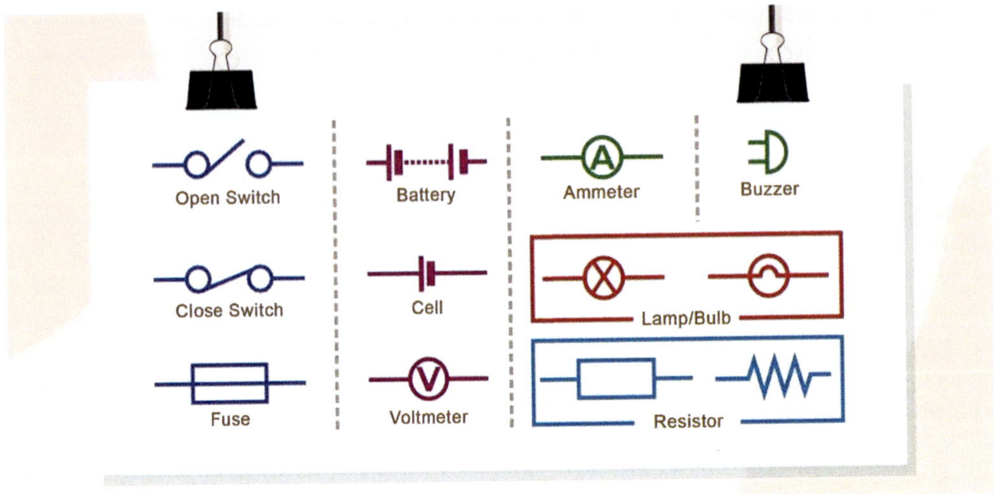

Figure 1.7. *Electric circuit symbols.*

Source: https://media.geeksforgeeks.org/wp-content/uploads/20230419183956/Electric-Circuit-2.webp.

1.1.4. Simple Circuit

A basic circuit includes a power source, conductors, switch, and load. It does not contain multiple components, and the Figure 1.8 displays an example of a simple circuit.

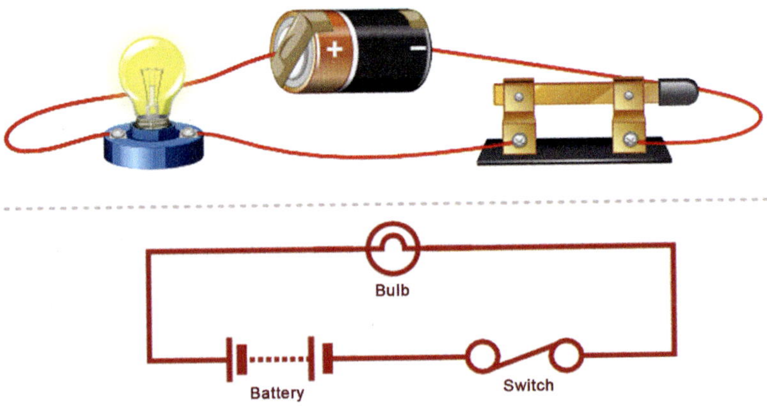

Figure 1.8. *A simple circuit.*

Source: https://media.geeksforgeeks.org/wp-content/uploads/20230227144647/Electric-Circuit-1.png.

- **Cell:** It is the power source that provides the potential difference in the electric circuit.
- **Load:** A resistor that consumes electrical power in a circuit is called the load. Generally, a load is a bulb.

- **Conductors:** Copper conductors that conduct electricity in a circuit are called conductors.
- **Switch:** A device that controls the incoming flow of current in an electric circuit is called a switch.

1.1.5. Electric Circuit Formula (Table 1.1)

Table 1.1. Various Formulas that are Used in Solving Electric Circuits

	Formula	Notations
Electric current	$I = Q/t$	• I is the current flowing • Q is the charge flowing • t is the time period
Voltage	$V = IR$	• V is the potential difference
Resistance	$R = \rho\, l/A$	• R is the resistance of the circuit • ρ is the specific resistivity per length of the wire • l is the length of the wire • A is the cross-sectional area
Power	$P = VI$ or E/t	• P is the power • E is the energy gain or loss • t is the time period
Series Resistance	$R_{eq} = R_1 + R_2 + \ldots + R_n$	• R_{eq} is the equivalent resistance of the resistors in series • $R_1, R_2 \ldots R_n$ are the individual resistors added in series
Parallel Resistance	$1/R_{eq} = 1/R_1 + 1/R_2 + \ldots + 1/R_n$	• R_{eq} is the equivalent resistance of the resistors in parallel • $R_1, R_2 \ldots R_n$ are the individual resistors added in parallel

Source: https://www.geeksforgeeks.org/electric-circuit/.

1.2. CIRCUIT ELEMENTS AND SOURCES

This gives an overview of the most frequently used elements in electric circuits, including the laws that govern current flow and voltage across these components, as well as the power supplied/dissipated and energy storage within this context. It also discusses the difference between ideal and non-ideal voltage and current sources, highlighting the importance of understanding sign conventions such as voltage polarity and current direction.

The concepts of current and voltage are initially presented as they are essential components in electronics and electrical engineering.

1.2.1. Current

Current is the movement of electric charge through a conductor. The unit of current is Ampere, while charge is measured in Coulombs.

- **Definition of an Ampere:** An Ampere is defined as the amount of total charge that flows through any random cross section of a conducting material in one second.

Mathematically;

$$I = \frac{Q}{t} or\, Q = It \tag{1}$$

Where; Q is the symbol of charge measured in Coulombs (C), I is the current in amperes (A) and t is the time in seconds (s).

The current can also be defined as the rate of charge passing through a point in an electric circuit i.e.,

$$i = \frac{dQ}{dt} \tag{2}$$

A constant current, represented by the symbol I, is also known as direct current or DC. On the other hand, a time-varying current, symbolized by i or i(t), is also known as alternating current or AC. Current is always measured across a circuit element. In Figure 1.9, an ampere-meter or ammeter is shown connected in series with a circuit element, R, to measure the current flowing through it (Naeem, 2009).

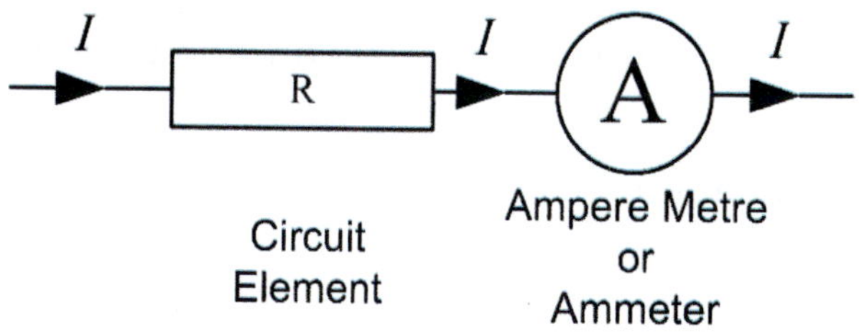

Figure 1.9. *An ammeter is connected in series to measure current, I, through the element, R.*

Source: https://sietm.com/wp-content/uploads/2015/03/concepts-in-electric-circuits.pdf.

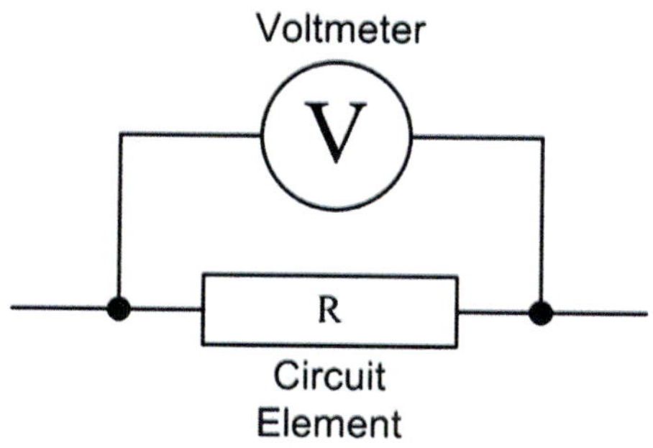

Figure 1.10. *A voltmeter is connected in parallel with the circuit element, R to measure the voltage across it.*

Source: https://sietm.com/wp-content/uploads/2015/03/concepts-in-electric-circuits.pdf.

Example

Determine the current in a circuit if a charge of 80 coulombs (C) passes a given point in 20 seconds (s).

$$Q = 80\ \mathrm{C}, t = 20\ \mathrm{s}, I = ?$$

$$I = \frac{Q}{t} = \frac{80}{20} = 4\ \mathrm{A} \tag{3}$$

1.2.2. Voltage or Potential Difference

If 1 joule of energy is used to transfer 1 coulomb of charge between two points in an electric circuit, the voltage or potential difference between those points is 1 volt.

Voltage is typically represented by the symbol V and measured in volts (V). It's worth noting that both the symbol and unit of voltage are denoted by the same letter, but this rarely leads to confusion.

The symbol V represents a constant voltage (DC), while a time-varying voltage (AC) is represented by the symbol v or v(t).

Voltage is consistently measured across a circuit component, as depicted in Figure 1.10 by Wasif Naeem in 2009.

An energy source, such as a voltage source, is necessary to provide the electromotive force needed for current flow. However, current will only flow if there is a potential difference and a clear path for it to follow.

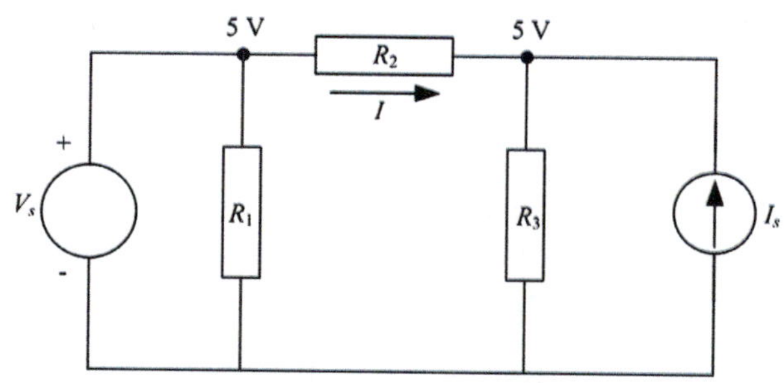

Figure 1.11. *The potential difference across R_2 is 0 V, hence the current I is 0 A where V_s and I_s are the voltage and current sources respectively.*

Source: https://sietm.com/wp-content/uploads/2015/03/concepts-in-electric-circuits.pdf.

If there is no potential difference, indicated by a voltage of 0 V between two points, then no current will flow. In the scenario shown in Figure 1.11, the current I is 0 A because there is no potential difference across R_2. This means that although there is a physical path for the current to travel, there is no potential difference driving it, essentially creating an open circuit (Wasif Naeem, 2009).

Table 1.2 summarizes the fundamental electric circuit quantities, their symbols and standard units.

1.2.3. Circuit Loads

The term "load" typically describes a device or item that is connected to the output of an electrical circuit. This could include various circuit elements such as resistors, capacitors, or inductors.

- Resistor (R)
- Inductor (L)
- Capacitor (C)

Table 1.2. Standard Quantities and Their Units Commonly Found in Electric Circuits

Quantity	Symbol	Unit
Voltage	V	Volts (V)
Current	I	Ampere (A)
Charge	Q	Coulomb (C)
Power	P	Watts (W)
Energy	W	Joules (J)
Time	t	seconds (s)

Source: https://sietm.com/wp-content/uploads/2015/03/concepts-in-electric-circuits.pdf.

Loads can be resistive, inductive, or capacitive, or a combination. For instance, a light bulb is purely resistive, while a transformer is both inductive and resistive. A circuit load is also known as a sink, as it consumes energy, while the voltage or current supply is a source.

Table 1.3 displays the fundamental circuit components, including their symbols and schematics, utilized in an electric circuit. The components R, L, and C are considered passive, meaning they do not produce their own electromotive force (emf). Contrastingly, the DC voltage and current sources are considered active elements.

Table 1.3. Common Circuit Elements and Their Representation in an Electric Circuit

Circuit Element	Symbol	Schematic
Resistor	R	 or 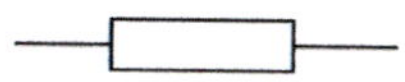
Inductor	L	
Capacitor	C	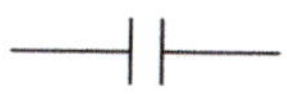
DC Voltage Source	V_s	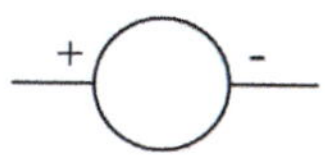 or 
DC Current Source	I_s	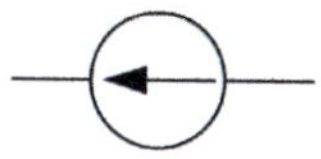

Source: https://sietm.com/wp-content/uploads/2015/03/concepts-in-electric-circuits.pdf.

1.2.4. Sign Convention

Many people believe that current is the movement of electrons, but the usual practice is to consider the movement of protons when determining the direction of the current.

The direction of current in a circuit is determined by the polarity of the source voltage. Current flows from the positive (high potential) side to the negative (low potential) side of the source, as illustrated in Figure 1.12(a). This diagram shows V_s as the source voltage, V_L as the voltage across the load, and I as the loop current moving in a clockwise direction (Naeem, 2009).

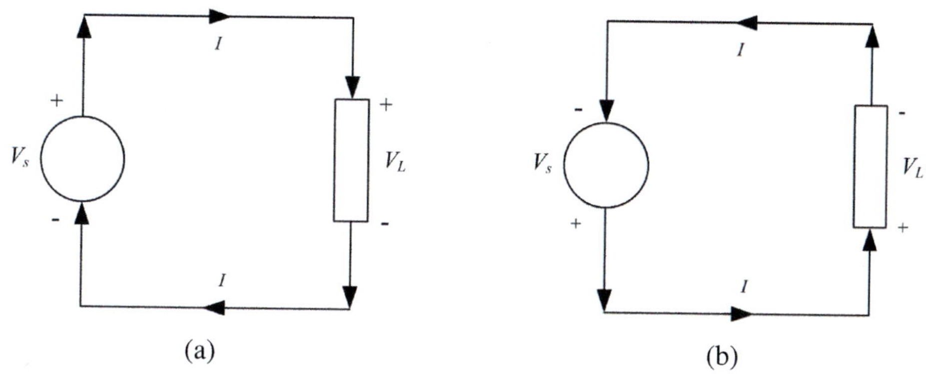

Figure 1.12. *Effect of reversing the voltage polarity on current direction.*

Source: https://sietm.com/wp-content/uploads/2015/03/concepts-in-electric-circuits.pdf.

Please observe that the voltage polarity and current direction in a sink is opposite to that of the source.

In Source current leaves from the positive terminal

In Load (Sink) current enters from the positive terminal

A reversal in source voltage polarity changes the direction of the current flow and vice versa as depicted in Figure 1.12(a) and 1.12(b).

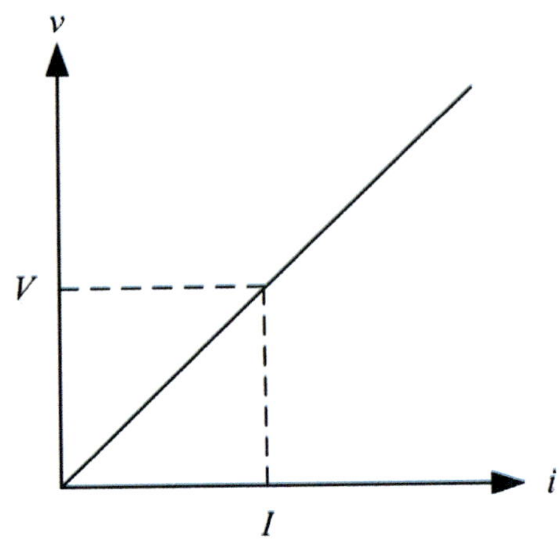

Figure 1.13. *V–I relationship for a resistor according to Ohm's law.*

Source: https://sietm.com/wp-content/uploads/2015/03/concepts-in-electric-circuits.pdf.

At any given point in the above graph, the ratio of voltage to current is always constant.

1.2.5. Passive Circuit Elements

- **Resistor:** In order to understand the resistance and characteristics of a resistor, it is crucial to define Ohm's law.
- **Ohm's Law:** The most basic principle utilized in circuit analysis is Ohm's Law, which offers a straightforward equation that explains the relationship between voltage and current in a conducting material.
  - **Statement:** The voltage across a conducting material is directly proportional to the current passing through it.
  - **Mathematically:**

$$V \propto I$$

$$V = RI \text{ or } I = \frac{V}{R} \text{ or } R = \frac{V}{I} \tag{4}$$

Where the constant of proportionality R is called the resistance or electrical resistance, measured in ohms (Ω). Graphically, the V–I relationship for a resistor according to Ohm's law is depicted in Figure 1.13 (Wasif Naeem, 2009).

Example

Find R if the voltage V and current I in Figure 1.13 are equal to 10 V and 5 A respectively.

$$V = 10 \text{ V}, I = 5 \text{ A}, R = ? \tag{5}$$

Using Ohm's Law

$$V = IR \text{ or } R = \frac{V}{I} = \frac{10}{5} \tag{6}$$

$$\therefore R = 2\,\Omega \tag{7}$$

A short circuit between two points represents a zero resistance whereas an open circuit corresponds to an infinite resistance as demonstrated in Figure 1.14 (Wasif Naeem, 2009).

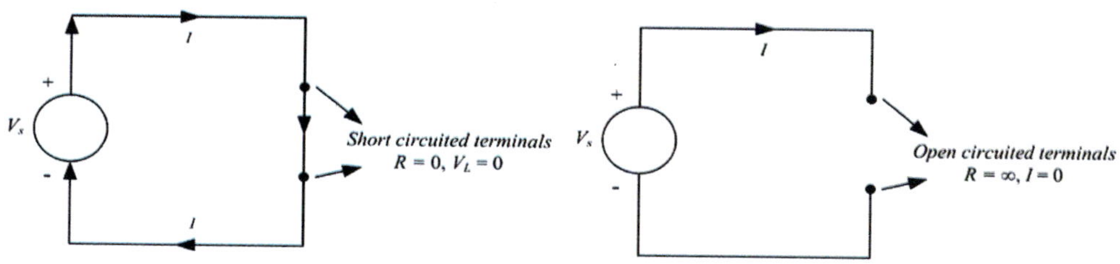

Figure 1.14. *Short circuit and open circuit resistance characteristics.*

Source: https://sietm.com/wp-content/uploads/2015/03/concepts-in-electric-circuits.pdf.

Using Ohm's law;

When R = 0 (short circuit), V = 0 V

When R = ∞ (open circuit), I = 0 A

- **Conductance:** Conductance (G) is the exact opposite of resistance. In mathematical terms;

$$G = \frac{1}{R}$$

$$\therefore I = \frac{V}{R} = VG \quad (8)$$

Where; G is measured in siemens (S) and sometimes also represented by the unit mho (℧) (upside-down omega).

- **Capacitor:** A capacitor is a passive component in circuits that can store charge in an electric field. It is commonly used in electric circuits as a filter. The relationship between voltage and current in a capacitor is determined by the following equation:

$$i = C\frac{dv}{dt} \text{ or } v = \frac{1}{C}\int_0^t idt + v(0) \quad (9)$$

Where; C is the capacitance measured in Farads (F) and v(0) is the initial voltage or initial charge stored in the capacitor

When; v = V (constant DC voltage), $\frac{dv}{dt} = 0$, and i = 0. Hence a capacitor acts as an open circuit to DC.

Example

For the circuit diagram shown in Figure 1.15, determine the current, I flowing through the 5 Ω resistance (Wasif Naeem, 2009).

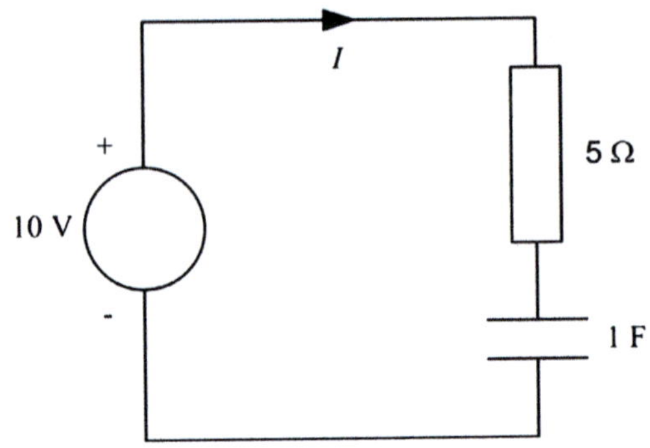

Figure 1.15. *In the circuit diagram determine the current.*

Source: https://sietm.com/wp-content/uploads/2015/03/concepts-in-electric-circuits.pdf.

Since the supply voltage is DC, therefore the capacitor will act as an open circuit. Hence no current can flow through the circuit regardless of the values of capacitor and resistor i.e.,

$$I = 0$$

- **Inductor:** An inductor is a conducting wire wound around a ferromagnetic core. While they can be used as filters like capacitors, their most common application is in AC transformers and power supplies to convert voltage levels.

In an inductor, the V – I relationship is given by the following differential equation:

$$v = L\frac{di}{dt} \text{ or } i = \frac{1}{L}\int_0^t v dt + i(0) \tag{10}$$

Where; L is the inductance in Henrys (H) and i(0) is the initial current stored in the magnetic field of the inductor.

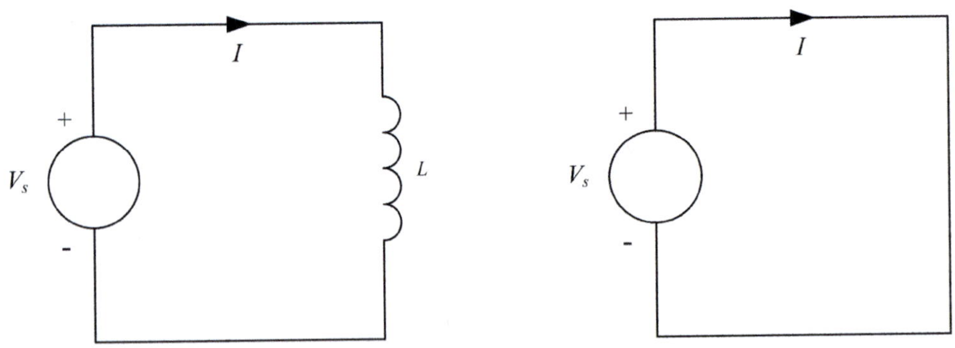

Figure 1.16. *An ideal inductor can be replaced by a short circuit when the supply voltage is DC.*

Source: https://sietm.com/wp-content/uploads/2015/03/concepts-in-electric-circuits.pdf.

When i = I (constant DC current), $\frac{di}{dt} = 0, v = 0$. Hence an inductor acts as a short circuit to DC. An ideal inductor is just a piece of conducting material with no internal resistance or capacitance. The schematics in Figure 1.16 are equivalent when the supply voltage is DC (Wasif Naeem, 2009).

A summary of the V–I relationships for the three passive circuit elements is provided in Table 1.4.

Table 1.4. V–I Relationships for a Resistor, Inductor and Capacitor

Circuit Element	Voltage	Current
Resistor	$V = IR$	$I = \frac{V}{R}$
Inductor	$v = L\frac{di}{dt}$, $v = 0$ for DC	$i = \frac{1}{L}\int_0^t vdt + i(0)$
Capacitor	$v = \frac{1}{C}\int_0^t idt + v(0)$	$i = C\frac{dv}{dt}$, $i = 0$ for DC

Source: https://sietm.com/wp-content/uploads/2015/03/concepts-in-electric-circuits.pdf.

1.2.6. DC Sources

In generally, there are two main types of DC sources:

1. Independent (Voltage and Current) Source
2. Dependent (Voltage and Current) Sources

Independent sources generate their own voltage and current through a chemical reaction and are not influenced by any other variables in the circuit. On the contrary, the output of a dependent source is impacted by changes in a specific parameter (voltage or current) in a circuit element. This discussion will focus solely on independent sources.

- **DC Voltage Source:** This can be further subcategorized into ideal and non-ideal sources.
  - **The Ideal Voltage Source:** According to Figure 1.17(a), an ideal voltage source maintains a consistent terminal voltage regardless of load changes. While the supply current may fluctuate, the terminal voltage (VL) remains steady. This property is illustrated in Figure 1.17(b) (Naeem, 2009).

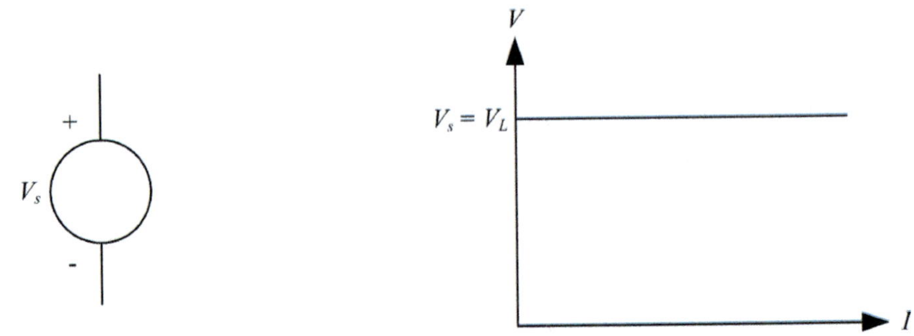

(a) An ideal voltage source.

(b) $V - I$ characteristics of an ideal voltage source.

Figure 1.17. *Schematic and characteristics of an ideal voltage source.*

Source: https://sietm.com/wp-content/uploads/2015/03/concepts-in-electric-circuits.pdf.

– **Non-Ideal or Practical Voltage Source:** In practical applications, the terminal voltage decreases as the load current increases. This relationship is illustrated in Figure 1.18(a).

This behavior can be represented by adding an internal resistance, Rs, in series with the source, as depicted in Figure 1.18(b) (Naeem, 2009).

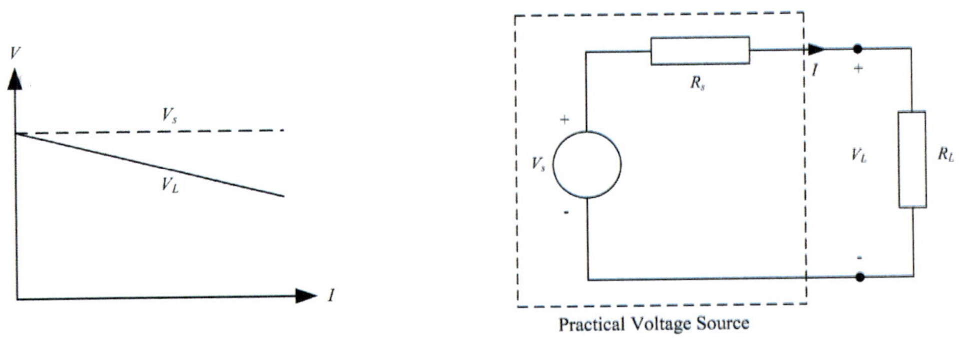

(a) $V - I$ characteristics of a practical voltage source

(b) A practical voltage source has an internal resistance connected in series with the source.

Figure 1.18. *Characteristics and model of a practical voltage source.*

Source: https://sietm.com/wp-content/uploads/2015/03/concepts-in-electric-circuits.pdf.

Where; R_L represents the load resistance.

The characteristic equation of the practical voltage source can be written as;

$$V_L = V_s - R_s I$$

For an ideal source, $R_s = 0$ and therefore $V_L = V_s$.

Example

The terminal voltage of a battery is 14 V at no load. When the battery is supplying 100 A of current to a load, the terminal voltage drops to 12 V. Calculate the source impedance.

$V_s = V_{NL} = 14.0$ V when $I = 0$ A (without load)

$V_L = 12.0$ V when $I = 100$ A (at full load)

$$\because V_L = V_s - R_s I$$

$$R_s = \frac{V_s - V_L}{I} = \frac{14 - 12}{100} = \frac{2}{100}$$

$$R_s = 0.02\ \Omega \qquad (11)$$

- **Voltage Regulation:** Voltage regulation (VR) is a critical indicator of the quality of a power source. It quantifies the change in terminal voltage from no load to full load conditions, as displayed in Figure 1.19 (Wasif Naeem, 2009).

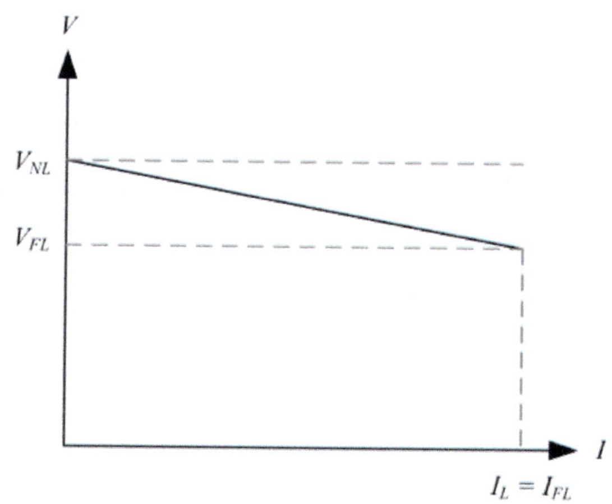

Figure 1.19. *No load and full load voltages specified on a V–I characteristic plot of a practical voltage source.*

Source: https://sietm.com/wp-content/uploads/2015/03/concepts-in-electric-circuits.pdf.

If V_{NL} and V_{FL} represents the no load and full load voltages, then the VR of a source is defined mathematically as:

$$VR = \frac{V_{NL} - V_{FL}}{V_{FL}} \times 100\% \qquad (12)$$

For an ideal source, there is no internal resistance and hence $V_{NL} = V_{FL}$ and

V R = 0%

Hence, the smaller the regulation, the better the source.

In the previous example, $V_{NL} = 14.0$ V and $V_{FL} = 12.0$ V, therefore;

$$VR = \frac{14 - 12}{12} \times 100 = 16.67\% \qquad (13)$$

- **DC Current Source**: A current source, in contrast to a DC voltage source, does not exist in physical form. Nonetheless, it is valuable in creating equivalent circuit diagrams for semiconductor devices like transistors. Current sources can be further classified as either ideal or non-ideal.
  - **The Ideal Current Source:** In Figure 1.20(a), an ideal current source is shown, which provides a constant current regardless of changes in load. This means that the current from an ideal current source remains constant regardless of the voltage across the load.
  - **Non-Ideal or Practical Current Source:** As the load or load voltage increases, the current provided by a practical current source decreases. This can be represented by adding a parallel resistance to the ideal current source, as illustrated in Figure 1.20(b). The internal

resistance of the current source is denoted by R_s, while R_L represents the load (Naeem, 2009).

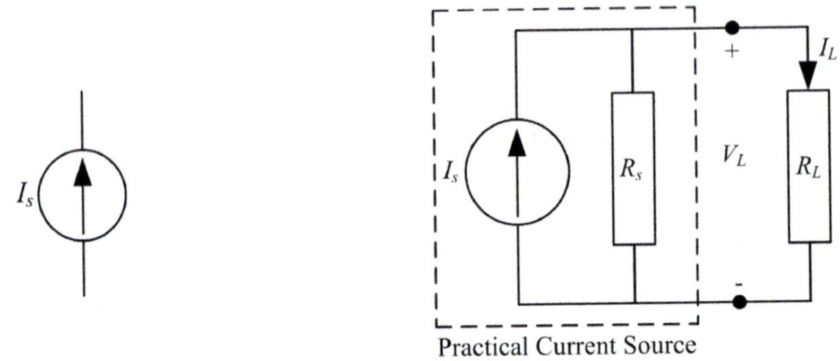

(a) An ideal current source

(b) A practical current source has an internal resistance connected in parallel with the source.

Figure 1.20. *Ideal and non-ideal current sources.*

Source: https://sietm.com/wp-content/uploads/2015/03/concepts-in-electric-circuits.pdf.

The characteristic equation of the practical current source can be written as;

$$I_L = I_s - \frac{V_L}{R_s} \qquad (14)$$

In an ideal current source, $R_s = \infty$ (open circuit), therefore $I_L = I_s$.

1.2.7. Power

Power can be determined by multiplying the magnitudes of V and I together, and it is measured in Watts (W).

Mathematically;

$$P = VI(\text{W}) \qquad (15)$$

Example

If the power dissipated in a circuit element is 100 W and a current of 10 A is flowing through it, calculate the voltage across and resistance of the element.

$P = 100\text{ W}, I = 10\text{ A}, V = ?, R = ?$

$$P = VI$$

$$V = \frac{P}{I} = \frac{100}{10} = 10\text{ V}$$

$$\text{Also, } R = \frac{V}{I} = \frac{10}{10} = 1\,\Omega \qquad (16)$$

- **Alternate Expressions for Power Using Ohm's Law:** Using Ohm's law i.e., V = IR, two more useful expressions for the power absorbed/delivered can be derived as follows:

$$P = VI = (IR)I = I^2R \qquad (17)$$

Also; $$I = \frac{V}{R} \qquad (18)$$

$$\therefore P = VI = V\frac{V}{R} = \frac{V^2}{R} \qquad (19)$$

Example

A Light bulb draws 0.5 A current at an input voltage of 230 V. Determine the resistance of the filament and also the power dissipated.

- From Ohm's Law

$$R = \frac{V}{I} = \frac{230}{0.5} = 460\ \Omega \qquad (20)$$

Since a bulb is a purely resistive load, therefore all the power is dissipated in the form of heat. This can be calculated using any of the three power relationships shown above.

$$\begin{aligned} P &= VI = 230 \times 0.5 = 115\ \text{W} \\ P &= I^2R = (0.5)^2 \times 460 = 115\ \text{W} \\ P &= \frac{V^2}{R} = \frac{(230)^2}{460} = 115\ \text{W} \end{aligned} \qquad (21)$$

1.2.8. Energy

Energy refers to the ability of a physical system to do work. In electric circuits, the relationship between energy (w) and power is as follows.

$$p = vi = \frac{dw}{dt} \qquad (22)$$

i.e., power is the rate of change of energy.

Voltage can also be written in terms of energy as the work done or energy supplied per unit charge (q) i.e.,

$$v = \frac{dw}{dq} \qquad (23)$$

In terms of the three passive circuit elements, R, L and C, the energy relationships can be derived as follows

- **Resistor:** When electrical power or energy is supplied to a resistor, it is converted and released as heat. This process is irreversible and often referred to as i^2R losses.

$$p = vi = i^2R = \frac{v^2}{R} \qquad (24)$$

- Inductor

$$p = vi = Li\frac{di}{dt} \qquad (25)$$

Total energy supplied from 0 to t is:

$$w = \int_0^t p\,dt = L\int_0^t i\frac{di}{dt}dt = L\int i\,di$$

$$\therefore \boxed{W = \frac{1}{2}LI^2} \qquad (26)$$

The magnetic field of the inductor stores this energy, and it can be returned to the circuit when the original source is taken away.

- Capacitor:

$$p = vi = Cv\frac{dv}{dt} \qquad (27)$$

Total energy supplied from 0 to t is:

$$w = \int_0^t p\,dt = C\int_0^t v\frac{dv}{dt}dt = C\int v\,dv$$

$$\therefore \boxed{W = \frac{1}{2}CV^2} \qquad (28)$$

This energy is stored in the electric field of the capacitor which is supplied back to the circuit when the actual source is removed.

1.3. CIRCUIT THEOREMS

This text discusses the essential laws and theorems needed to analyze and solve electric circuits. It also covers the relationships between different laws and techniques for writing equations through inspection. Numerous examples are included to illustrate how these laws are applied. Furthermore, it addresses situations where applying the concepts directly may not be possible, and offers solutions for overcoming these challenges.

1.3.1. Kirchoff's Laws

Kirchhoff's voltage and current laws are widely regarded as the most common and fundamental set of laws for analyzing electric circuits. These laws form the basis for deriving many other useful relationships in circuit analysis.

- **Kirchoff's Voltage Law (KVL):** The total sum of voltages in a closed loop, including rises and drops, is zero. This means that the total increase in voltage is equal to the total decrease in voltage in the closed loop.

In Figure 1.21, consider mesh BEDB, then according to KVL

$$V_3 = V_4 + V_5 \quad (29)$$

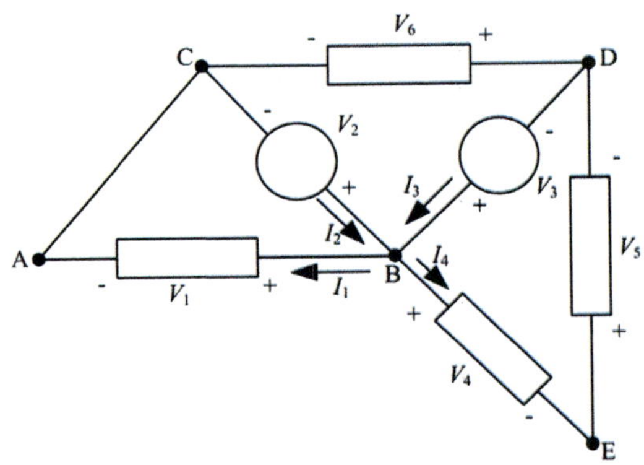

Figure 1.21. *An electric network showing nodes, branches, elements and loops.*

Source: https://sietm.com/wp-content/uploads/2015/03/concepts-in-electric-circuits.pdf.

Example

In each of the circuit diagrams in Figure 1.22, write the mesh equations using KVL (Wasif Naeem, 2009).

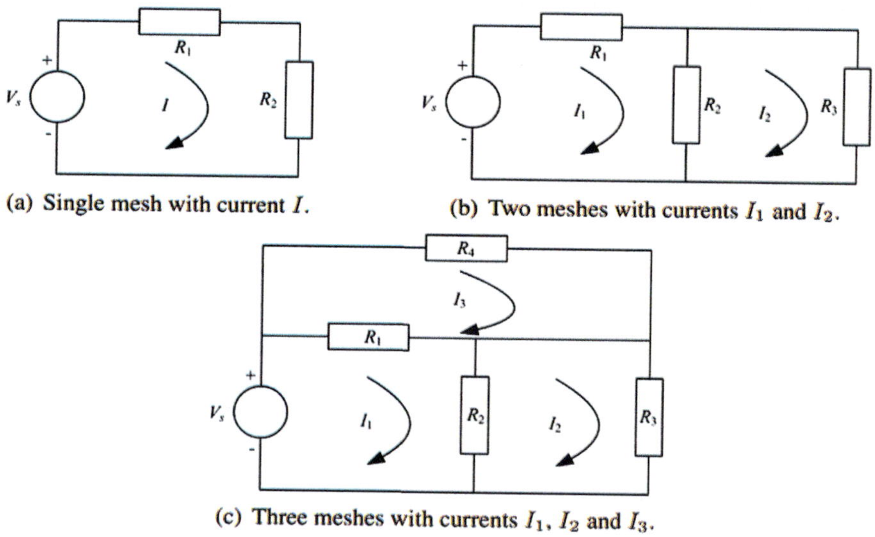

Figure 1.22. *Circuit diagrams to demonstrate the application of KVL in the above example.*

Source: https://sietm.com/wp-content/uploads/2015/03/concepts-in-electric-circuits.pdf.

Figure 1.22(a) contains a single loop hence a single current, I is flowing around it. Therefore, a single equation will result as given below:

$$V_s = IR_1 + IR_2 \quad (30)$$

If V_s, R_1 and R_2 are known, then I can be found.

Figure 1.22(b) contains two meshes with currents I_1 and I_2 hence there will be two equations as shown below. Note that the branch containing R_2 is common to both meshes with currents I_1 and I_2 flowing in opposite directions.

- Left Loop

$$\begin{aligned} V_s &= I_1R_1 + (I_1 - I_2)R_2 \\ V_s &= (R_1 + R_2)I_1 - R_2I_2 \end{aligned} \quad (31)$$

- Right Loop

$$\begin{aligned} 0 &= (I_2 - I_1)R_2 + I_2R_3 \\ 0 &= -R_2I_1 + (R_2 + R_3)I_2 \end{aligned} \quad (32)$$

Given V_s, R_1, R_2 & R_3, Equations can be solved simultaneously to evaluate I_1 and I_2.

In Figure 1.22(c) circuit diagram, three equations must be written. It is important to note that loops 2 and 3 do not share any circuit element, so I_2 and I_3 are independent variables.

- Left Bottom Loop

$$\begin{aligned} V_s &= (I_1 - I_3)R_1 + (I_1 - I_2)R_2 \\ V_s &= (R_1 + R_2)I_1 - R_2I_2 - R_1I_3 \end{aligned} \quad (33)$$

- Right Bottom Loop

$$\begin{aligned} 0 &= (I_2 - I_1)R_2 + I_2R_3 \\ 0 &= -R_2I_1 + (R_2 + R_3)I_2 \end{aligned} \quad (34)$$

- Upper Loop

$$\begin{aligned} 0 &= (I_3 - I_1)R_1 + I_3R_4 \\ 0 &= -R_1I_1 + (R_1 + R_4)I_3 \end{aligned} \quad (35)$$

If V_s and resistors' values are known, the mesh currents can be evaluated by solving.

- **Resistors in Series:** Figure 1.23 illustrates a circuit containing one voltage source and two resistors connected in series to create a single mesh with current I (Naeem, 2009).

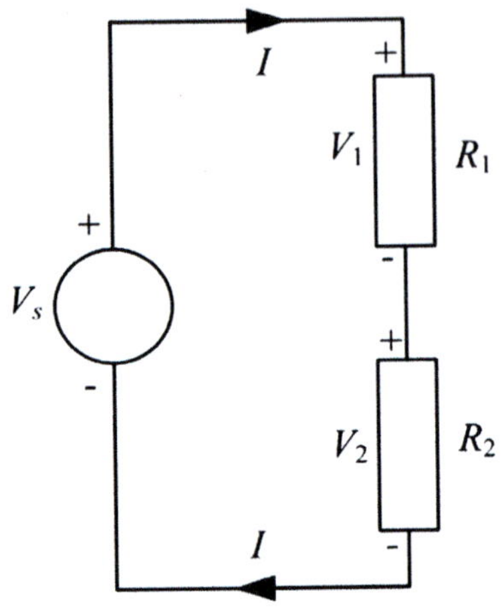

Figure 1.23. *Series combination of two resistors.*

Source: https://sietm.com/wp-content/uploads/2015/03/concepts-in-electric-circuits.pdf.

According to KVL

$$V_s = V_1 + V_2 \quad (36)$$

Using Ohm's law (V = IR);

$$\begin{aligned} IR_{eq} &= IR_1 + IR_2 \\ R_{eq} &= R_1 + R_2 \end{aligned} \quad (37)$$

Where; R_{eq} is the combined or equivalent resistance of the series network. Hence the equivalent resistance of two or more resistors connected in series is given by the algebraic sum of all the resistances. In general, for n number of serial resistors, R_{eq} is given by:

$$\boxed{R_{eq} = R_1 + R_2 + R_3 + \cdots + R_n} \quad (38)$$

- **Voltage Divider Rule (VDR):** The Voltage divider rule is a handy formula that can be used to calculate the voltage across a specific resistor when multiple resistors are connected in series to a voltage source. Figure 1.23 shows how the voltage across each resistor can be expressed based on the supply voltage and the resistance values of the individual resistors.

$$V_1 = V_s \frac{R_1}{R_1 + R_2} \quad (39)$$

$$V_2 = V_s \frac{R_2}{R_1 + R_2} \quad (40)$$

In general, for n number of resistors connected in series, the voltage across the ith resistor can be specified as:

$$\boxed{V_i = V_s \frac{R_i}{R_1 + R_2 + \cdots + R_i + \cdots + R_n}} \quad (41)$$

1.3.2. Kirchoff's Current Law (KCL)

The total current entering or leaving a node in an electric circuit must be zero. This implies that the sum of currents entering a node is equal to the sum of currents leaving the node. For instance, if we consider node B in Figure 1.21, this principle can be applied according to Kirchhoff's Current Law (KCL).

$$I_1 + I_4 = I_2 + I_3 \quad (42)$$

- **Resistors in Parallel:** Examine Figure 1.24 featuring a single current source and two resistors connected in parallel. In parallel circuits, all elements have the same voltage, V across them, meaning $V_1 = V_2 = V$ (Naeem, 2009).

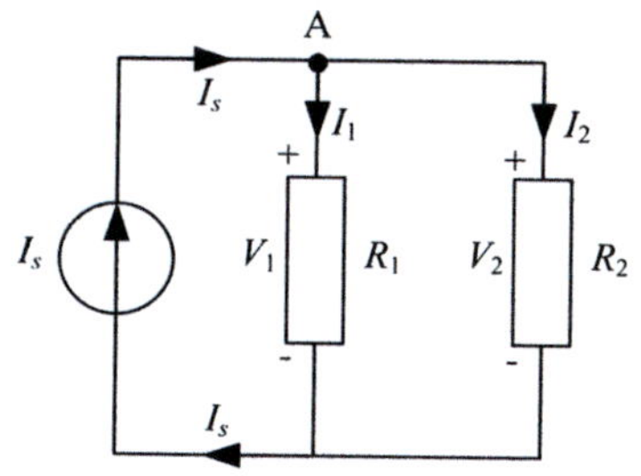

Figure 1.24. *Parallel connection of resistors.*

Source: https://sietm.com/wp-content/uploads/2015/03/concepts-in-electric-circuits.pdf.

Applying KCL at node A

$$I_s = I_1 + I_2 \quad (43)$$

Using Ohm's law $(I = \frac{V}{R})$

$$\frac{V}{R_{eq}} = \frac{V}{R_1} + \frac{V}{R_2}$$

$$\boxed{\frac{1}{R_{eq}} = \frac{1}{R_1} + \frac{1}{R_2}} \quad (44)$$

Where; R_{eq} is the equivalent resistance of the parallel network of resistors. For n number of resistors connected in parallel, the combined resistance is given by:

$$\boxed{\frac{1}{R_{eq}} = \frac{1}{R_1} + \frac{1}{R_2} + \frac{1}{R_3} + \cdots + \frac{1}{R_n}} \quad (45)$$

or in terms of conductance, G

$$\boxed{G_{eq} = G_1 + G_2 + G_3 + \cdots + G_n} \quad (46)$$

- **Current Divider Rule (CDR):** The current divider rule is a helpful way to calculate the current flow through individual components connected in parallel. In the circuit shown in Figure 1.24, the current through each resistor can be determined using the following equations.

$$\begin{aligned} I_1 &= I_s\frac{R_2}{R_1 + R_2} \\ I_2 &= I_s\frac{R_1}{R_1 + R_2} \end{aligned} \quad (47)$$

Observe the contrast between VDR and CDR (for two resistors) in relation to the resistor value in the numerators. To extend CDR to a general case of n resistors, the conductance parameter is employed. Therefore, to determine the current through the ith of n resistors connected in parallel, the following equation can be utilized.

$$I_i = I_s\frac{G_i}{G_1 + G_2 + \cdots + G_i + \cdots + G_n} \quad (48)$$

The equation shown above is structurally similar to the generalized VDR, but with the variable R substituted for G and voltages replaced with current variables.

1.3.3. Electric Circuits Analysis

The main tools for analyzing DC electric circuits are Kirchhoff's Voltage Law (KVL), Kirchhoff's Current Law (KCL), and Ohm's Law. Nodal analysis is typically used when applying KCL to electric circuits, while loop or mesh analysis is used with KVL for problem solving.

The mesh and nodal analysis methods described are highly organized and typically yield a solution for a given problem. Yet, they can be quite complex computationally. An easier alternative may involve circuit reduction techniques like combining resistors in series/parallel or using voltage division ratio/current division ratio methods.

- Mesh Analysis: The mesh analysis technique consists of the following steps
  - Transform all currents sources to voltage sources, if possible.
  - Identify and assign a current

to each mesh of the network (preferably in the same direction).

- Write mesh equations using KVL and simplify them.
- Solve the simultaneous system of equations.
- Number of equations is equal to number of meshes in the network.

- Nodal Analysis: The following steps describe the nodal analysis method
  - Transform all voltage sources to current sources, if possible
  - Identify and assign an arbitrary voltage to each node including the reference node in the network assuming all currents leaving the node. The reference node is normally assumed to be at zero potential.
  - Write nodal equations using KCL and simplify them.
  - Solve the simultaneous system of equations.
  - Number of equations is N – 1 where N is the number of nodes in the network including the reference node.

1.3.4. Superposition Theorem

The Superposition theorem is a valuable tool for analyzing electric circuits with multiple active sources. It involves evaluating the current or voltage of a component by considering each source individually, and then summing up the effects of all sources acting independently. This theorem can be formally expressed as:

In a linear circuit with multiple sources, the total current through or voltage across any element is the sum of the individual currents or voltages from each source, with all other sources removed from the circuit.

Linearity is a necessary condition for the theorem to apply. Fortunately, the v, i relationship for R, L and C are all linear.

The sources can be removed using the following methodology:

1. Ideal voltage sources are short-circuited
2. Ideal current sources are open-circuited

In general, practical sources are replaced by their internal resistances.

Example

Find the voltage V_L using Superposition theorem in the circuit diagram of Figure 1.25 (Wasif Naeem, 2009).

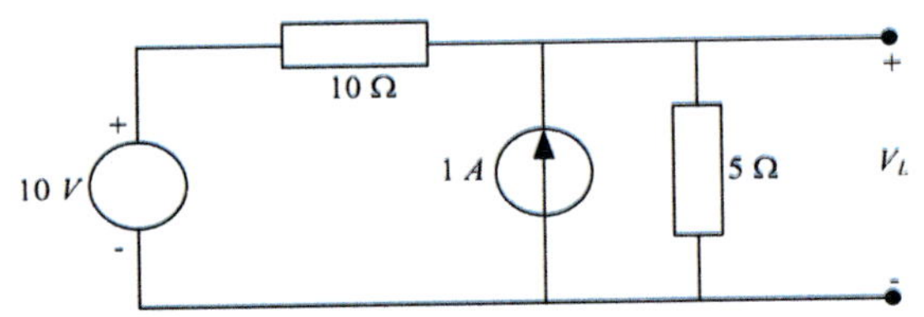

Figure 1.25. *An electric circuit containing multiple sources.*

Source: https://sietm.com/wp-content/uploads/2015/03/concepts-in-electric-circuits.pdf.

- **Step 1: Suppressing the 1 A Current Source by Replacing it With an Open Circuit:** This will result in the circuit diagram of Figure 1.26(a). The output voltage, now called V'_L, is the voltage across the 5 Ω resistance. Since both 10 Ω and 5 Ω are in series, VDR can be applied here

$$\therefore V_L' = 10\frac{5}{10+5} = 3.33\ \text{V} \tag{49}$$

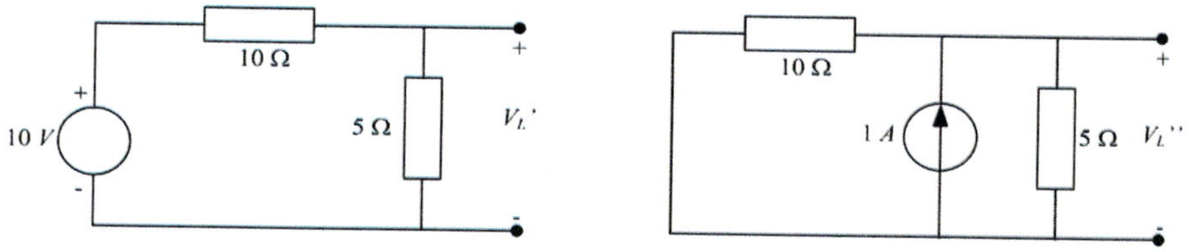

(a) Superposition (step 1) - suppressing the current source.

(b) Superposition (step 2) - suppressing the voltage source.

Figure 1.26. *Application of superposition theorem to figure 1.25.*

Source: https://sietm.com/wp-content/uploads/2015/03/concepts-in-electric-circuits.pdf.

- **Step 2: Suppressing the 10 V Voltage Source by Replacing it with a Short Circuit:** The resulting circuit diagram is shown in Figure 1.26(b). The output voltage is now represented by V_L'' and has the same polarity as VL. It is obvious that both the resistors are in parallel and therefore CDR can be applied to determine the current flowing through 5 Ω resistor.

$$\begin{aligned} I_{5\Omega} &= 1\frac{10}{10+5} &= 0.67\ \text{A} \\ \therefore V_L'' &= I_{5\Omega} \times 5 &= 3.33\ \text{V} \end{aligned} \tag{50}$$

The total voltage, according to the superposition theorem is then given by the sum of V_L' and V_L'' i.e.,

$$V_L = V_L' + V_L'' = 3.33 + 3.33 = 6.66\ \text{V} \tag{51}$$

- **Superposition Theorem – Power Calculation:** Although the superposition theorem can be used to find the voltage across and current through a circuit element by adding the responses due to each source acting alone, the power (as a nonlinear relationship) cannot be evaluated until the net voltage or current is found. This is because:

$$\begin{aligned} P &= (I_1 + I_2)^2 R \neq I_1^2 R + I_2^2 R \\ \text{or } P &= \frac{(V_1 + V_2)^2}{R} \neq \frac{V_1^2}{R} + \frac{V_2^2}{R} \end{aligned} \tag{52}$$

1.3.5. Thevenin's Theorem

Thevenin's theorem is a beneficial tool for simplifying the analysis of complex and large electric circuits by converting them into a single voltage source connected in series with a resistor. This method is especially useful when dealing with circuits that have a variable resistor or load.

Formally, the Thevenin's theorem can be stated as;

A linear electric circuit with resistors and sources can be simplified by replacing it with a single voltage source and resistor connected in series across the load.

The current IL passing through the load resistance RL is consistent in the circuit diagrams depicted in Figure 1.27, indicating equivalence between the circuits with regards to the load resistor RL (Naeem, 2009).

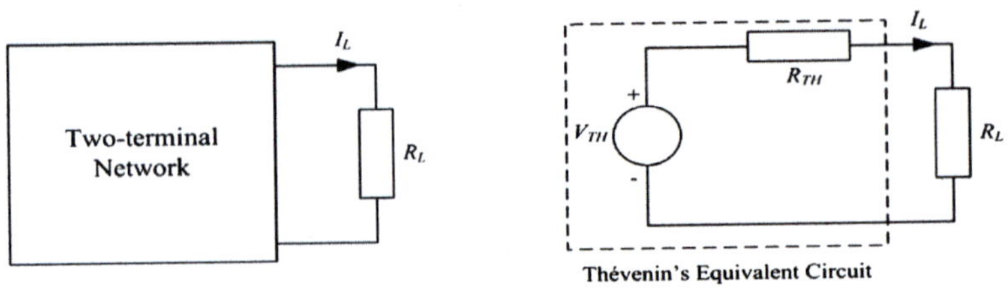

Figure 1.27. *Illustration of Thevenin's theorem.*

Source: https://sietm.com/wp-content/uploads/2015/03/concepts-in-electric-circuits.pdf.

The following steps outline the procedure to simplify an electric circuit using Thevenin's theorem where V_{TH} and R_{TH} are the Thevenin's voltage and Thevenin's resistance, respectively.

- Remove the load resistance R_L.
- V_{TH} is the open circuit (OC) voltage across the load terminals and
- R_{TH} is the resistance across the load terminals with all sources replaced by their internal resistances.

Alternatively, measure the OC voltage across, and the short circuit (SC) current through the load terminals. Then

$$V_{TH} = V_{oc} \text{ and } R_{TH} = \frac{V_{oc}}{I_{sc}} \quad (53)$$

1.3.6. Norton's Theorem

Thevenin's equivalent circuit can be represented as a practical voltage source, while Norton's equivalent circuit can be represented as a practical current source. In other words, any linear two-terminal circuit consisting of resistors and sources can be replaced by a single current source connected in parallel with a resistor.

To determine Norton's equivalent circuit, Norton current, I_N, and Norton resistance, R_N, are required. The following steps outline the procedure required.

- Remove the load resistance, R_L.
- I_N is the SC current through the load terminals and
- R_N is the resistance across the load terminals with all sources replaced by their internal resistances. Clearly $R_N = R_{TH}$.

CASE STUDY

NEW MAINS DISTRIBUTION, CABLE MANAGEMENT, MACHINE CIRCUITS, LIGHTING CIRCUITS AT SUSSEX ENGINEERING CO.

About:

Based in Rustington, West Sussex, IEC Engineering Ltd delivers uniquely integrated precision engineering solutions. Part of the Nasmyth Group of companies, IEC Engineering manufacture high quality components and working assemblies with specialist experience in machining high quality complex components using multi-axis CNC machines. They supply solutions for safety critical markets including Commercial Aerospace, Submarine, Medical, Surveillance, Mobility, Instrumentation, and nuclear industries.

Summary:

IEC Engineering has been upgrading and updating their factory units by installing new mains distribution systems, galvanized steel tray cable management systems, and new machine circuits.

These works have been carried out while the factory is in full production and whilst the office and factory staff are working. The aim was to minimize disruption and also make sure that productions levels remained effective as much as possible.

These machines have been designated new positions and all required new feed cables isolators, localized fuse box and distribution circuits. This also included new lighting supplies, fittings and localized power sockets in the new areas.

Brief:

To work with the maintenance department in design and structure for the new layout. This included the rearrangement of existing machines to a new more efficient layout, with space for increased production levels of machining areas.

To include completely new wiring for each unit in turn (4 units in total). This ranged from:

- MCCB mains distribution systems with galvanized steel cable tray management systems fixed from suspended ceiling fixings
- New sub main cables to feed localized distribution system

- New fuse boxes close to new machine positions
- Individual circuits for each machine with means of isolation and small power circuits for socket outlets where required

Delivery

Completed over several months to ensure minimal disruption to the client and production. The project was completed fully with all the client's objectives met:

- New three-phase mains distribution boards (MCCB) panel boards
- New main galvanized steel cable tray cable management system
- New three-phase Sub main circuits feeding local fuse box / distribution boards
- New single-phase sub main circuits feeding local fuse box / distribution boards
- New machine circuits for each new position of existing machine
- New isolation for each machine
- New power circuits from new sub main fuse boxes for socket outlets in new positions
- New lighting circuits wired from new lighting fuse boxes and switch positions
- New emergency lighting system designed and installed
- New main earthing and earth bonds to main water, gas and steel
- Testing and NICEIC certification upon completion

CLASS ACTIVITY

Students will apply Ohm's Law to calculate the expected current and voltage drops, comparing their theoretical calculations with actual measurements. To extend the activity, students can modify the circuit by adding additional resistors in series or parallel and observe how these changes affect the overall circuit behavior, reinforcing concepts like total resistance and voltage division.

SUMMARY

- An electric circuit is a representation of a real circuit that is used to connect different devices.
- A circuit in which the switch is open and current does not flow through the circuit is called an open circuit.
- Whenever the current in the circuit follows a shorter path rather than following its usual path short circuit happens.
- Electric current is passed to our house at 220 volts through a circuit called a Fuse. If the supplied voltage varies a lot as compared to the allowed voltage the fuse wire gets to break and stopped the flow of current in the circuit.
- Current can be defined as the motion of charge through a conducting material. The unit of current is Ampere whilst charge is measured in Coulombs.
- Voltage or potential difference between two points in an electric circuit is 1 V if 1 J (Joule) of energy is expended in transferring 1 C of charge between those points.
- A circuit load can also be referred to as a sink since it dissipates energy whereas the voltage or current supply can be termed as a source.
- The voltage or potential difference across a conducting material is directly proportional to the current flowing through the material.
- A capacitor is a passive circuit element that has the capacity to store charge in an electric field. It is widely used in electric circuits in the form of a filter.
- An independent source produces its own voltage and current through some chemical reaction and does not depend on any other voltage or current variable in the circuit.
- A current source, unlike the DC voltage source, is not a physical reality. However, it is useful in deriving equivalent circuit models of semiconductor devices such as a transistor.
- Voltage divider rule provides a useful formula to determine the voltage across any resistor when two or more resistors are connected in series with a voltage source.
- Current divider rule provides a useful relationship for determining the current through individual circuit elements that are connected in parallel.

REVIEW QUESTIONS

1. What is electric circuit? Explain the types of electric circuits.
2. What is simple circuit?
3. What are the different circuit elements? Explain each of them.
4. Elaborate the Kirchoff's Current Law (KCL).
5. What is electric circuits analysis?

REFERENCES

1. Cathey, J., & Nasar, S. (1997). *Basic Electrical Engineering* (2nd ed.). McGraw-Hill.
2. Dorf, R. C., & Svoboda, J. A. (2014). *Introduction to Electric Circuits* (9th ed.). John Wiley & Sons, Inc.
3. Fitzgerald, A. E., Higginbotham, D. E., & Grabel, A. (1975). *Basic Electrical Engineering*. McGraw-Hill.
4. Hambley, A. R. (2019). *Electrical Engineering: Principles and Applications* (7th ed., global ed.). Pearson Education Limited. Available from: https://ebookcentral.proquest.com/lib/gla/detail.action?docID=5573664 (accessed on 1 July 2024).
5. Kuphaldt, T. (2021). Using the Spice circuit simulation program. In *Lessons in electricity: Reference* (Vol. 5, Ch. 7). Retrieved from: https://www.ibiblio.org/kuphaldt/electricCircuits/Ref/REF.pdf.
6. Naeem, W. (2009). *Concepts in Electric Circuits*. ISBN 978-87-7681-499-1.
7. Nahvi, M., & Edminister, J. (2014). *Schaum's Outlines Electric Circuits* (6th ed.). McGraw-Hill Education. Available from: http://www.vlebooks.com/vleweb/product/openreader?id=GlasgowUni&isbn=9780071830461 (accessed on 1 July 2024).
8. Nilsson, J. W., & Riedel, S. A. (2015). *Electric Circuits* (10th ed.). Pearson.
9. Nilsson, J., & Riedel, S. (1999). *Electric Circuits* (2nd ed.). Prentice Hall.
10. Robbins, A., & Miller, W. (2000). *Circuit Analysis: Theory and Practice* (6th ed.). Delmar.
11. Spiegel, M. R., & Stephens, L. J. (2018). *Schaum's Outlines: Statistics* (6th ed.). McGraw-Hill Education. Available from: https://ebookcentral.proquest.com/lib/gla/detail.action?docID=6255552 (accessed on 1 July 2024).

CHAPTER 2

Resistive Network Analysis

LEARNING OBJECTIVES

After studying this chapter, you will be able to:

- Understand about series circuit
- Learn about parallel circuits
- Identify a parallel resistive circuit
- Determine the voltage across each branch in a parallel circuit
- Apply Kirchhoff's current law
- Determine total parallel resistance
- Determine the total effect of current sources in parallel
- Use a parallel circuit as a current divider
- Determine power in a parallel circuit
- Describe some basic applications of parallel circuits
- Troubleshoot parallel circuits
- Demonstrate the node voltage method
- Provide a step-by-step introduction to using the mesh current method

KEY TERMS FROM THIS CHAPTER

Branch	Branch current	Circuit
Current divider	Current flow	Current source
Currents	Electrical circuit	Electrical network
Kirchhoff's current law	Kirchhoff's voltage law	Loop
Mesh	Millman's theorem	Network analysis
Node	Norton's theorem	Ohm's Law

INTRODUCTION

Many circuit elements connected in a necessary way and powered by one or more voltage or current sources make up a network. The computation of circuit voltage and current is a component of network analysis. Calculating the potential difference between any two points in the network or the current passing through a specific circuit element may be necessary. For this purpose, various techniques are available. When performing calculations, an expert will employ the best method. The following list includes some fundamental terms related to electrical networks.

- **Network elements**: An electrical network will have interconnected components such as resistors, inductors, capacitors, and diode. The network will also have voltage source or current source or both.
- **Branch**: An electrical network will have a number of branches. Each branch will have two terminals.
- **Node**: In a network, two or more branches meet at certain points. The junction point of two or more branches in a network is called node.
- **Loop or Mesh**: A closed path is called a loop, and the current that travels along the closed path after a loop is known as loop current. Any closed path in a network is called a loop. The smallest loop that exists in a network is called a mesh. Any loop devoid of any other loops is called a mesh. The terms mentioned above are illustrated in Figure 2.1.

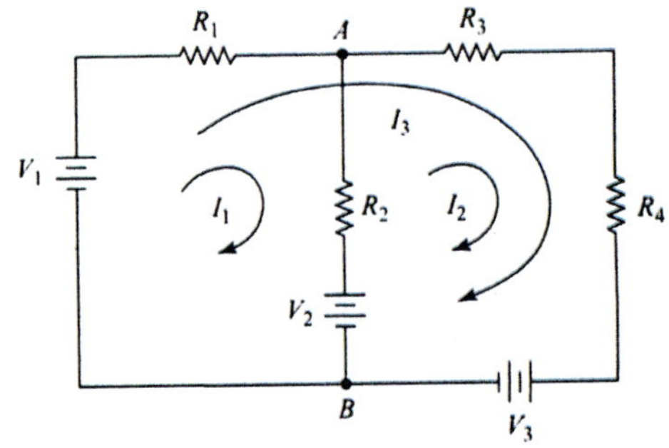

Figure 2.1. *Distinction between loop and mesh illustrated.*

Source: S. K. Bhattacharya. Basic Electrical Engineering. Pearson. ISBN: 9788131763261.

Voltage sources and resistances make up the network elements. There are nodes at A and B. Current flows through three loops, denoted as I_1, I_2, and I_3, respectively. Current I_1 and I_2, respectively, are flowing through two independent loops. The smallest loops are these ones. We refer to them as mesh.

The voltage equations for each loop can be written by using Kirchhoff's voltage law. We can compute the loop current and, in turn, the branch currents I by solving the equations. e. every component in the circuit has current flowing through it.

2.1. SERIES CIRCUIT

A series circuit is a type of electrical circuit arrangement in which all of the parts are connected end to end to create a single path for current flow. Consequently, the circuit's current will pass through each component one after the other. As the number of resistors connected in a series circuit increases, the total resistance—which is the sum of the individual resistances of the circuit's components—increases.

- This is because the current goes on through each resistor, one after the other, then the resistance it will face is the sum of the resistances of each resistor.
- This is because the voltage drop in each element on which current is flowing is constant and proportional to the element's resistance.
- A key characteristic of the series circuit is that if one part or component fails or is removed, the circuit would be broken so that no current will flow through it.

Four resistors are shown in series with a voltage source in Figure 2.2. All points in a series circuit have the same amount of current. This is due to the fact that the current flowing into and out of any given point in this circuit must always be equal. Kirchhoff's voltage law (KVL) states that the total voltage source in a closed path in a circuit should equal the sum of all the voltage drops surrounding it.

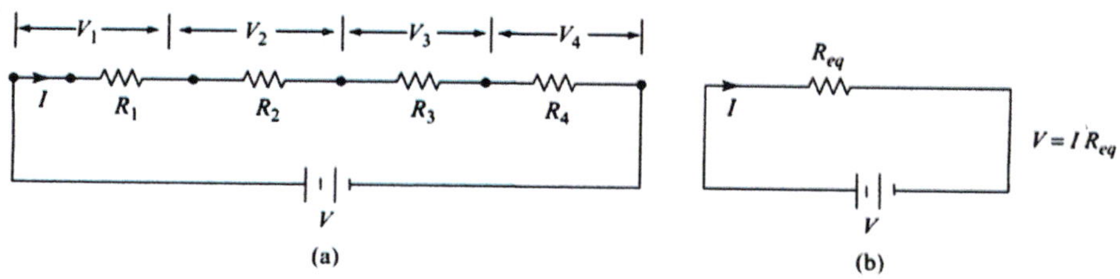

Figure 2.2. *DC series circuit.*

Source: S. K. Bhattacharya. Basic Electrical Engineering. Pearson. ISBN: 9788131763261.

The voltage drops across the resistances are V_1, V_2, V_3, and V_4, respectively. Since the same current is flowing through all the resistances, we can write;

$$V_1 = IR_1,\ V_2 = IR_2,\ V_3 = IR_3,\ \&\ V_4 = IR_4 \qquad (1)$$

Again, the total voltage applied is V is equal to the sum of the voltage drops across the resistances. Thus, we can write;

$$V = V_1 + V_2 + V_3 + V_4 \qquad (2)$$

To find the value of equivalent resistance of a number of resistances connected in series, we equate the voltage, V of the two equivalents in units are shown in Figure 2.2(a) and Figure 2.2(b) as;

$$IR_{eq} = IR_1 + IR_2 + IR_3 + IR_4 \quad (3)$$

or

$$R_{eq} = R_1 + R_2 + R_3 + R_4 \quad (4)$$

Assuming

R_{eq} equal to R, $R = R_1 + R_2 + R_3 + R_4$ (5)

Resistances connected in series can therefore be understood to have an equivalent resistance across the supply that is equal to the sum of the individual resistances.

2.1.1. Current in a Series Circuit

Every component in a circuit that is connected in series receives the same amount of current. This is due to the fact that the current flow has only one path. Because electric charge moves through conductors like marbles through a tube, there must always be an equal marble speed or flow rate throughout the circuit (tube) at all times.

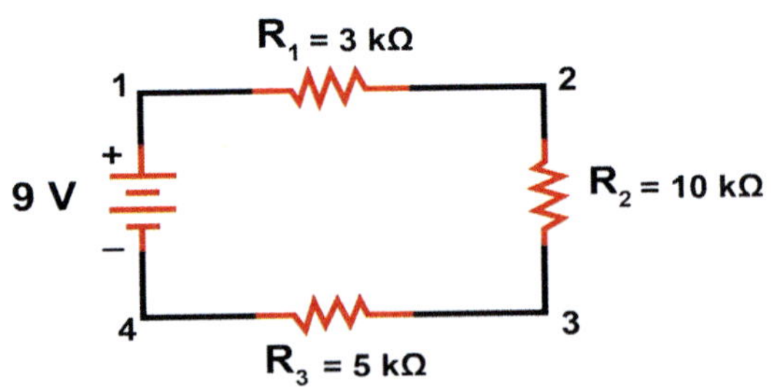

Figure 2.3. *Series circuit with a battery and three resistors.*

Source: https://www.allaboutcircuits.com/textbook/direct-current/chpt-5/simple-series-circuits/.

The requirement that all quantities (voltage, current, resistance, and power) relate to one another in terms of the same two points in a circuit is a crucial qualification to Ohm's law. Let's look at this idea for a circuit with a single resistor before moving on to the more intricate series circuit shown in Figure 2.3.

2.1.1.1. Using Ohm's Law in a Single Resistor Circuit

In this preliminary analysis, we will assess the voltage and current for the circuit with a single resistor shown in Figure 2.4.

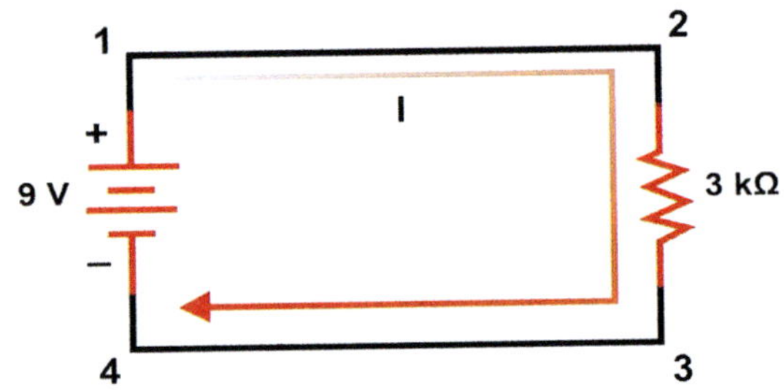

Figure 2.4. *Series circuit with a battery and a single resistor.*

Source: https://www.allaboutcircuits.com/textbook/direct-current/chpt-5/simple-series-circuits/.

We can state that point 1 is electrically common to point 2 and that point 3 is electrically common to point 4 because points 1 and 2 as well as points 3 and 4 are connected together with a wire of negligible resistance.

The circuit must drop 9 V between points 2 and 3 (directly across the resistor) since it has 9 V of electromotive force between points 1 and 4 (directly across the battery). This is due to the requirement of Kirchhoff's Voltage Law, which states that the total of all loop voltages must equal zero.

Since we are aware of the voltage (V) across the resistor and its resistance (R), we can use Ohm's Law (I = V/R) to calculate the current flowing through it. We can apply Ohm's law formula without any reservations because all terms (V, I, and R) relate to the same two points in the circuit as well as that resistor:

$$I = \frac{V}{R} \quad (6)$$

$$I = \frac{9\ \text{V}}{3\ \text{k}\Omega} = 3\ \text{mA} \quad (7)$$

2.1.1.2. Using Ohm's Law for Series Circuits with Multiple Resistors

Looking back at the circuit in Figure 2.3, we can observe that the 9 V battery's polarity will once more cause a current, I, to flow from point 1 to point 2 to point 3 to point 4 and back to point 1. Figure 2.5 provides an illustration of this idea.

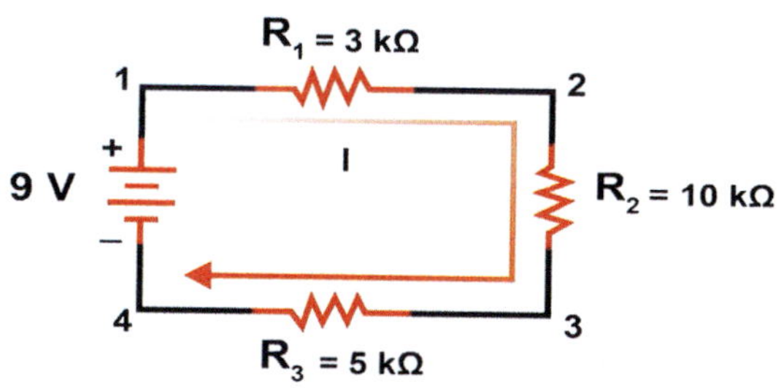

Figure 2.5. *Current in a series circuit with a battery and three resistors.*

Source: https://www.allaboutcircuits.com/textbook/direct-current/chpt-5/simple-series-circuits/.

Three resistances and one voltage source, though, are what we have. From there, we could think about applying Ohm's law in this case.

The electromotive force driving the current through the series combination of R_1, R_2, and R_3 in the three-resistor example circuit of Figure 2.5 is 9 V, which is present between points 1 and 4. To attempt to determine the current value, we are unable to divide the 9 V value by 3 kΩ, 10 kΩ, or 5 kΩ since we are unsure of the voltage across each of those resistors separately.

While the values of 3 kΩ, 10 kΩ, or 5 kΩ correspond to the individual resistors, the voltage value of 9 V represents the total quantity for the entire circuit. The result would not precisely correspond to any quantity in the actual circuit if we were to enter a value for individual resistance and a value for total voltage into an Ohm's law equation.

For R_1, Ohm's law will relate the amount of voltage across R_1 with the current through R_1, given R_1's resistance of 3 kΩ:

$$I_{R1} = \frac{V_{R1}}{3\ \text{k}\Omega} \quad (8)$$

$$E_{R1} = I_{R1} \ldots 3\ \text{k}\Omega \quad (9)$$

However, we are unable to perform any calculations using either formula because we do not know the voltage across R_1 (only the total voltage supplied by the battery across the three-resistor series combination) or the current flowing through R_1. For R_2 and R_3, the same holds true: we can only use Ohm's law equations if every term represents the quantity between the same two points in the circuit. We know the resistance of each resistor and the voltage of the source (9 V) applied across the series combination of R_1, R_2, and R_3. So, what can we do? We are unable to calculate the circuit current using Ohm's law, though, because those quantities aren't in the same context. We could use our value for total voltage (I = V/R) to calculate the total current if we only knew the circuit's total resistance.

2.1.2. How to Calculate Total Resistance in a Series Circuit

This leads us to the second principle of series circuits: the sum of the individual resistances in a series circuit equals the total resistance of the circuit.

This should make intuitive sense; in essence, the more resistors in series the current must pass through, the more challenging the current will be to flow.

In the example problem, we had a 3 kΩ, 10 kΩ, and 5 kΩ resistors in series, giving us a total resistance of 18 kΩ:

$$R_{total} = R_1 + R_2 + R_3 \quad (10)$$

$$R_{total} = 3\ k\Omega + 10\ k\Omega + 5\ k\Omega \quad (11)$$

$$R_{total} = 18\ k\Omega \quad (12)$$

Essentially, what we have determined is the total equivalent resistance of R_1, R_2, and R_3 taken together. With this knowledge, we can recreate the circuit (Figure 2.6) using just one equivalent resistor to stand in for the R_1, R_2, and R_3 series combination.

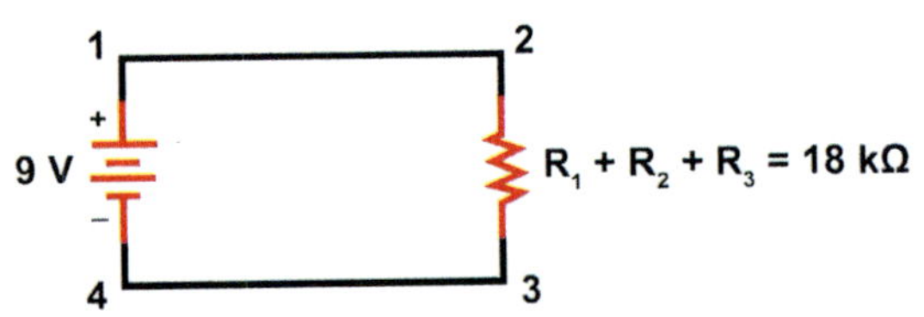

Figure 2.6. *Circuit showing total equivalent resistance of series resistors.*

Source: https://www.allaboutcircuits.com/textbook/direct-current/chpt-5/simple-series-circuits/.

2.1.2.1. Using Ohm's Law to Calculate Circuit Current in a Series Circuit

After completing all of those computations, we now have the voltage between points 1 and 4 (9 V) and the resistance between points 1 and 4 (18 kΩ), which allows us to compute the circuit current for Figure 2.6.

$$I_{total} = \frac{V_{total}}{R_{total}} \quad (13)$$

$$I_{total} = \frac{9\ V}{18\ k\Omega} = 500\ \mu A \quad (14)$$

2.1.3. How to Calculate Voltage Drop in a Series Circuit

We can refer back to our initial circuit schematic of Figure 2.6 and observe the current through each component, as shown in Figure 2.7, knowing that current flows equally through all of a series circuit's components (we have just found the current through the battery).

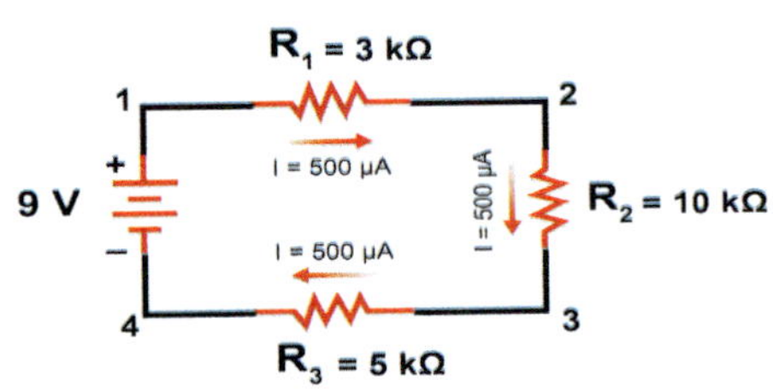

Figure 2.7. *Calculated current for the series circuit.*

Source: https://www.allaboutcircuits.com/textbook/direct-current/chpt-5/simple-series-circuits/.

We can use Ohm's law to calculate the voltage drop across each resistor now that we know how much current flows through them (using Ohm's law appropriately):

$$V_{R1} = I_{R1} \cdot R_1 = (500\ \mu A) \cdot (3\ k\Omega) = 1.5 \quad (15)$$

$$V_{R2} = I_{R2} \cdot R_2 = (500\ \mu A) \cdot (10\ k\Omega) = 5.0 \quad (16)$$

$$V_{R3} = I_{R3} \cdot R_3 = (500\ \mu A) \cdot (5\ k\Omega) = 2.5 \quad (17)$$

Notice that sum of the voltage drops (1.5 + 5.0 + 2.5 = 9.0 V) is equal to the battery (supply) voltage of 9 V.

This is the third principle of series circuits—the total voltage drop in a series circuit equals the sum of the individual voltage drops.

2.1.4. Applications of Series Circuit

Series circuits have various applications, including:

- **Lighting Circuits:** The lighting boards have a series circuit because the bulbs are connected in series there. Any incident, like one lightbulb burning out, will cause an electrical circuit, which will break the connection and turn off all of the bulbs. Older holiday light strings frequently have this feature.
- **Voltage Dividers:** When using a voltage divider circuit, which has an output voltage that is a fraction of the input voltage, a series circuit is frequently used. Using a couple of resistors to create a series circuit that produces a voltage across a single resistor, this technique is achieved.
- **Temperature Sensors:** Temperature sensors are a practical illustration of a series circuit; the true temperature of a temperature-variable resistor (thermistor) is measured by measuring the change in resistance of the resistor. The circuit contains two thermistors. The thermistor is connected to the voltage divider circuit and the fixed resistor in series.
- **Battery Packs:** One tool for raising the battery's overall output voltage is a series circuit. As you connect the batteries in series, the voltage gradually rises. In this manner, a high output voltage can be obtained, which has numerous uses, such as in mobile electronics and electric vehicles.
- **Electrical Safety Devices:** Circuit breakers and fuses are examples of series circuits. The latter are circuit-protecting devices that, upon detection of an overload or short circuit, interrupt the circuit to preserve its constituent parts.

2.2. PARALLEL CIRCUITS

Electric current can travel through several paths in a parallel circuit. The current can flow through it in two or more paths. One of the two circuit configurations that give the impression of being in parallel is a parallel circuit, where all of the devices share an end point.

Each resistor offers a distinct path for current when they are connected in parallel and a voltage is applied across the parallel circuit. As more resistors are connected in parallel, the total resistance of the circuit is lowered. The voltage applied across the parallel circuit as a whole is equal to the voltage across each parallel resistor.

When current flows from a node to several paths that eventually converge at a common node, the circuit is said to be parallel. The name "parallel circuits" comes from the path's apparent parallel appearance caused by the branching. Each branch has a different current flow because of branching, but it's crucial to remember that every branch has the same potential difference—that is, the potential drop between the two node points.

The current in these circuits is first branched out before being recombined at the common point. Elements in a parallel circuit aren't connected end to end.

2.2.1. Key Principles of Parallel Circuit

Here are the three key principles of laws of the parallel circuit:

- **Voltage**: In a parallel circuit, voltage is significant. The voltage is the same across all parts. This indicates that there is always the same voltage drop in each branch of a parallel circuit.
- **Resistance**: Reduced resistances add up to a smaller overall resistance. This indicates that as more components are added, the circuit's overall resistance decreases, leading to a smaller overall resistance.
- **Current**: A higher total current is obtained by adding the branch currents. This indicates that the total current in the circuit increases when we add the current of each individual branch. Using Ohm's Law for Parallel Circuits to Determine Current.

2.2.2. Working of Parallel Circuit

The parallel circuit looks like any other circuit with the addition of branching as shown below:

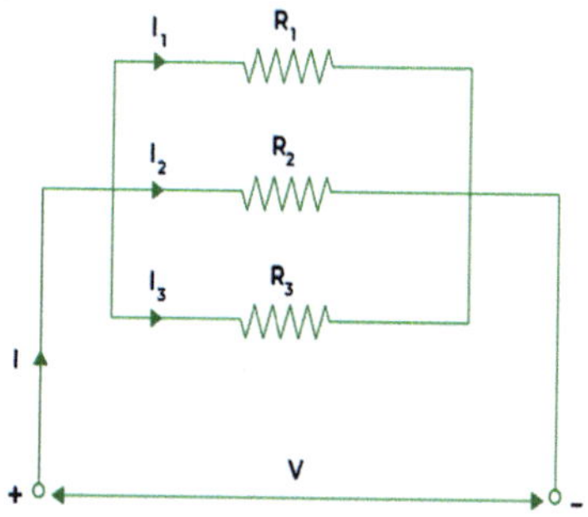

Figure 2.8. *Parallel circuits.*

Source: https://www.geeksforgeeks.org/parallel-circuits/.

In a parallel circuit, the current flowing in the circuit is equal to the sum of current in the individual branch. We will apply this in the above circuit.

Hence; $I = I_1 + I_2 + I_3$

Since the voltage across each branch is the same we write $I = V/R_i$

$$I = V/R_1 + V/R_2 + V/R_3 \quad (18)$$

Let R_{eq} be the equivalent Resistance of the circuit then

$$V/R_{eq} = V/R_1 + V/R_2 + V/R_3 \quad (19)$$

$$1/R_{eq} = 1/R_1 + 1/R_2 + 1/R_3 \quad (20)$$

Hence equivalent resistance of the circuit is;

$$1/R_{eq} = 1/R_1 + 1/R_2 + 1/R_3 \quad (21)$$

2.2.3. Characteristics of Parallel Circuits

Some characteristics of parallel circuits are mentioned below:

- Instead of being connected end-to-end, each device joins the same two spots i.e., there are two common nodes where each branch starts and ends. Between these two common points, there can be multiple branches
- Voltage is the same across each branch of the parallel circuit equal to the potential drop between the two nodes.
- The sum of the currents through each branch is equal to the total current that flows from the source. This means current $I = I_1 + I_2 + ... I_n$.
- The Total circuit resistance decreases as the number of branches increases. This can be shown mathematically, equivalent resistance of a parallel circuit is less than the smallest individual resistance in the circuit.
- Each component in a parallel circuit works on its own i.e., it operates independently. If one component gets damaged or is removed, the others will continue to function without having any effect.

2.2.4. Advantages of Parallel Circuits

The advantages of parallel circuits are:

- **Independent Components:** In a parallel circuit, all components operate independently. Since the components have an independent voltage they work even when a component in any other branch is not working.
- **Different Current Distribution:** In the real world, every appliance has its current requirement depending on its rating. In a parallel circuit, every branch may have a different

current flowing through it thereby allowing different current distribution.

- **Stable Voltage:** Each device in a parallel circuit receives constant voltage, this stable running voltage is the same across each branch ensuring a stable circuit.
- **Low Resistance:** In a parallel circuit, the equivalent resistance is less than the smallest individual resistance. As we know current is inversely proportional to resistance. This can result in increased current flow, hence being useful when a high current is required.
- **Low Complexity:** Parallel circuits are easy to design due to their low complexity and are reliable circuits.

2.2.5. Disadvantages of Parallel Circuits

The disadvantages of parallel circuits are:

- **Additional Cost:** Due to additional components in parallel circuits as compared to series circuits, parallel circuits have an additional cost associated with purchasing components.
- **Power Consumption:** Due to more components in parallel circuits than series circuits, the power consumption of these circuits can increase drastically. This is common when multiple components are operating simultaneously.
- **Complex Diagnosis:** In a parallel circuit, identifying the source of an error can be more difficult as compared to a series circuit because there are multiple paths for current to flow hence multiple sources of error.

2.2.6. Applications of Parallel Circuits

There is a need for Parallel circuits because they have various applications in different fields some of which are given below:

- Parallel circuits form the major circuit of housing outlets. Since there are different appliances in the household, a parallel circuit ensures each appliance is getting an equal and stable voltage supply.
- Parallel circuits are commonly used in a factory. If someone shuts off a conveyor belt, it's not going to impact the performance of any other piece of equipment on the factory floor like rollers will still be working. This way we can run the industry in case one equipment fails.
- Security systems also utilize parallel circuits. Earlier security cameras were installed using series circuits but parallel circuits have overpowered them now due to the errors in series circuits. One major issue with a series circuit was that if one part of the circuit fails, the other parts will fail as well resulting in a failed security system. Thereby we use parallel circuits for this.
- Parallel circuits are also used within automobile units, specifically for wiring car headlights. The effectiveness of the other lights is unaffected by a single damaged tail light, ensuring safety.

2.2.7. Difference Between Series and Parallel Circuit

The Table 2.1 represents the difference between series and parallel circuits

Table 2.1. Difference Between Series and Parallel Circuits

Parameter	**Series Circuit**	**Parallel Circuit**
Voltage distribution	Voltage across each component may not necessarily same	The voltage across each branch is necessarily the same
Current distribution	Current across each component is necessarily the same	Current is divided so current across each component may not be necessarily the same
Overall Resistance	Total Resistance has a larger value than the maximum resistance	Overall Resistance has less value than the minimum resistance
Alignment	In this electrical circuit, components are arranged in a line	In this electrical circuit, components are arranged parallel to each other
Dependency	If one component in the Circuit breaks down, the whole circuit will get damaged.	Other components will function even if one component breaks down since components are independent
Example	An example of a series circuit is a string of Diwali lights. If any one of the bulbs gets damaged, no current will flow and none of the lights will go on.	Parallel circuits are like the smaller veins that divide into branches from our heart and then connect to other parts to return blood to the heart.

2.3. RESISTORS IN PARALLEL

In a circuit, resistors are connected in parallel when two or more of them are connected individually between two different points, or nodes. Multiple paths for current flow are provided by a parallel circuit.

A parallel circuit is one that has multiple branches, each of which is referred to as a current path. Figure 2.9(a) depicts two resistors connected in parallel. When the current leaves the source (I_T) and reaches point A, it splits, as part (b) illustrates. I_1 passes through R_1, and I_2 passes via R_2. More current paths are available between point A and point B if additional resistors are connected in parallel with the first two, as seen in Figure 2.9(c). Every point along the top, represented by the color blue, is electrically equivalent to point A, and every point along the bottom, represented by the color green, is electrically equivalent to point B.

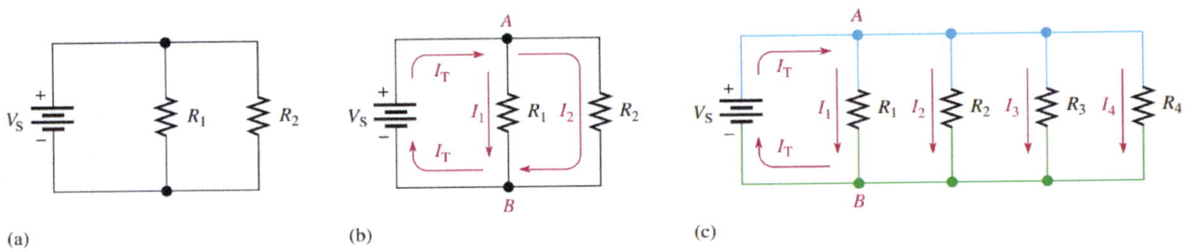

Figure 2.9. *Resistors in parallel.*

Source: Thomas L. Floyd. Principles of Electric Circuits. Pearson. ISBN: 9781292025667.

It is clear that the resistors are connected in parallel in Figure 2.9. In real circuit diagrams, the parallel relationship is frequently not as obvious. It is imperative that you acquire the ability to identify parallel circuits, irrespective of their drawing style.

A rule for identifying parallel circuits is as follows:

There is a parallel circuit between two points if there are multiple current paths, or branches, connecting them and if the voltage across each branch is present between the two points.

Parallel resistors drawn in various configurations between two distinct points, A and B, are shown in Figure 2.10. Notably, there are two routes for the current to travel from point A to point B in each instance, and the voltage across each branch is the same. There can be any number of resistors in parallel, even though these examples only display two parallel paths.

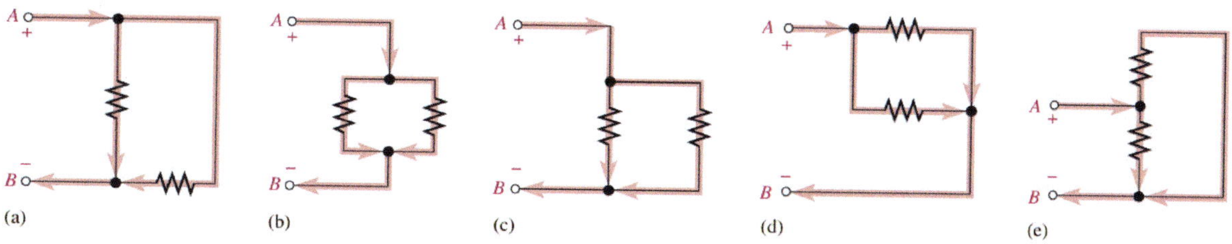

Figure 2.10. *Examples of circuits with two parallel paths.*

Source: Thomas L. Floyd. Principles of Electric Circuits. Pearson. ISBN: 9781292025667.

Example 1

The figure below illustrates the placement of five resistors on a protoboard. Align all of the resistors in parallel between A and B using the wiring that is needed. Put the values of the resistors on each label and draw a schematic.

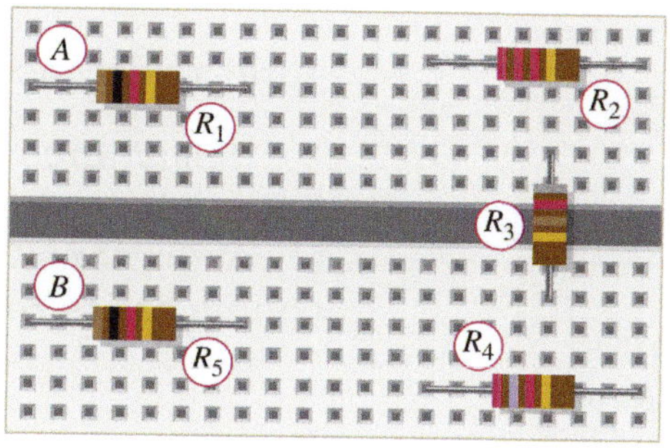

Source: Thomas L. Floyd. Principles of Electric Circuits. Pearson. ISBN: 9781292025667.

- **Solution:** The wiring diagram in figure (a) below illustrates how the wires are connected. Figure below (b) displays the schematic. Remember that the resistor arrangement as it is in reality need not always be depicted in the schematic. The schematic illustrates the electrical connections between the parts.

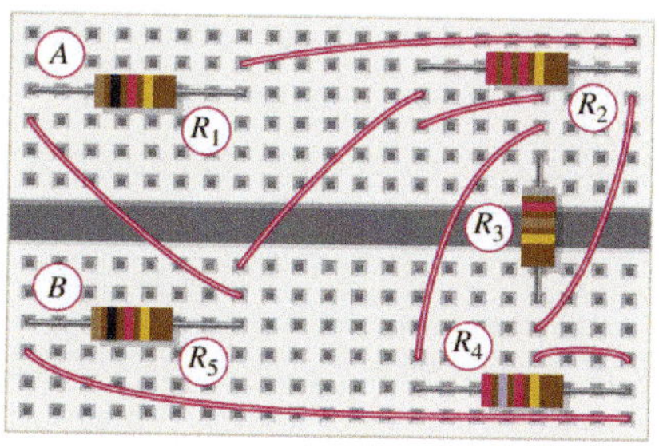

(a) Assembly wiring diagram

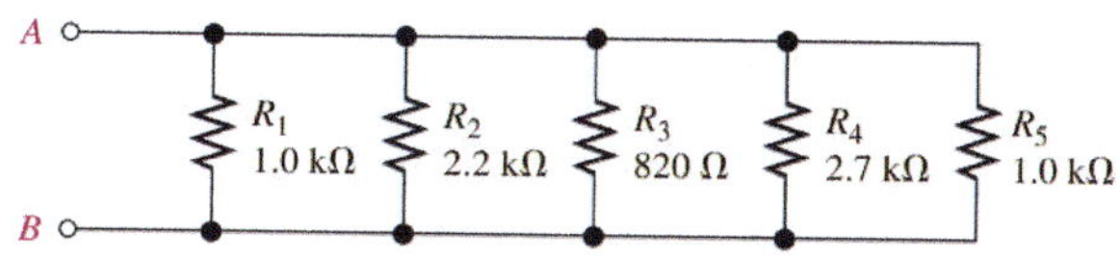

(b) Schematic

Source: Thomas L. Floyd. Principles of Electric Circuits. Pearson. ISBN: 9781292025667.

Example 2

The following are tools to calculate the ohm value and tolerance based on resistor color codes, the total resistance of a group of resistors in parallel or in series, and the resistance of a conductor based on size and conductivity.

Resistor color code calculator: Use this calculator to find out the ohm value and tolerance based on resistor color codes.

Result

52 MΩ ±5% (J)

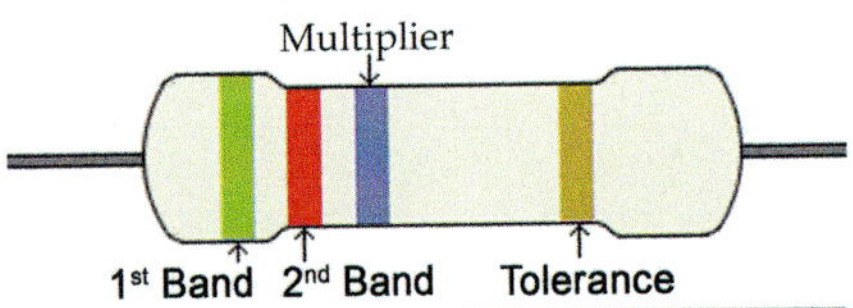

Color	1st 2nd Band Significant Figures	Multiplier	Tolerance
Black	0	x 1	
Brown	1	x 10	±1% (F)
Red	2	x 100	±2% (G)
Orange	3	x 1K	±0.05% (W)
Yellow	4	x 10K	±0.02% (P)
Green	5	x 100K	±0.5% (D)
Blue	6	x 1M	±0.25% (C)
Violet	7	x 10M	±0.1% (B)
Grey	8	x 100M	±0.01% (L)
White	9	x 1G	
Gold		x 0 1	±5% (J)
Silver		x 0.01	±10% (K)

Source: https://www.calculator.net/resistor-calculator.html

Parallel Resistor Calculator: Provide all of the resistance values in parallel, separated by a comma "," and click the "Calculate" button to determine total resistance.

Result

Total resistance in parallel is 1.2703136679494.

Resistors in Series Calculator: Provide all of the resistance values in series separated by a comma "," and click the "Calculate" button to determine total resistance.

Result

Total resistance in series is 155.

Resistance of a Conductor: Use the following to calculate the resistance of a conductor. This calculator assumes the conductor is round.

Result

Total resistance of the conductor is 0.40371174065481 ohm (Ω).

2.4. VOLTAGE IN A PARALLEL CIRCUIT

Any branch in a parallel circuit has a voltage across it that is equal to the voltage across all the other branches in the circuit. As you are aware, in a parallel circuit, every current path is referred to as a branch. Let's look at Figure 2.11(a) to see voltage in a parallel circuit illustrated. Because the voltage is the same along this line, points A, B, C, and D on the left side of the parallel circuit are electrically identical. You can imagine that a single wire connects each of these points to the battery's negative terminal. The voltage at each of the points E, F, G, and H on the right side of the circuit is the same as the voltage at the source's positive terminal. As a result, the voltage across every parallel resistor is the same and equal to the voltage at the source. Take note of how Figure 2.11's parallel circuit resembles a ladder.

The circuit shown in Part (a) of Figure 2.11(b) is the same, but it is drawn slightly differently. Here, the negative battery terminal serves as the single point to which the left side of each resistor is connected. Every resistor has a single point of connection, the positive battery terminal, on its right side. All of the resistors across the source are still connected in parallel.

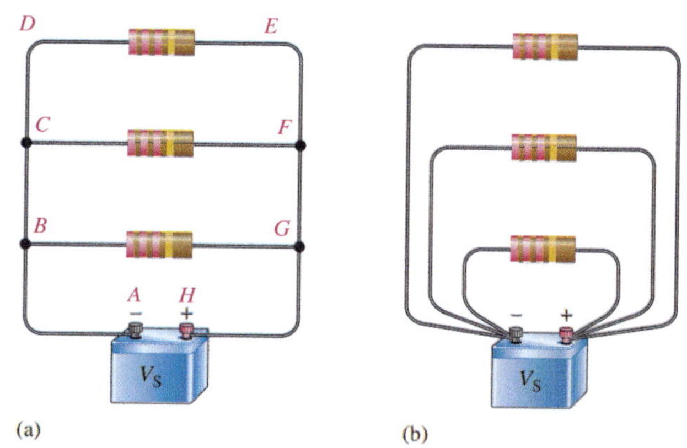

Figure 2.11. *Voltage across parallel branches is the same.*

Source: Thomas L. Floyd. Principles of Electric Circuits. Pearson. ISBN: 9781292025667.

Three parallel resistors are connected across a 12 V battery in Figure 2.12. The values obtained from measuring the voltage across the battery and then across each resistor are identical. As you can see, in a parallel circuit, every branch exhibits the same voltage.

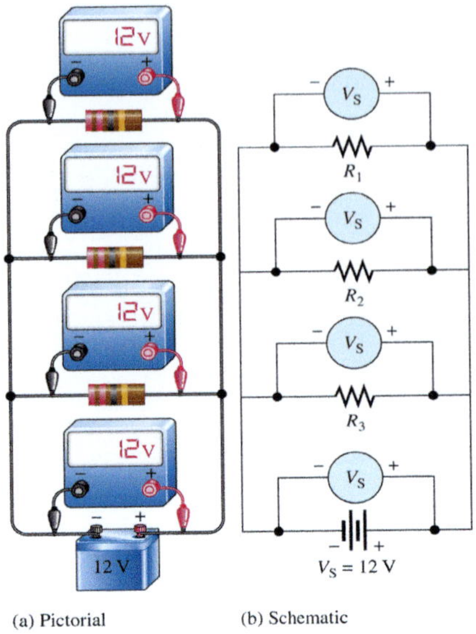

Figure 2.12. *The same voltage appears across each resistor in parallel.*

Source: Thomas L. Floyd. Principles of Electric Circuits. Pearson. ISBN: 9781292025667.

Example 3

Determine the voltage across each resistor in figure below.

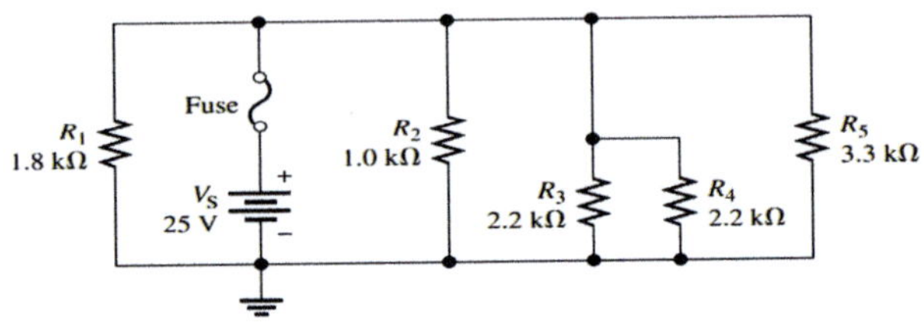

Source: Thomas L. Floyd. Principles of Electric Circuits. Pearson. ISBN: 9781292025667.

- **Solution:** The five resistors are in parallel, so the voltage across each one is equal to the applied source voltage. There is no voltage across the fuse. The voltage across the resistors is;

$$V_1 = V_2 = V_3 = V_4 = V_5 = V_S = 25 \text{ V} \quad (22)$$

2.5. KIRCHHOFF'S CURRENT LAW

Regarding voltages in a single closed path, Kirchhoff's voltage law applies. Kirchhoff's law of currents in multiple paths is applicable (Athavale Prashant, 2018).

Kirchhoff's current law, often abbreviated KCL, can be stated as follows:

A node's total current in (total current in) and total current out (total current out) are the same.

Any junction or point in a circuit where two or more components are connected is called a node. A node, also known as a junction, is where the parallel branches of a circuit come together. Point A and Point B are two examples of nodes in the circuit shown in Figure 2.13. Let's follow the current and begin at the source's positive terminal. Node A is receiving all of the source's current IT. This is where the current divides into the three branches as shown. Node A is the source of all three branch currents (I_1, I_2, and I_3). Kirchhoff's current law says that the total current into node A is equal to the total current out of node A; that is;

$$I_T = I_1 + I_2 + I_3 \qquad (23)$$

The currents in Figure 2.13 now show that they re-join at node B after passing through the three branches. Nodes I_1, I_2, and I_3 are receiving currents, while node B is receiving IT. Therefore, at node B, Kirchhoff's current law formula is the same as it is at node A.

$$I_T = I_1 + I_2 + I_3 \qquad (24)$$

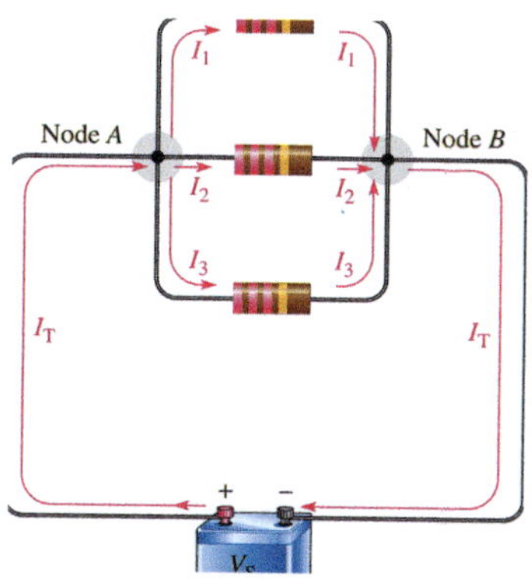

Figure 2.13. *Kirchhoff's current law: The current into a node equals the current out of that node.*

Source: Thomas L. Floyd. Principles of Electric Circuits. Pearson. ISBN: 9781292025667.

Figure 2.14 illustrates the general case of Kirchhoff's current law and can be written as a mathematical relationship:

$$I_{IN(1)} + I_{IN(2)} + I_{IN(3)} + \cdots + I_{IN(n)} = I_{OUT(1)} + I_{OUT(2)} + I_{OUT(3)} + \cdots + I_{OUT(m)} \quad (25)$$

Moving the terms on the right side to the left side and changing the sign results in the following equivalent equation:

$$I_{IN(1)} + I_{IN(2)} + I_{IN(3)} + \cdots + I_{IN(n)} - I_{OUT(1)} - I_{OUT(2)} - I_{OUT(3)} - \cdots - I_{OUT(m)} = 0 \quad (26)$$

This equation shows that all current into and out of the junction sums to zero and can be stated as;

The algebraic sum of all of the currents entering and leaving a node is equal to zero.

Kirchhoff's current law can be written in an equivalent manner using mathematical summation shorthand.

$$\sum_{i=1}^{n} I_i = 0 \quad (27)$$

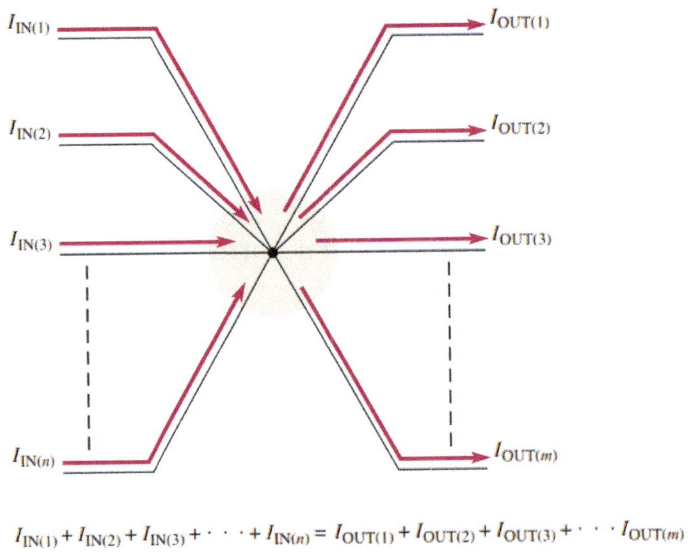

Figure 2.14. *Generalized circuit node illustrating Kirchhoff's current law.*

Source: Thomas L. Floyd. Principles of Electric Circuits. Pearson. ISBN: 9781292025667.

This method of expressing Kirchhoff's current law involves giving each current a sequential subscript (1, 2, 3, and so on), regardless of whether the current is entering or exiting the node. Positive currents enter the node, and negative currents exit the node. As shown in Figure 2.15, you can confirm Kirchhoff's current law by connecting a circuit and measuring the current flowing through each branch as well as the total current coming from the source. The total current will equal the sum of the branch currents when they are added together. Any number of branches are covered by this rule (Oldham, Kalil T. Swain (2008).

The following four examples illustrate use of Kirchhoff's current law.

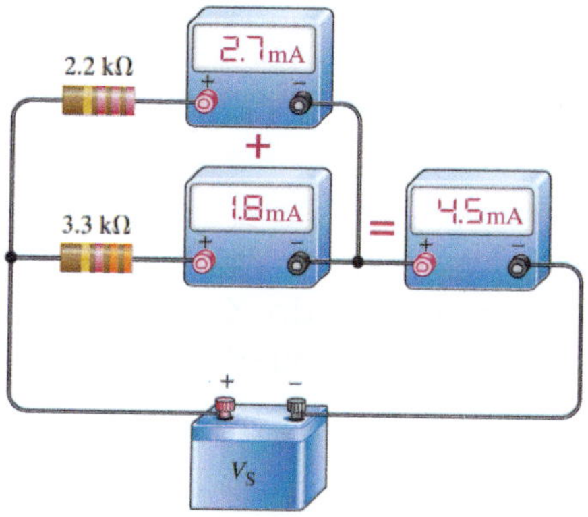

Figure 2.15. *An illustration of Kirchhoff's current law.*

Source: Thomas L. Floyd. Principles of Electric Circuits. Pearson. ISBN: 9781292025667.

Example 4

The branch currents are shown in the circuit of figure below. Determine the total current entering node A and the total current leaving node B.

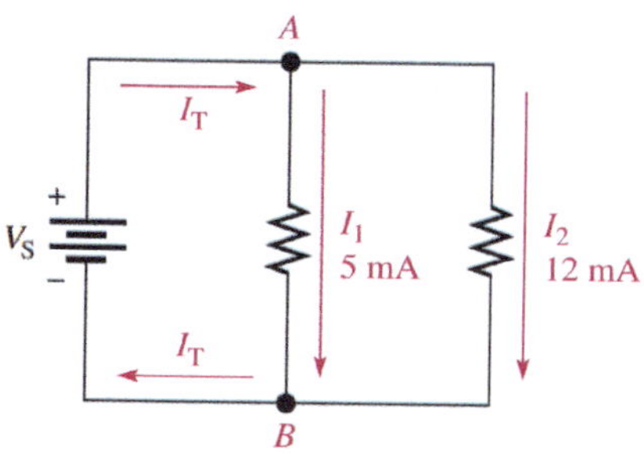

Source: Thomas L. Floyd. Principles of Electric Circuits. Pearson. ISBN: 9781292025667.

Solution

The total current out of node A is the sum of the two branch currents. So the total current into node A is;

$$I_T = I_1 + I_2 = 5 \text{ mA} + 12 \text{ mA} = 17 \text{ mA} \quad (28)$$

The total current entering node B is the sum of the two branch currents. So, the total current out of node B is;

$$I_T = I_1 + I_2 = 5 \text{ mA} + 12 \text{ mA} = 17 \text{ Ma} \quad (29)$$

Example 5

Determine the current I_2 through R_2 in figure below.

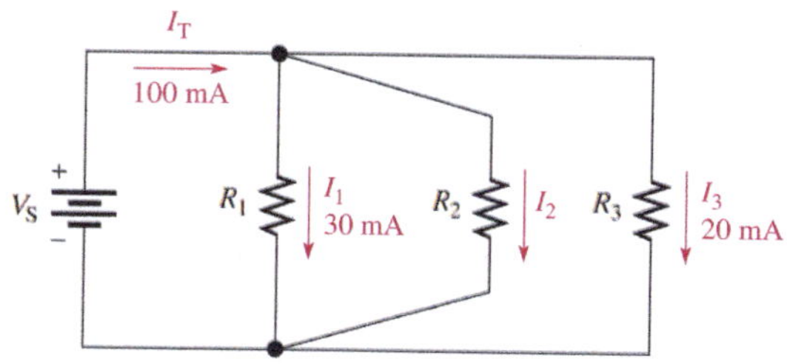

Source: Thomas L. Floyd. Principles of Electric Circuits. Pearson. ISBN: 9781292025667.

Solution

The total current into the junction of the three branches is $I_T = I_1 + I_2 + I_3$. From this figure, you know the total current and the branch currents through R_1 and R_3. Solve for I_2 as follows:

$$I_2 = I_T - I_1 - I_3 = 100 \text{ mA} - 30 \text{ mA} - 20 \text{ mA} = 50 \text{ mA} \quad (30)$$

Example 6

Use Kirchhoff's current law to find the current measured by ammeters A3 and A5 in figure below.

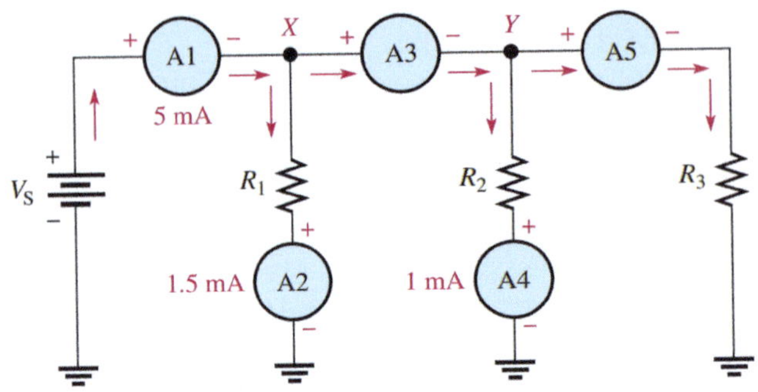

Source: Thomas L. Floyd. Principles of Electric Circuits. Pearson. ISBN: 9781292025667.

Solution

The total current into node X is 5 mA. Two currents are out of node X: 1.5 mA through resistor R_1 and the current through A3. Kirchhoff's current law applied at node X gives;

$$5 \text{ mA} - 1.5 \text{ mA} - I_{A3} = 0 \tag{31}$$

Solving for I_{A3} yields

$$I_{A3} = 5 \text{ mA} - 1.5 \text{ mA} = 3.5 \text{ mA} \tag{32}$$

The total current into node Y is I_{A3} = 3.5 mA. Two currents are out of node Y: 1 mA through resistor R_2 and the current through A5 and Kirchhoff's current law applied at node Y gives

$$3.5 \text{ mA} - 1 \text{ mA} - I_{A5} = 0 \tag{33}$$

Solving for I_{A5} yields

$$I_{A5} = 3.5 \text{ mA} - 1 \text{ mA} = 2.5 \text{ mA} \tag{34}$$

Example 7

Using Kirchhoff's current law, find the current in each of two taillights given that the total current from the battery is 8.0 A and together the headlights draw 5.6 A.

Solution

$$I_{BAT} - I_{T(HEAD)} - I_{T(TAIL)} = 0 \tag{35}$$

Rearranging and changing the signs,

$$I_{T(TAIL)} = I_{BAT} - I_{T(HEAD)} = 8.0 \text{ A} - 5.6 \text{ A} = 2.4 \text{ A} \tag{36}$$

Because the taillights are identical, each draws 2.4 A/2 = 1.2 A.

2.6. TOTAL PARALLEL RESISTANCE

The circuit's overall resistance drops when resistors are connected in parallel. In a parallel circuit, the total resistance is always less than the value of the smallest resistor. For instance, the total resistance is less than 10 Ω when two resistors, 10 Ω and 100 Ω, are connected in parallel.

As you are aware, there are multiple paths for the current to travel when resistors are connected in parallel. The number of parallel branches and the number of current paths are equal.

Because the circuit in Figure 2.16(a) is a series circuit, there is only one current path. A specific amount of current, I_1, flows through R_1. As seen in Figure 2.16(b), when resistor R_2 is connected in parallel with R_1, more current, I_2, flows through R_2. With the parallel resistor added, the source's overall current has increased. According to Ohm's law, an increase in the total current coming from the source indicates a decrease in the total resistance, assuming that the source voltage remains constant. Further parallel connections between resistors will lower resistance even more while raising total current.

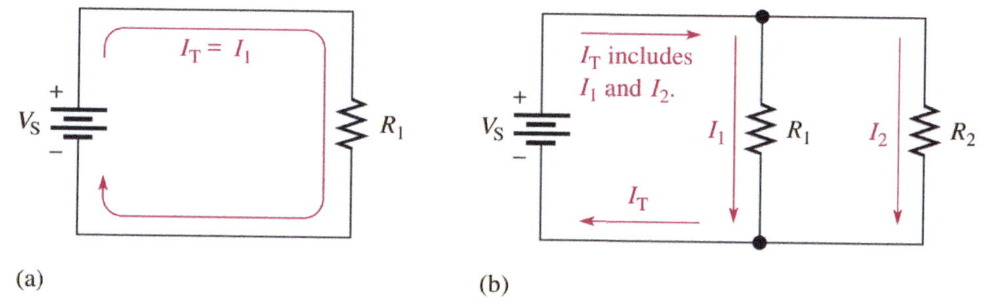

Figure 2.16. *Addition of resistors in parallel reduces total resistance and increases total current.*

Source: Thomas L. Floyd. Principles of Electric Circuits. Pearson. ISBN: 9781292025667.

2.6.1. Formula for Total Parallel Resistance

A general example of n resistors connected in parallel is shown in the circuit in Figure 2.17 (n can be any integer greater than 1). From Kirchhoff's current law, the equation for current is;

$$I_T = I_1 + I_2 + I_3 + \cdots + I_n \tag{37}$$

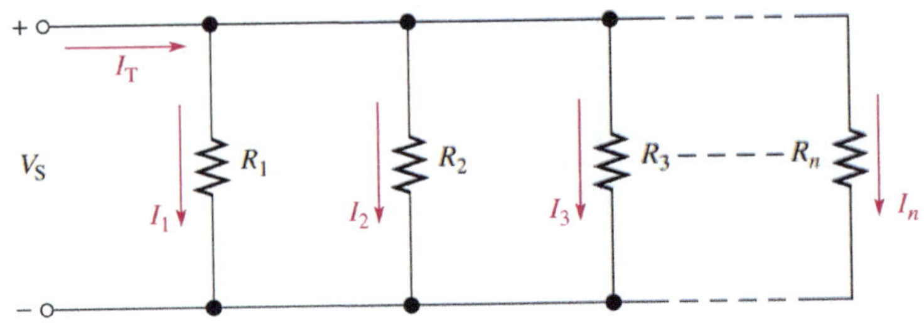

Figure 2.17. *Circuit with n resistors in parallel.*

Source: Thomas L. Floyd. Principles of Electric Circuits. Pearson. ISBN: 9781292025667.

Since V_S is the voltage across each of the parallel resistors, by Ohm's law, $I_1 = V_S/R_1$, $I_2 = V_S/R_2$, and so on. By substitution into the equation for current;

$$\frac{V_S}{R_T} = \frac{V_S}{R_1} + \frac{V_S}{R_2} + \frac{V_S}{R_3} + \cdots + \frac{V_S}{R_n} \tag{38}$$

Only the resistance terms remain after factoring out the term V_S from the right side of the equation and canceling it with V_S on the left.

$$\frac{1}{R_T} = \frac{1}{R_1} + \frac{1}{R_2} + \frac{1}{R_3} + \cdots + \frac{1}{R_n} \tag{39}$$

Remember that conductance, represented by G, is the reciprocal of resistance (1/R). The Siemen is the unit of conductance (S). The equation for $1/R_T$ can be expressed in terms of conductance as

$$G_T = G_1 + G_2 + G_3 + \cdots + G_n \tag{40}$$

Solve for R_T by taking the reciprocal of (that is, by inverting) both sides of the equation for $1/R_T$.

$$R_T = \frac{1}{\left(\frac{1}{R_1}\right) + \left(\frac{1}{R_2}\right) + \left(\frac{1}{R_3}\right) + \cdots + \left(\frac{1}{R_n}\right)} \tag{41}$$

Eqn. (41) demonstrates that you must add up all of the 1/R (or conductance, G) terms and then take the reciprocal of the total to determine the total parallel resistance.

$$R_T = \frac{1}{G_T} \tag{42}$$

Example 8

Calculate the total parallel resistance between points A and B of the circuit in figure below.

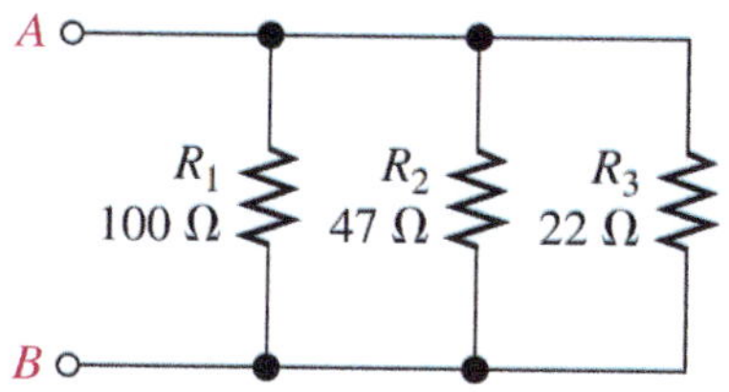

Source: Thomas L. Floyd. Principles of Electric Circuits. Pearson. ISBN: 9781292025667.

Solution

To calculate the total parallel resistance when you know the individual resistances, first find the conductance, which is the reciprocal of the resistance, of each of the three resistors.

$$G_1 = \frac{1}{R_1} = \frac{1}{100\ \Omega} = 10\text{ mS}$$
$$G_2 = \frac{1}{R_2} = \frac{1}{47\ \Omega} = 21.3\text{ mS}$$
$$G_3 = \frac{1}{R_3} = \frac{1}{22\ \Omega} = 45.5\text{ mS} \quad (43)$$

Next, calculate R_T by adding G_1, G_2, and G_3 and taking the reciprocal of the sum.

$$R_T = \frac{1}{G_T} = \frac{1}{10\text{ mS} + 21.3\text{ mS} + 45.5\text{ mS}} = \frac{1}{76.8\text{ mS}} = \mathbf{13.0\ \Omega} \quad (44)$$

For a quick accuracy check, notice that the value of R_T (13.0 Ω) is smaller than the smallest value in parallel, which is R_T (22 Ω) as it should be.

2.6.2. Calculator Tip

Eqn. (41) can be used to quickly solve the parallel-resistance formula on a calculator. Typically, you would enter R_1's value and then use the x^{-1} key to get its reciprocal. On certain calculators, the reciprocal is a secondary function. Afterwards, hit the + key. After entering R_2's value and finding its reciprocal with the x^{-1} key, hit the + key once more. Once all of the resistor values have been entered, repeat this process and hit ENTER. To obtain R_T, the last step is to press the x^{-1} and ENTER keys. On the display at this point is the total parallel resistance. Depending on the calculator in question, the display format may change. For example, the steps required for a typical calculator solution of Example 2.7 are as follows:

- Enter 100. Display shows 100.
- Press x^{-1} (or 2nd then x^{-1}). Display shows 100^{-1}.
- Press +. Display shows 100^{-1} + .
- Enter 47. Display shows $100^{-1} + 47$.
- Press x^{-1} (or 2nd then x^{-1}). Display shows $100^{-1} + 47^{-1}$.
- Press +. Display shows $100^{-1} + 47^{-1}$ + .
- Enter 22. Display shows $100^{-1} + 47^{-1} + 22$.
- Press x^{-1} (or 2nd then x^{-1}). Display shows $100^{-1} + 47^{-1} + 22^{-1}$.
- Press ENTER. Display shows a result of $76.7311411992_E{-3}$.
- Press x^{-1} (or 2nd then x^{-1}) and then ENTER. Display shows a result of 13.0325182758_E0.

The number displayed in Step 10 is the total resistance in ohms. Round it to 13.0 Ω.

2.6.3. The Case of Two Resistors in Parallel

A general formula for calculating the total resistance of any number of resistors connected in parallel is Eqn. (41). In practice, it is common to combine two resistors in parallel. As an alternative method of determining the R_T, resistors in parallel of any number can also be divided into pairs. Based on Eqn. (41), the formula for the total resistance of two resistors in parallel is:

$$R_T = \frac{1}{\left(\frac{1}{R_1}\right) + \left(\frac{1}{R_2}\right)} \quad (45)$$

Combining the terms in the denominator yields

$$R_T = \frac{1}{\left(\frac{R_1 + R_2}{R_1 R_2}\right)} \quad (46)$$

Which can be rewritten as follows:

$$R_T = \frac{R_1 R_2}{R_1 + R_2} \quad (47)$$

The total resistance for two resistors connected in parallel is equal to the product of the resistors divided by the total of the resistors, according to Eqn. (47).

The "product over the sum" formula is another name for this equation.

Example 9

Calculate the total resistance connected to the voltage source of the circuit in figure below.

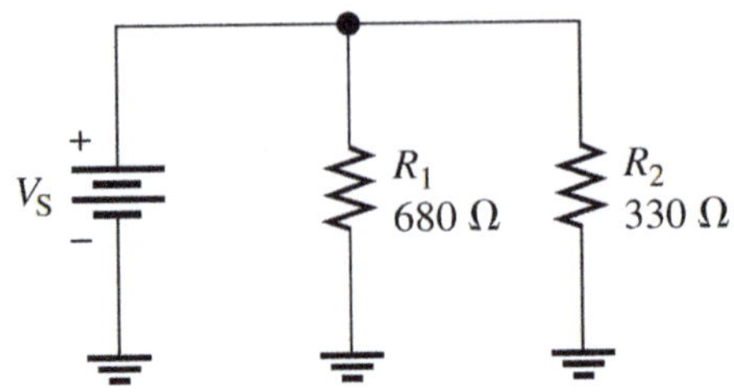

Source: Thomas L. Floyd. Principles of Electric Circuits. Pearson. ISBN: 9781292025667.

Solution

Use Eqn. (47).

$$R_T = \frac{R_1 R_2}{R_1 + R_2} = \frac{(680\ \Omega)(330\ \Omega)}{680\ \Omega + 330\ \Omega} = \frac{224{,}400\ \Omega^2}{1010\ \Omega} = 222\ \Omega \quad (48)$$

2.6.4. The Case of Equal-Value Resistors in Parallel

The parallel connection of multiple resistors with the same resistance value is another special case of a parallel circuit. When this situation arises, there is a quick way to calculate R_T.

If several resistors in parallel have the same resistance, they can be assigned the same symbol R. For example, $R_1 = R_2 = R_3 = \cdots = R_n = R$. Starting with Eqn. (41), you can develop a special formula for finding R_T.

$$R_T = \frac{1}{\left(\frac{1}{R}\right) + \left(\frac{1}{R}\right) + \left(\frac{1}{R}\right) + \cdots + \left(\frac{1}{R}\right)} \quad (49)$$

Observe that the same term, 1/R, is added n times to the denominator (n being the number of parallel equal-value resistors). Therefore, the formula can be written as:

$$R_T = \frac{1}{n/R} \quad (50)$$

or

$$R_T = \frac{R}{n} \quad (51)$$

The resistance divided by the number of resistors in parallel, or R_T, is what happens when any number of resistors (n) with the same resistance (R) are connected in parallel, according to Eqn. (51).

Example 10

Four 8 Ω speakers are connected in parallel to the output of an amplifier. What is the total resistance across the output of the amplifier?

Solution

There are four 8 Ω resistors in parallel. Use Equation 51 as follows:

$$R_T = \frac{R}{n} = \frac{8\,\Omega}{4} = 2\,\Omega \quad (52)$$

2.6.5. Determining an Unknown Parallel Resistor

There are situations when you need to figure out the resistor values that go together to create the desired total resistance. To get a known total resistance, for instance, you can use two resistors in parallel. Eqn. (47) for two parallel resistors can be used to determine the second resistor value if you know or choose at random the value of the first resistor. The formula for determining the value of an unknown resistor R_x is developed as follows:

$$\frac{1}{R_T} = \frac{1}{R_A} + \frac{1}{R_x}$$

$$\frac{1}{R_x} = \frac{1}{R_T} - \frac{1}{R_A}$$

$$\frac{1}{R_x} = \frac{R_A - R_T}{R_A R_T}$$

$$R_x = \frac{R_A R_T}{R_A - R_T} \quad (53)$$

Where; R_x is the unknown resistor and R_A is the known or selected value.

Example 11

Suppose that you wish to obtain a resistance as close to 150 Ω as possible by combining two resistors in parallel. There is a 330 Ω resistor available. What other value do you need?

Solution

R_T = 150 Ω and R_A = 330 Ω. Therefore;

$$R_x = \frac{R_A R_T}{R_A - R_T} = \frac{(330\,\Omega)(150\,\Omega)}{330\,\Omega - 150\,\Omega} = 275\,\Omega \quad (54)$$

The closest standard value is 270 Ω.

2.6.6. Notation for Parallel Resistors

Two parallel vertical marks are sometimes used to identify parallel resistors for ease of use. For instance, R_1 and R_2 can be written as $R_1 \parallel R_2$. Additionally, this notation can be applied when multiple resistors are connected in parallel. For example;

$$R_1 \| R_2 \| R_3 \| R_4 \| R_5 \quad (55)$$

Indicates that R_1 through R_5 are all in parallel.

This notation is also used with resistance values. For example;

$$10\,\text{k}\Omega \| 5\,\text{k}\Omega \quad (56)$$

Means that a 10 kΩ resistor is in parallel with a 5 kΩ resistor.

2.7. CURRENT SOURCES IN PARALLEL

An energy source that supplies a steady current to a load regardless of the load's resistance is known as a current source. In electronic circuits, current sources are crucial because a transistor can serve as a current source.

Generally speaking, the algebraic sum of the individual current sources equals the total current generated by current sources operating in parallel. When combining the sources in parallel, the algebraic sum requires you to take the current's direction into account. For instance, the three parallel current sources in figure below (a) supply current in the same direction (into node A). So the total current into node A is;

$$I_T = 1\ A + 2\ A + 2\ A = 5\ A \qquad (57)$$

In figure below (b), the 1 A source provides current in a direction opposite to the other two. The total current into node A in this case is;

$$I_T = 2\ A + 2\ A - 1\ A = 3\ A \qquad (58)$$

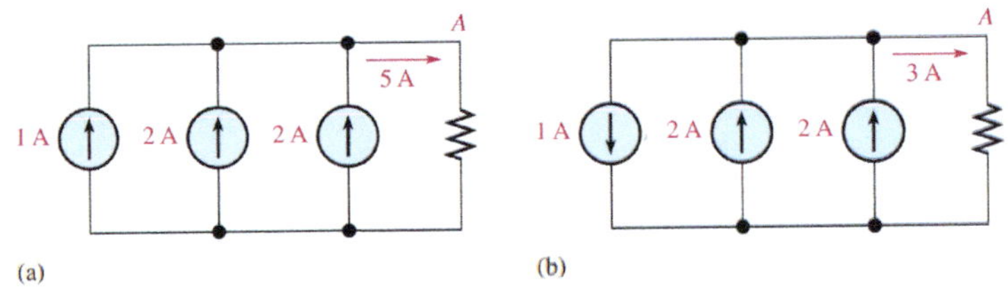

Source: Thomas L. Floyd. Principles of Electric Circuits. Pearson. ISBN: 9781292025667.

Example 12

Determine the current through R_L in figure below.

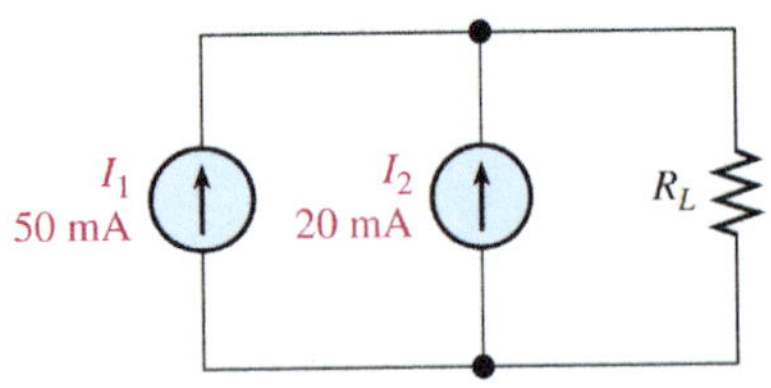

Source: Thomas L. Floyd. Principles of Electric Circuits. Pearson. ISBN: 9781292025667.

Solution

The two current sources are in the same direction; so, the current through R_L is:

$$I_{R_L} = I_1 + I_2 = 50\ mA + 20\ mA = \mathbf{70\ mA} \qquad (59)$$

2.8. CURRENT DIVIDERS

Because the current entering the junction of parallel branches "divides" up into several separate branch currents, a parallel circuit functions as a current divider. Total current flowing into the junction of parallel branches in a circuit splits among the branches. Therefore, a current divider is a parallel circuit. For a two-branch parallel circuit, where part of the total current I_T passes through R_1 and part through R_2, Figure 2.18 illustrates the current-divider principle.

The branch currents are inversely proportional to the resistor values because each resistor in parallel is experiencing the same voltage. For instance, if R_2 is twice as valuable as R_1, then I_2 is half as valuable as I_1. Stated differently, the total current splits into currents inversely proportional to the resistance values among the parallel resistors.

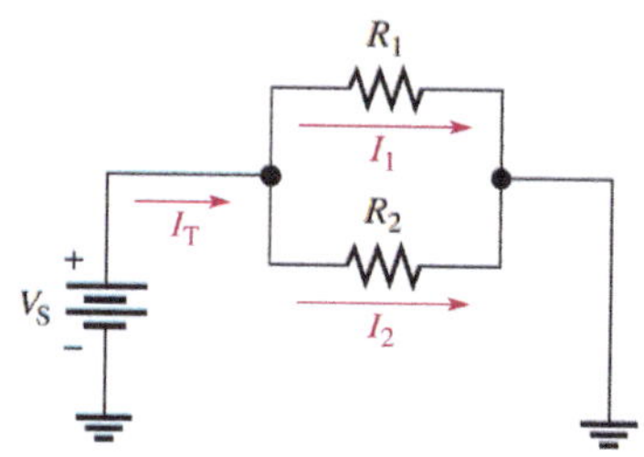

Figure 2.18. *Total current divides between the two branches.*

Source: Thomas L. Floyd. Principles of Electric Circuits. Pearson. ISBN: 9781292025667.

Ohm's law states that current flows more through branches with lower resistance and less through branches with higher resistance. Branch currents are equal if there is a uniform resistance across all branches.

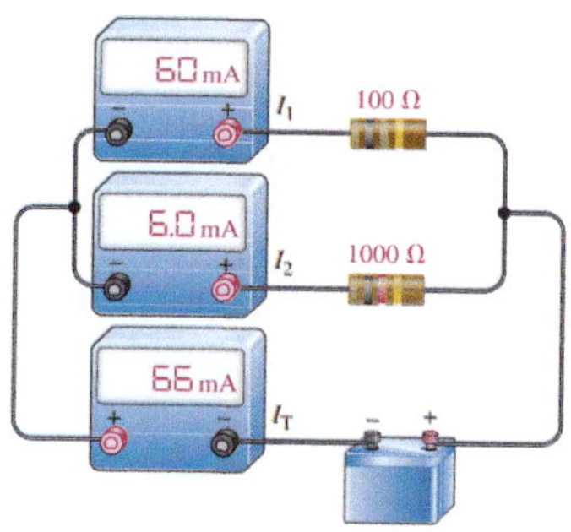

Figure 2.19. *The branch with the lower resistance has more current, and the branch with the higher resistance has less current.*

Source: Thomas L. Floyd. Principles of Electric Circuits. Pearson. ISBN: 9781292025667.

The values in Figure 2.19 illustrate how the currents split based on the branch resistances. It is noticeable that in this instance, the upper branch current is ten times greater than the lower branch current, despite the upper branch resistance being one-tenth that of the lower branch.

2.8.1. Current-Divider Formula

As seen in Figure 2.20, you can create a formula to find the ratio of currents among any number of parallel resistors, where n is the total number of resistors.

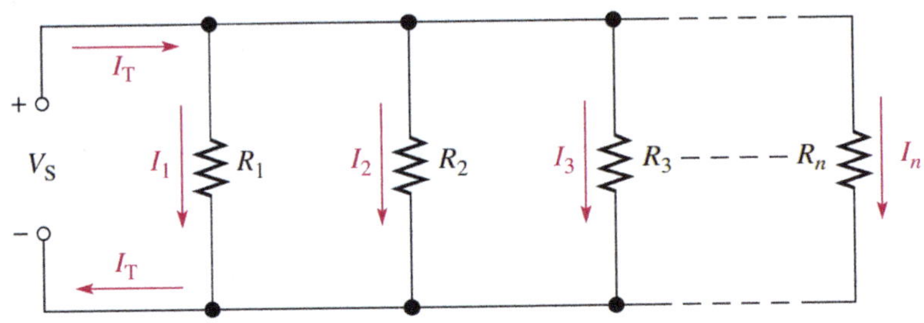

Figure 2.20. *A parallel circuit with n branches.*

Source: Thomas L. Floyd. Principles of Electric Circuits. Pearson. ISBN: 9781292025667.

Any resistor in the parallel configuration has a current of Ix, where x is the resistor number (1, 2, 3, and so on). The current flowing through any resistor in Figure 2.20 can be expressed as follows using Ohm's law:

$$I_x = \frac{V_S}{R_x} \tag{60}$$

Each parallel resistor has the source voltage, V_S, across it, and any one of the resistors is represented by R_x. The total current multiplied by the total parallel resistance equals the total source voltage, or V_S.

$$V_S = I_T R_T \tag{61}$$

Substituting $I_T R_T$ for V_S in the expression for I_x results in;

$$I_x = \frac{I_T R_T}{R_x} \tag{62}$$

Rearranging terms yields

$$I_x = \left(\frac{R_T}{R_x}\right) I_T \tag{63}$$

where x = 1, 2, 3, etc. The general current-divider formula, which can be applied to any number of branches in a parallel circuit, is Eqn. (63).

The total current (I_T) into the junction of parallel branches multiplied by the total parallel resistance (R_T) divided by the resistance (R_x) of that branch yields the current (I_x) through that branch.

Example 13

Determine the current through each resistor in the circuit of figure below.

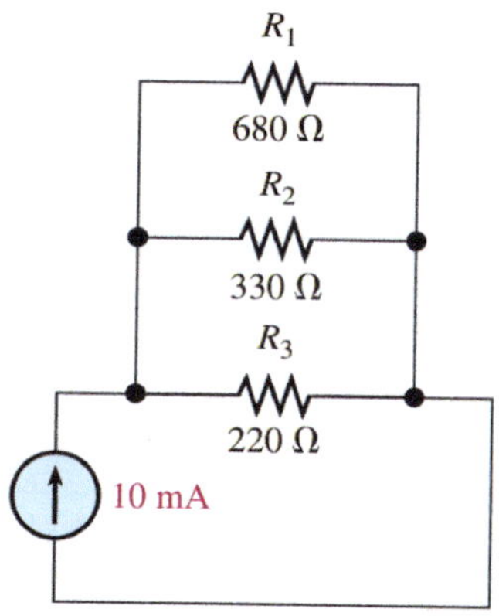

Source: Thomas L. Floyd. Principles of Electric Circuits. Pearson. ISBN: 9781292025667.

Solution

First calculate the total parallel resistance.

$$R_T = \frac{1}{\left(\frac{1}{R_1}\right) + \left(\frac{1}{R_2}\right) + \left(\frac{1}{R_3}\right)} = \frac{1}{\left(\frac{1}{680\ \Omega}\right) + \left(\frac{1}{330\ \Omega}\right) + \left(\frac{1}{220\ \Omega}\right)} = 111\ \Omega \tag{64}$$

The total current is 10 mA. Use Eqn. (63) to calculate each branch current.

$$\begin{aligned} I_1 &= \left(\frac{R_T}{R_1}\right)I_T = \left(\frac{111\ \Omega}{680\ \Omega}\right)10\ \text{mA} = \mathbf{1.63\ mA} \\ I_2 &= \left(\frac{R_T}{R_2}\right)I_T = \left(\frac{111\ \Omega}{330\ \Omega}\right)10\ \text{mA} = \mathbf{3.36\ mA} \\ I_3 &= \left(\frac{R_T}{R_3}\right)I_T = \left(\frac{111\ \Omega}{220\ \Omega}\right)10\ \text{mA} = \mathbf{5.05\ mA} \end{aligned} \tag{65}$$

- **Current-Divider Formulas for Two Branches:** Practical circuits typically have two parallel resistors, as the figure below illustrates. As you know from Eqn. (47);

$$R_T = \frac{R_1 R_2}{R_1 + R_2} \tag{66}$$

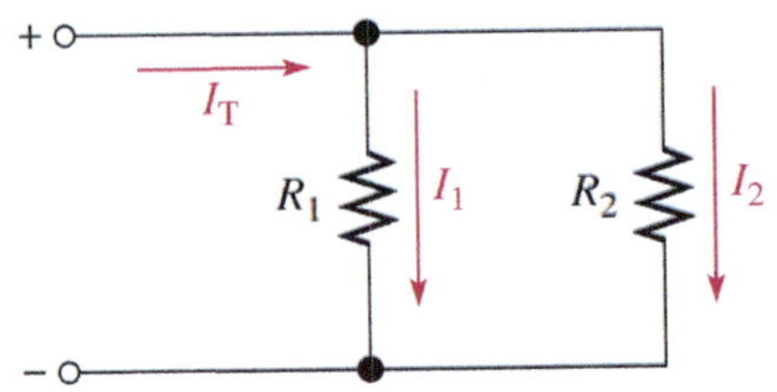

Source: Thomas L. Floyd. Principles of Electric Circuits. Pearson. ISBN: 9781292025667.

The general current-divider formula found in Eqn. (63) can be used to write the I_1 and I_2 formulas as follows:

$$I_1 = \left(\frac{R_T}{R_1}\right)I_T \quad \text{and} \quad I_2 = \left(\frac{R_T}{R_2}\right)I_T \tag{67}$$

Substituting $R_1R_2/(R_1 + R_2)$ for R_T and canceling terms result in:

$$I_1 = \frac{\left(\frac{\cancel{R_1}R_2}{R_1 + R_2}\right)}{\cancel{R_1}}I_T \quad \text{and} \quad I_2 = \frac{\left(\frac{R_1\cancel{R_2}}{R_1 + R_2}\right)}{\cancel{R_2}}I_T \tag{68}$$

Therefore, the current-divider formulas for the special case of two branches are:

$$I_1 = \left(\frac{R_2}{R_1 + R_2}\right)I_T \tag{69}$$

$$I_2 = \left(\frac{R_1}{R_1 + R_2}\right)I_T \tag{70}$$

Keep in mind that the current in one of the branches in Eqns. (69) and (70) is equal to the resistance of the opposite branch divided by the total of the two resistors, multiplied by the total current. You need to know the total current flowing into the parallel branches in order to use the current-divider equations in any application.

2.9. POWER IN PARALLEL CIRCUITS

The sum of the powers of each individual resistor in a parallel circuit yields the total power, just as in series circuits.

Eqn. (71) provides a concise formula for calculating total power for any number of resistors connected in parallel.

$$P_T = P_1 + P_2 + P_3 + \cdots + P_n \tag{71}$$

Where; P_T is the total power and P_n is the power in the last resistor in parallel. As you can see, the powers are additive, just as in a series circuit.

Parallel circuits can be directly addressed by power formulas. The total power P_T is computed using the following formulas:

$$P_T = VI_T$$
$$P_T = I_T^2 R_T$$
$$P_T = \frac{V^2}{R_T} \tag{72}$$

Where; V is the voltage across the parallel circuit, I_T is the total current into the parallel circuit, and R_T is the total resistance of the parallel circuit.

2.10. PARALLEL CIRCUIT APPLICATIONS

Almost every electronic system has some sort of parallel circuit. For the time being, let's examine a few typical and well-known uses for parallel circuits.

2.10.1. Automotive

The fact that the other branches remain unaffected when one branch opens gives a parallel circuit an advantage over a series circuit. For instance, a simplified diagram of a car lighting system can be found in Figure 2.21. Due to their parallel design, an automobile's headlights do not automatically go out when one goes out.

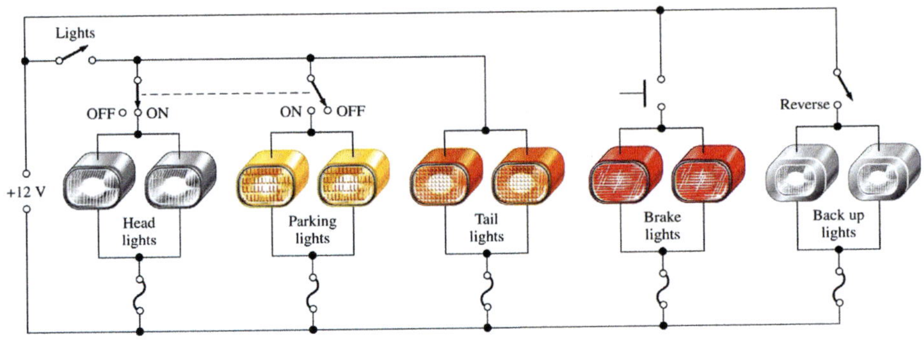

Figure 2.21. *Simplified diagram of the exterior light system of an automobile.*

Source: Thomas L. Floyd. Principles of Electric Circuits. Pearson. ISBN: 9781292025667.

Take note of the brake lights' separate on/off switches from the headlights and taillights. They only turn on when the driver depresses the brake pedal to close the brake light switch. The headlights and taillights are both on when the lights switch is in the closed position. A dashed line connects the switches; when the headlights are on, the parking lights are off, and vice versa. There is still current flowing through all of the lights in the event that one burns out or opens. Engaging the reverse gear causes the backup lights to come on.

The rear window defroster in an automobile is another device that uses parallel resistances. You are aware that resistance causes power to be lost as heat. When power is applied, a set of parallel resistance wires in the defroster warms the glass. Across the window, a standard defroster can dissipate more than 100 W. Resistance heating is easy to use and reasonably priced in situations such as this, despite its inferior efficiency compared to other heat sources.

2.10.2. Residential

In residential electrical systems, parallel circuits are also frequently used. In a house, every light and appliance is connected in parallel. A typical room wiring configuration with two switch-controlled lights and three parallel wall outlets is depicted in Figure 2.22.

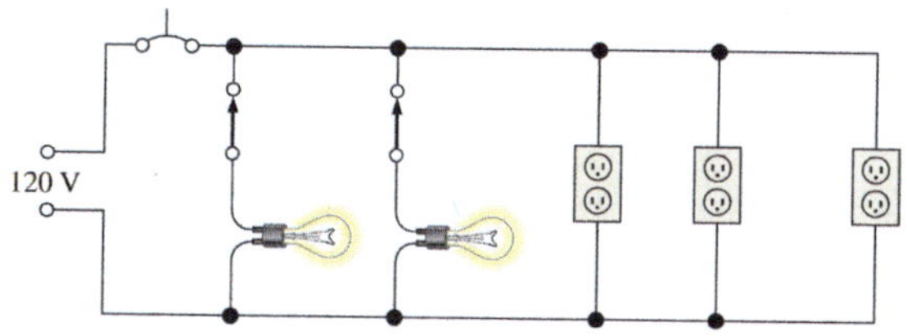

Figure 2.22. *Example of a parallel circuit in residential wiring.*

Source: Thomas L. Floyd. Principles of Electric Circuits. Pearson. ISBN: 9781292025667.

2.10.3. Control Circuits

A production line is one example of an industrial process that many control systems monitor and control using parallel circuits or equivalent circuits. The majority of sophisticated control applications are run on programmable logic controllers (PLCs), which are specialized computers. An analogous parallel circuit is shown by a PLC and appears on a computer screen. Internally, the circuit might only exist as computer code that is written in a programming language for computers.

On the other hand, the circuit that is shown can theoretically be built using hardware. These circuits can be represented as a ladder, with the two main nodes represented by the rails and the load (and source) represented by the rungs. (Note, for instance, how the lights and wall outlets in Figure 2.22 resemble a ladder on its side). Ladder diagrams are used in parallel control circuits, but additional control components (switches, relays, timers, and more) are added. Ladder logic is the term for the logical diagram that is produced when control elements are added. Because ladder logic is simple to comprehend, it is frequently used to illustrate control logic in industrial settings like factories or food processing facilities.

2.10.4. Analog Ammeters

The analog (needle-type) ammeter and milliammeter use parallel circuits. Even though they are less common than they formerly were, analog millimeters and panel meters are still made and can be used in some situations. Because parallel circuits enable the selection of multiple ranges for the purpose of measuring a wide range of current values, they are a crucial component of analog ammeter operation.

The meter movement in an ammeter is the mechanism that moves the pointer in proportion to the current. For the time being, knowing that a particular meter movement has a maximum current and a specific resistance is sufficient.

The pointer moves all the way to the end of the scale as a result of this maximum current, also known as the full-scale deflection current. For instance, a particular meter movement has a full-scale deflection current of 1 mA and a resistance of 50 Ω.

This specific movement of the meter allows it to measure currents of 1 mA or less, as shown in Figure 2.23(a) and (b). As shown in part (c), currents higher than 1 mA will cause the pointer to "peg"—or stop—slightly past the full scale mark, potentially damaging the meter.

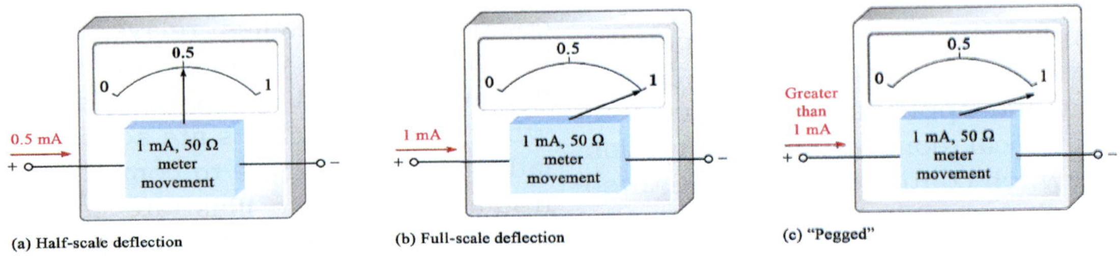

Figure 2.23. *A 1 mA analog ammeter.*

Source: Thomas L. Floyd. Principles of Electric Circuits. Pearson. ISBN: 9781292025667.

A basic ammeter with a resistor in parallel with the 1 mA meter movement is depicted in Figure 2.24; this resistor is known as a shunt resistor. Its goal is to increase the amount of current that can be measured by bypassing a portion of the current that flows around the meter movement. In particular, the figure displays 1 mA through the meter movement and 9 mA through the shunt resistor. So, it is possible to measure up to 10 mA. Just multiply the scale reading by ten to determine the current value.

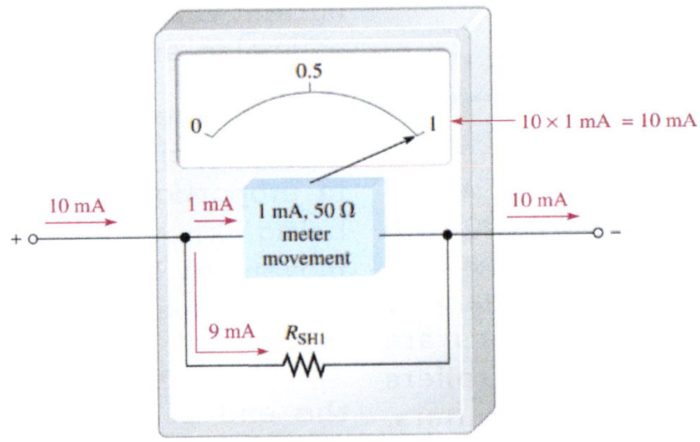

Figure 2.24. *A 10 mA analog ammeter.*

Source: Thomas L. Floyd. Principles of Electric Circuits. Pearson. ISBN: 9781292025667.

A range switch on a multiple-range ammeter allows the user to choose between multiple full-scale current settings. A parallel resistor bypasses a specific amount of current in each switch position, which is based on the resistance value. The current flowing through the movement in our example is never more than 1 mA.

A meter with three ranges—1 mA, 10 mA, and 100 mA—is shown in Figure 2.25. All of the current entering the meter passes through the meter movement when the range switch is in the 1 mA position. Up to 9 mA can flow through R_{SH1} and up to 1 mA can flow through the movement when using the 10 mA setting. Up to 99 mA can pass through R_{SH2} when the 100 mA setting is used, and the movement can still only have 1 mA at full scale.

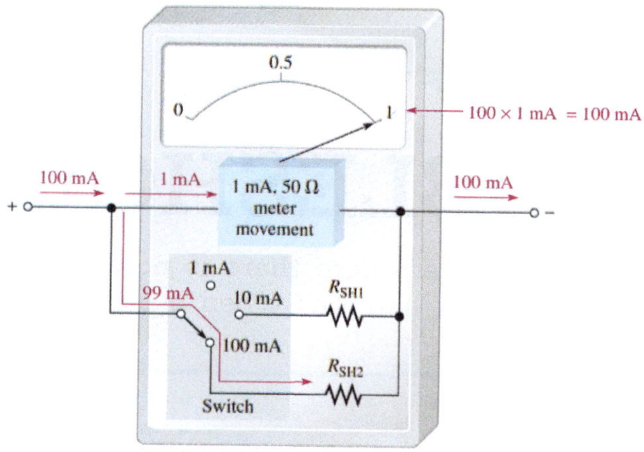

Figure 2.25. *An analog ammeter with three ranges.*

Source: Thomas L. Floyd. Principles of Electric Circuits. Pearson. ISBN: 9781292025667.

The range setting determines how the scale reading should be interpreted. For instance, in Figure 2.25, the needle points at 0.5 on the scale when 50 mA of current is being measured; to find the current value, multiply 0.5 by 100. In this case, 49.5 mA are through R_{SH2}, and 0.5 mA are through the movement (half-scale deflection).

2.11. TROUBLESHOOTING

Remember that an open circuit is one where there is no current and the current path is broken. This section looks at what happens when a parallel circuit branch opens.

2.11.1. Open Branches

As seen in Figure 2.26, a switch can create an open or closed path if it is connected in a branch of a parallel circuit. As shown in Figure 2.26(a), R_1 and R_2 are in parallel when the switch is closed. Two parallel 100 Ω resistors make up the total resistance of 50 Ω. With both resistors, current flows. R_1 is essentially taken out of the circuit and there is 100 Ω of resistance overall if the switch is opened, as shown in Figure 2.26(b). Right now, only R_2 is being used. Generally speaking, when a parallel branch opens, all of the remaining parallel paths continue to carry the same current while total resistance and total current both rise.

The amount of current that was previously in the open branch is equal to the decrease in total current. There is no change in the other branch currents.

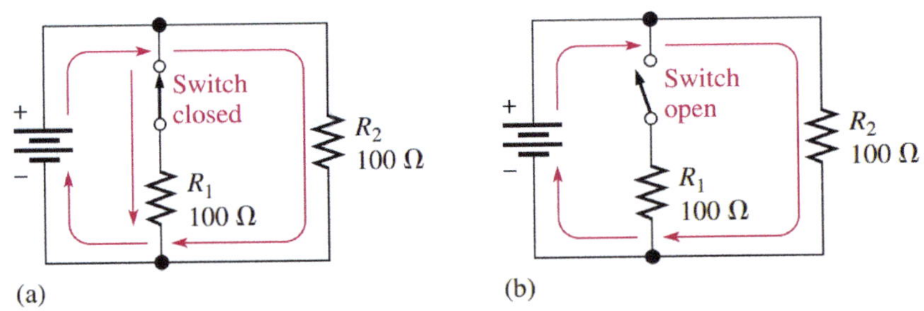

Figure 2.26. *When switch opens, total current decreases and current through R_2 remains unchanged.*

Source: Thomas L. Floyd. Principles of Electric Circuits. Pearson. ISBN: 9781292025667.

Examine the lamp circuit depicted in Figure 2.27. A 12 V source is connected in parallel to four bulbs. Every bulb in section (a) has current flowing through it. Now imagine that one of the lightbulbs burns out, leaving an exposed path as Figure 2.27(b) illustrates. There is no current flowing through the open path, so this light will go out. Observe, nevertheless, that all of the other parallel bulbs are still receiving current and are still glowing. The current flowing through each branch stays constant, and the open branch has no effect on the voltage across the parallel branches, which stays at 12 V.

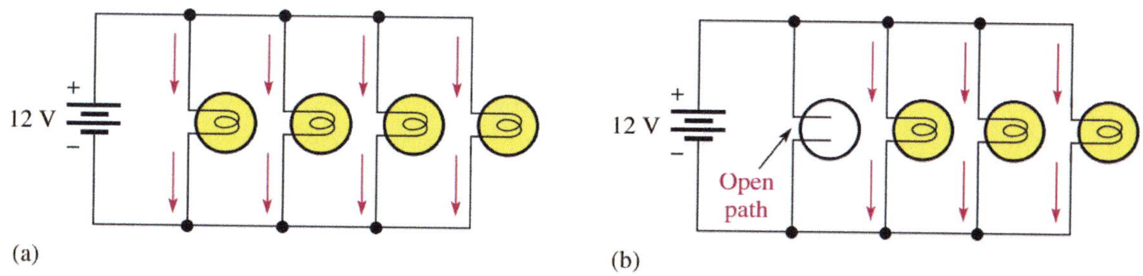

Figure 2.27. *When one lamp opens, total current decreases and other branch currents remain unchanged.*

Source: Thomas L. Floyd. Principles of Electric Circuits. Pearson. ISBN: 9781292025667.

When it comes to lighting systems, it is evident that a parallel circuit has an advantage over a series circuit since the remaining parallel bulbs will continue to function even in the event that one or more burn out. When a single bulb in a series circuit burns out, the current path is totally disrupted, which also causes the other bulbs to burn out.

Because the voltage across all of the branches in a parallel circuit is the same, it is impossible to locate the open resistor by measuring the voltage across the branches. Therefore, merely measuring voltage will not reveal which resistor is open. As shown in Figure 2.28, the good resistors will always have the same voltage as the open resistor (note that the middle resistor is open).

Current measurements must be used to locate the open resistor if a visual inspection is unable to identify it. In actuality, measuring current requires inserting the ammeter in series, making it more challenging than measuring voltage. To connect the ammeter in series, therefore, one must cut or disconnect a wire or a PC board connection, or lift one end of a component off the circuit board. Naturally, when voltage measurements are taken, this process is not necessary because the meter leads are just connected across a component.

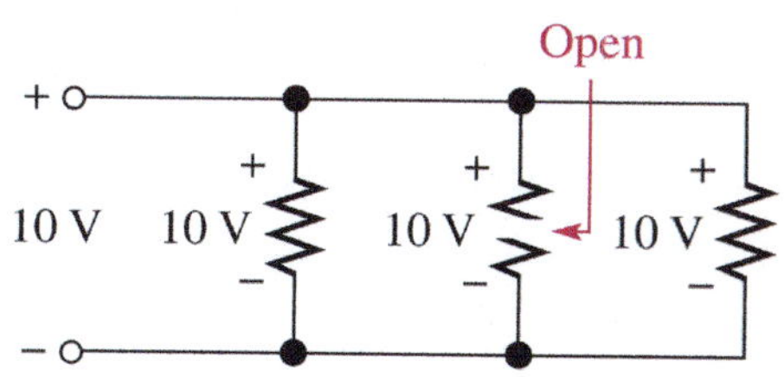

Figure 2.28. *Parallel branches (open or not) have the same voltage.*

Source: Thomas L. Floyd. Principles of Electric Circuits. Pearson. ISBN: 9781292025667.

2.11.2. Finding an Open Branch by Current Measurement

The total current in a parallel circuit with a possible open branch can be measured to identify the open. The total current, I_T, is always less than its typical value when a parallel resistor opens. When all the resistors have different resistance values, you can calculate which resistor is open by knowing I_T and the voltage across the branches.

Examine the circuit with two branches shown in Figure 2.29(a). The total current will equal the current in the good resistor if one of the resistors opens. You can quickly determine the recommended current in each resistor by using Ohm's law.

$$I_1 = \frac{50\ \text{V}}{560\ \Omega} = 89.3\ \text{mA}$$

$$I_2 = \frac{50\ \text{V}}{100\ \Omega} = 500\ \text{mA}$$

$$I_T = I_1 + I_2 = 589.3\ \text{mA} \qquad (73)$$

As shown in Figure 2.29(b), the total current when R_2 is open is 89.3 mA. As shown in Figure 2.29(c), the total current is 500 mA if R_1 is open.

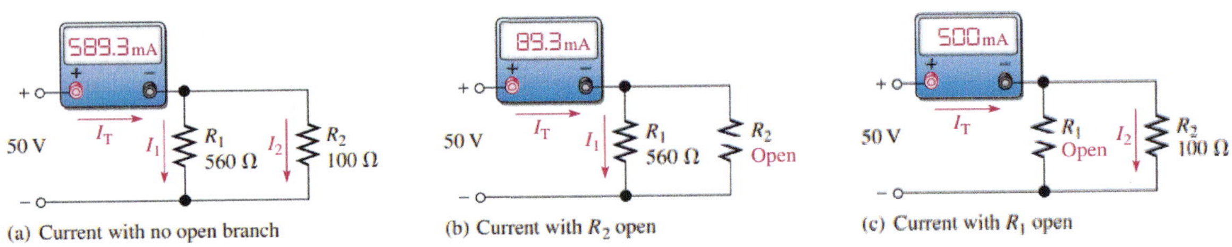

Figure 2.29. *Finding an open path by current measurement.*

Source: Thomas L. Floyd. Principles of Electric Circuits. Pearson. ISBN: 9781292025667.

You can repeat this process with any number of branches that have different resistances. The current in each branch must be examined if the parallel resistances are all the same until a branch with no current is found. This resistor is open.

2.11.3. Finding an Open Branch by Resistance Measurement

An open branch can be found by measuring the total resistance of the parallel circuit to be checked, provided that it can be disconnected from both its voltage source and any other circuit to which it might be connected.

Remember that the unit of conductance, G, is the Siemen (S), and that it is the reciprocal of resistance (1/R). The total of all the resistors' conductances is the circuit's conductance in parallel.

$$G_T = G_1 + G_2 + G_3 + \cdots + G_n \tag{74}$$

To locate an open branch, do the following steps:

1. Calculate what the total conductance should be using the individual resistor values.

$$G_{T(calc)} = \frac{1}{R_1} + \frac{1}{R_2} + \frac{1}{R_3} + \cdots + \frac{1}{R_n} \tag{75}$$

2. Measure the total resistance with an ohmmeter and calculate the total measured conductance.

$$G_{T(meas)} = \frac{1}{R_{T(meas)}} \tag{76}$$

3. Subtract the measured total conductance (Step 2) from the calculated total conductance (Step 1). The result is the conductance of the open branch and the resistance is obtained by taking its reciprocal (R = 1/G).

$$R_{open} = \frac{1}{G_{T(calc)} - G_{T(meas)}} \tag{77}$$

2.11.4. Shorted Branches

A fuse or circuit breaker will blow when a branch in a parallel circuit shorts, resulting in an excessive current surge. This makes it challenging to isolate the shorted branch, which makes troubleshooting the issue challenging.

To locate shorts in a circuit, two common tools are pulsers and current tracers. They work well in any kind of circuit and are not limited to use in digital circuits. A pen-shaped device known as a pulser applies pulses to a specific point in a circuit, shorting out the path and causing current pulses to flow through it. Another pen-shaped device that detects current pulses is the current tracer.

2.12. NODE VOLTAGE METHOD

A DC network analysis method called the node voltage method is based on a methodical application of Kirchhoff's current law. Using this technique, resistances are swapped out for equivalent conductances and voltage sources are changed to current sources.

The Kirchhoff's current law (KCL) system of equations is used in the node voltage method of DC network analysis to solve for unknown voltages at circuit nodes. Because it substitutes equal current sources for voltage sources, this analysis appears odd. Additionally, equivalent conductances ($G = 1/R$) with units of siemens (S) are used in place of resistor values in ohms. Once grasped, the node voltage method can offer a straightforward method for rapidly resolving a variety of intricate circuits.

2.12.1. Node Voltage Method Steps

Let's begin by looking at a circuit with traditional voltage sources, as shown in Figure 2.30.

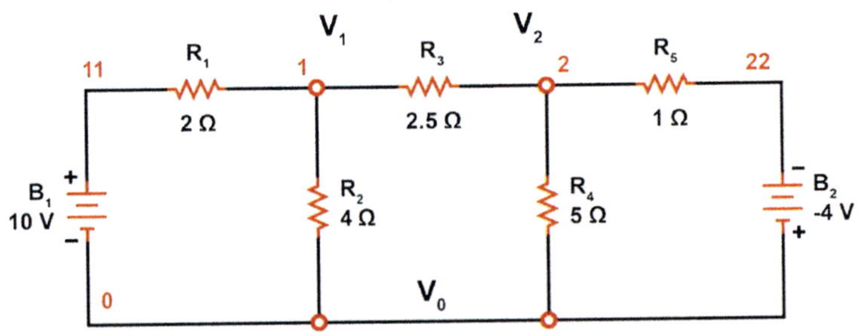

Figure 2.30. *Circuit schematic for demonstration of the node voltage method of circuit analysis.*

Source: https://www.allaboutcircuits.com/textbook/direct-current/chpt-10/node-voltage-method/.

- **Step 1: Label All Unique Voltage Nodes:** The ground node in the circuit is usually designated as the common node, or V_0, for reference purposes. This point will be used to calculate the other unknown node voltages in our Figure 2.30 circuit, V_1 and V_2.
- **Step 2: Replace All Voltage Sources and Series Resistors with Current Sources and Parallel Resistors:** The next step is to swap out all of the voltage sources

and series resistors for equal current sources and parallel resistors. The current source value is calculated as:

$$I_{source} = \frac{V_{source}}{R_{series}} \tag{78}$$

The voltage source, B_1, in Figure 2.30 is in series with resistor, R_1, which has a current source value of 5 A:

$$I_1 = \frac{V_{B1}}{R_1} = \frac{10\ \text{V}}{2\ \Omega} = 5\ \text{A} \tag{79}$$

The series resistance and the parallel resistance have the same value. With the resistance and current source values in hand, we can sketch the analogous parallel circuit, as seen in Figure 2.31.

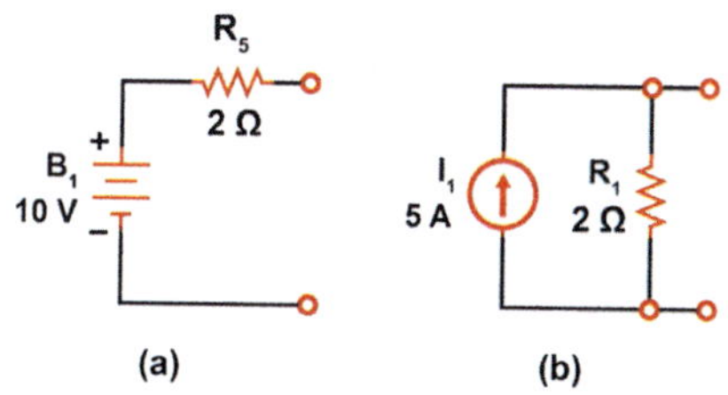

Figure 2.31. *Conversion of a voltage source with series resistance to a current source with parallel resistance.*

Source: https://www.allaboutcircuits.com/textbook/direct-current/chpt-10/node-voltage-method/.

For battery B_2, we must convert its voltage source to current source using the same series resistor R_5 from our circuit shown in Figure 2.30.

$$I_2 = \frac{V_{B2}}{R_5} = \frac{-4\ \text{V}}{1\ \Omega} = -4\ \text{A} \tag{80}$$

This allows us to redraw our circuit, as seen in Figure 2.32, adding two current sources and parallel resistors.

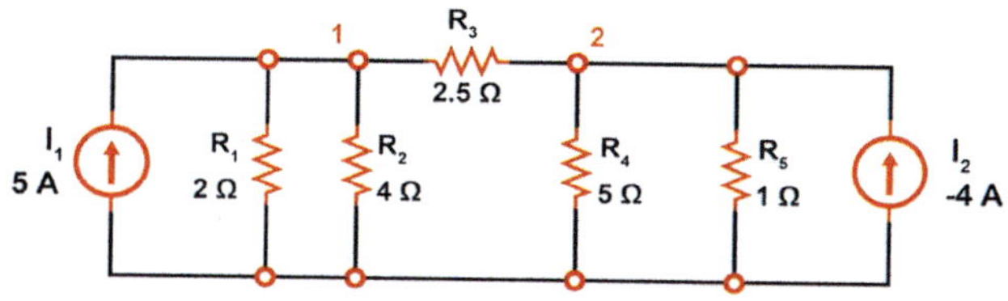

Figure 2.32. *New circuit with current sources and parallel resistors.*

Source: https://www.allaboutcircuits.com/textbook/direct-current/chpt-10/node-voltage-method/.

- **Step 3: Replace Resistor Resistance in Ohms with Conductance in Siemens:** We now need to change the resistance value in ohms for every resistor in our circuit to its reciprocal conductance value in siemens. The current unit of conductance is the siemens, which took the place of the antiquated mho (inverse of an ohm) unit.

$$G_1 = \frac{1}{R_1} = \frac{1}{2\,\Omega} = 0.5\ \text{S} \tag{81}$$

$$G_2 = \frac{1}{R_2} = \frac{1}{4\,\Omega} = 0.25\ \text{S} \tag{82}$$

$$G_3 = \frac{1}{R_3} = \frac{1}{2.5\,\Omega} = 0.4\ \text{S} \tag{83}$$

$$G_4 = \frac{1}{R_4} = \frac{1}{5\,\Omega} = 0.2\ \text{S} \tag{84}$$

$$G_5 = \frac{1}{R_4} = \frac{1}{1\,\Omega} = 1\ \text{S} \tag{85}$$

These calculated conductance values allow us to update our drawing (Figure 2.33).

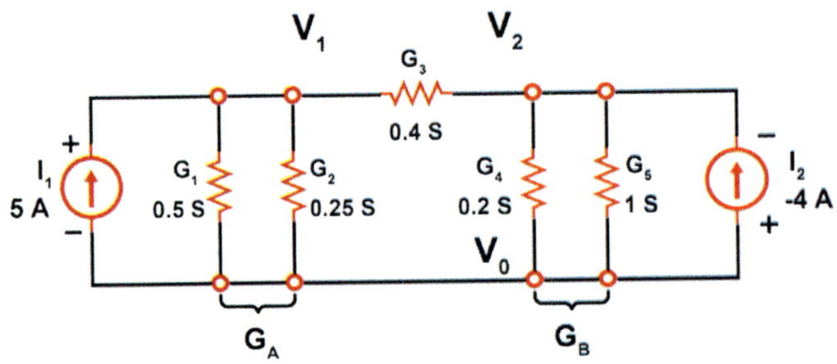

Figure 2.33. *All resistance values are replaced by equivalent conductance.*

Source: https://www.allaboutcircuits.com/textbook/direct-current/chpt-10/node-voltage-method/.

At this stage, adding parallel conductors to the circuit will also help to simplify it. For this, we just use addition. Two sets of parallel conductors are present in our circuit in Figure 2.33.

$$G_A = G_1 + G_2 = 0.5 + 0.25 = 0.75\ \text{S} \tag{86}$$

$$G_B = G_4 + G_5 = 0.2 + 1 = 1.2\ \text{S} \tag{87}$$

Figure 2.34 depicts our simplified circuit, which is prepared for use with the node voltage method.

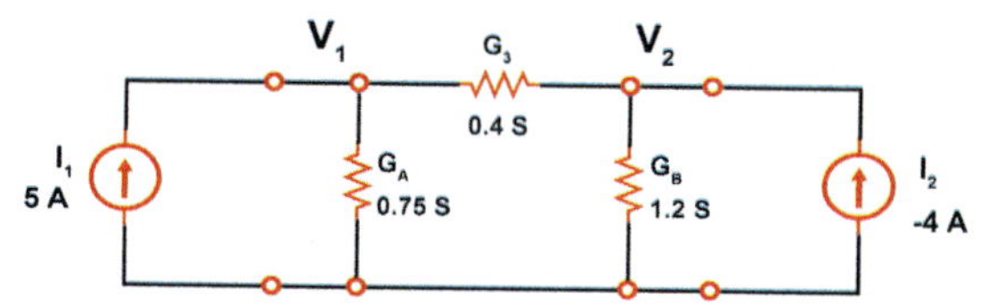

Figure 2.34. *Simplified circuit.*

Source: https://www.allaboutcircuits.com/textbook/direct-current/chpt-10/node-voltage-method/.

- **Step 4: Write KCL Equations for Each Unknown Voltage Node:** We now write KCL equations for all unknown voltage nodes using the node voltage method. There are just two unknown node voltages in our example circuit: V_1 and V_2.

$$I_1 = G_AV_1 + G_3(V_1 - V_2) \quad (88)$$

$$I_2 = G_BV_2 + G_3(V_2 - V_1) \quad (89)$$

To show how to write these equations by inspection, let's rewrite these two equations.

$$I_1 = (G_A + G_3)\ V_1 - G_3V_2 \quad (90)$$

$$I_2 = (G_B + G_3)\ V_2 - G_3V_1 \quad (91)$$

Take note of how these two equations are similar. The positive coefficient of the first voltage (V_1) at the first KCL node is equal to the sum of the conductances connected to it ($G_A + G_3$). Similarly, the positive coefficient of the voltage, V_2, is equal to the total of the conductances linked to that KCL node. The conductances between additional non-common nodes are represented by the other negative coefficients.

The left-hand side of each equation equals the corresponding current source that is attached to the node. We can quickly write the equations by looking at this pattern. Through the use of conductances and a relatively simple pattern to write the KCL equations, the node voltage method can be used to analyze a large number of circuits quickly.

- **Step 5: Solve the Set of Simultaneous KCL Equations for the Unknown Voltages:** From here, we can solve for the two unknown voltages in our circuit using the two equations. We can replace the unknown conductances and currents with our own and reorder them to make sense.

$$I_1 = 5 = (0.75 + 0.4)\ V_1 - 0.4V_2 = 1.15V_1 - 0.4V_2 \quad (92)$$

$$I_2 = -4 = (1.2 + 0.4)\ V_2 - 0.4V_1 = -0.4V_1 + 1.6V_2 \quad (93)$$

Solving these simultaneous equations, we get the following results for our node voltages:

$$V_1 = 3.8095\ \text{V} \quad (94)$$

$$V_2 = -1.5476\ \text{V} \quad (95)$$

2.12.2. Verification of the Node Voltage Method Using SPICE

The original schematic diagram with voltage sources can then be used to verify the solution using SPICE. On the other hand, it was also possible to simulate the circuit containing the current sources.

```
* Node Voltage Method circuit simulation
V1 11 0 DC 10
V2 22 0 DC -4
r1 11 1 2
r2 1 0 4
r3 1 2 2.5
r4 2 0 5
```

```
r5 2 22 1
.DC V1 10 10 1 V2 -4 -4 1
.print DC V(1) V(2)
.end
v(1)   v(2)
3.809524e+00 – 1.547619e+00
```

2.12.3. The Node Voltage Method Applied to an Unbalanced Wheatstone Bridge

Using the node voltage method, let's examine another example of circuit analysis (Paul, Clayton R. 2001). Together with the common voltage node, V_0, the circuit of Figure 2.35 is referred to as an unbalanced Wheatstone bridge. It consists of three unknown voltage nodes.

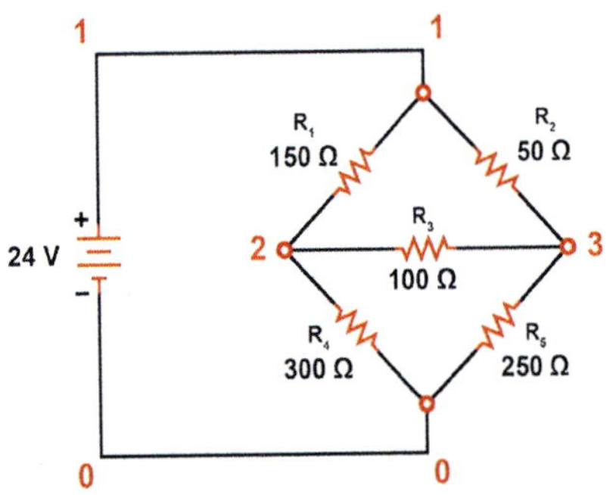

Figure 2.35. *Wheatstone bridge for node voltage method application.*

Source: https://www.allaboutcircuits.com/textbook/direct-current/chpt-10/node-voltage-method/.

It's crucial to remember that the schematic diagram does not include a list of conductances. But $G_1 = 1/R_1$, and so on.

Writing equations for the three nodes (V_1, V_2, and V_3) can be done by visual inspection. Recall that for conductors connected to the node for which we are creating the KCL equation, the conductance coefficients are positive. There are no other negative conductance coefficients.

For the time being, the subscripted G's are listed as the coefficients because we are too lazy to compute the conductances for the resistors on the diagram.

$$I_1 = 0.136092 = (G_1 + G_2)\,V_1 - G_1V_2 - G_2V_3 \quad (96)$$

$$I_2 = 0 = (G_1 + G_3 + G_4)\,V_2 - G_1V_1 - G_3V_3 \quad (97)$$

$$I_3 = 0 = (G_2 + G_3 + G_5)\,V_3 - G_2V_1 - G_3V_2 \quad (98)$$

Next, we can reorder to align the unknown voltages to make it easier to solve the simultaneous equations:

$$0.136092 = (G_1 + G_2)\,V_1 - G_1V_2 - G_2V_3 \quad (99)$$

$$0 = -G_1V_1 + (G_1 + G_3 + G_4)\,V_2 - G_3V_3 \quad (100)$$

$$0 = -G_2V_1 - G_3V_2 + (G_2 + G_3 + G_5) \quad (101)$$

Once more, because we are so indolent, we will solve this using GNU Octave. We will enter reciprocal resistances and their sums into the Octave "A" matrix and then allow Octave to compute the conductance matrix after "A =."

The first entry line was divided into three rows due to its length. The starting and ending square brackets define the entered "A" matrix. Elements in columns are separated by spaces. Rows are divided into "new lines." There is no need to use semicolons and commas as separators. However, a column vector of currents is produced by semicolon-separating the current vector at "b."

```
octave:12> A = [1/150+1/50 -1/150 -1/50
> -1/150 1/150+1/100+1/300 -1/100
```

```
> -1/50 -1/100 1/50+1/100+1/250]

A =

0.0266667 -0.0066667 -0.0200000

-0.0066667 0.0200000 -0.0100000

-0.0200000 -0.0100000 0.0340000

octave:13> b = [0.136092;0;0]

b =

0.13609

0.00000

0.00000

octave:14> x=A\b

x =

24.000

17.655

19.310
```

The diagonal coefficients in the "A" matrix are positive, whereas the other coefficients in the matrix must all be negative due to the requirements of our node voltage method equation writing rules.

The voltage vector solution is given in "x":

- V_1 = 24.000 V
- V_2 = 17.655 V
- V_3 = 19.310 V

These three voltages match the outcomes we got from the mesh current analysis and the unbalanced Wheatstone bridge SPICE simulation. This is not by accident; in that problem, the 0.13609 A current source was selected on purpose to produce the 24 V needed as a voltage source.

2.13. MESH CURRENT METHOD

In the mesh current method, Kirchhoff's voltage law (KVL) and Ohm's law are applied methodically to solve for the unknown currents and voltages after mesh (or loop) current directions are arbitrarily assigned.

An overview of the mesh current method, also called the loop current method, for electrical circuit analysis is given in this section. To ascertain unknown currents in a network, the mesh current method applies Kirchhoff's voltage law (KVL), Ohm's law, and simultaneous equations.

The branch current method and the mesh current method are fairly similar. But it doesn't apply Kirchhoff's current law (KCL), and it can typically solve a circuit with fewer simultaneous equations and unknown variables.

2.13.1. How to Use the Mesh Current Method

Before we dive into the mesh current method, it's crucial to remember that we'll be introducing other network analysis methods using the same example circuit (Figure 2.36):

- Branch current
- Superposition theorem
- Thevenin's theorem
- Norton's theorem
- Millman's theorem

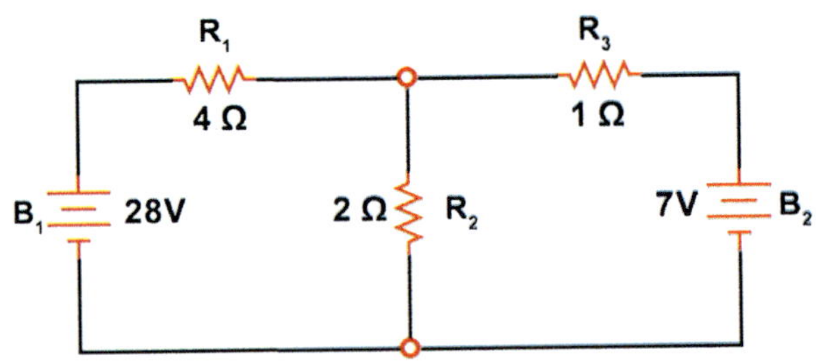

Figure 2.36. *Circuit schematic for explaining the mesh current method.*

Source: https://www.allaboutcircuits.com/textbook/direct-current/chpt-10/mesh-current-method/.

- **Step 1: Identify and Label the Current Loops:** The mesh current method starts with locating and labeling the circuit's current "loops." We need to identify at least one loop current flowing through each component in the circuit in order to accomplish this.

In Figure 2.37, our example circuit, the loop composed of B_1, R_1, and R_2 will be the first loop, and the loop composed of B_2, R_2, and R_3 will be the second.

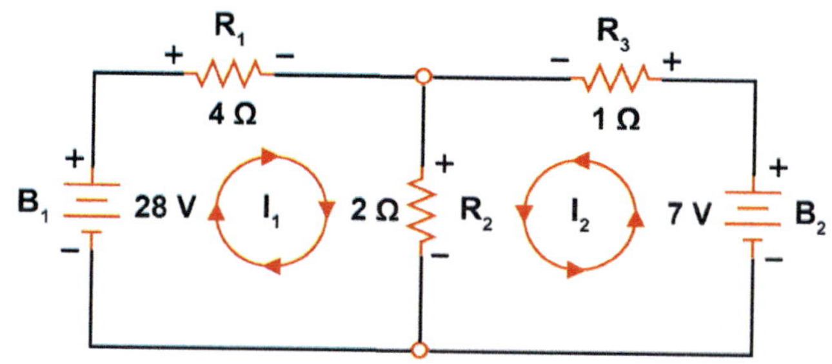

Figure 2.37. *Identify and label the current loops.*

Source: https://www.allaboutcircuits.com/textbook/direct-current/chpt-10/mesh-current-method/.

The most peculiar aspect of the mesh current approach is the idea of having currents flowing through each loop. The idea that these currents mesh together between loops like a set of rotating gears is actually how this technique gets its name.

The direction of each current loop can be chosen at will, but if the currents flow through components with multiple current loops in the same direction, the resulting equations are frequently simpler to solve. As an illustration, observe how currents I_1 and I_2 "down" or "mesh" at resistor R_2, where they intersect. The answer for a mesh current will be negative if the assumed direction of the current is incorrect.

- **Step 2: Label the Voltage Drop Polarities:** As indicated in Figure 2.38, the next step is to label all voltage drop polarities across resistors in accordance with the assumed directions of the mesh currents.

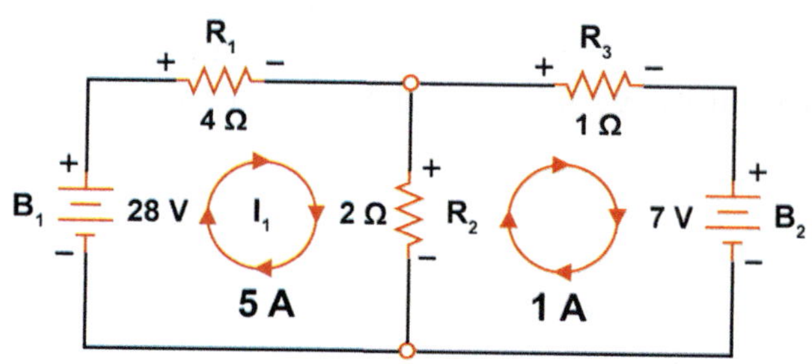

Figure 2.38. *Label the voltage drop polarities.*

Source: https://www.allaboutcircuits.com/textbook/direct-current/chpt-10/mesh-current-method/.

Recall that a resistor's "upstream" end is always going to be positive and its "downstream" end is always going to be negative. This occurs as a result of the resistor acting as a load, lowering voltage when current passes through it.

Note that battery polarities may or may not "agree" with resistor polarities and assumed loop current directions depending on their symbol orientations in the diagram.

- **Step 3: Apply Kirchhoff's Voltage Law to Each Loop:** We can now step around each of these loops using Kirchhoff's voltage law to create equations that represent the component voltage drops and polarities. Similar to the branch current method, the voltage drop of a resistor will be expressed as the product of its mesh current (which is currently unknown) and resistance (measured in ohms). The resistor current is the total of the two meshing currents, and we will write the term in the equation where two currents mesh together.

Both the beginning point and the direction of the voltage drop tracing around each loop are randomly chosen. Let's start at the lower-left corner of the circuit's left loop and work our way around it in a clockwise manner, counting polarity as if we were holding a voltmeter: red lead at the point forward, black lead at the point behind. For the left loop with current I_1, we get the following equation:

$$28 - R_1I_1 - R_2(I_1 + I_2) = 28 - 4I_1 - 2(I_1 + I_2) = 0 \qquad (102)$$

It is important to note that the current through resistor R_2 in the middle term of the equation is determined by adding the mesh currents I_1 and I_2. This is because mesh currents I_1 and I_2 complement one another because they are flowing through R_2 in the same direction.

We can distribute the coefficient of 2 to the I_1 and I_2 terms and then combine the I_1 terms to simplify the equation as:

$$28 - 4I_1 - 2I_1 - 2I_2 = 28 - 6I_1 - 2I_2 = 0 \qquad (103)$$

We currently have one equation that has two unknowns. We need two equations in order to be able to solve for two unknown mesh currents.

Let's now repeat the procedure for the circuit's right loop using current I2. This will give us an additional KVL equation. We can find the currents using just two equations and the two unknown currents. Due to habit, we'll begin at the lower-left corner of the right loop once more and work our way clockwise:

$$R_2(I_1 + I_2) + R_3I_2 - 7 = 2(I_1 + I_2) + 1I_2 - 7 = 0 \qquad (104)$$

Simplifying the equation as before, we end up with:

$$2I_1 + 2I_2 + 1I_2 - 7 = 2I_1 + 3I_2 - 7 = 0 \qquad (105)$$

- **Step 4: Solve the Simultaneous Equations for the Unknown Currents:** Now that we have two equations, we can solve for the unknown currents I_1 and I_2 mathematically using a variety of techniques. First, let's rearrange the two equations for an easier solution:

$6I_1 + 2I_2 = 28$ (106)

$2I_1 + 3I_2 = 7$ (107)

Now, solving for the currents, we get:

$I_1 = 5\text{ A}$ (108)

$I_2 = -1\text{ A}$ (109)

- **Step 5: Redraw the Mesh Currents and Determine the Branch Currents:** We must refer back to our diagram to see how these solutions fit together to provide currents through all components, keeping in mind that these values correspond to mesh currents rather than branch currents (Figure 2.39).

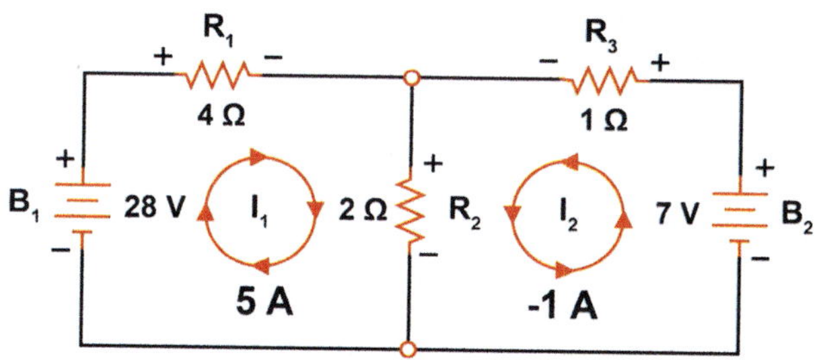

Figure 2.39. *Circuit with calculated mesh current values.*

Source: https://www.allaboutcircuits.com/textbook/direct-current/chpt-10/mesh-current-method/.

Our initial assumption about the current direction was incorrect, as indicated by the solution of -1 A for I_2. Actually, I_2 moves at a value of +1 A in a clockwise manner. Let's redraw our circuit using this correction, altering the resistor R_3's voltage drop and the I_2's current flow direction, as seen in Figure 2.40.

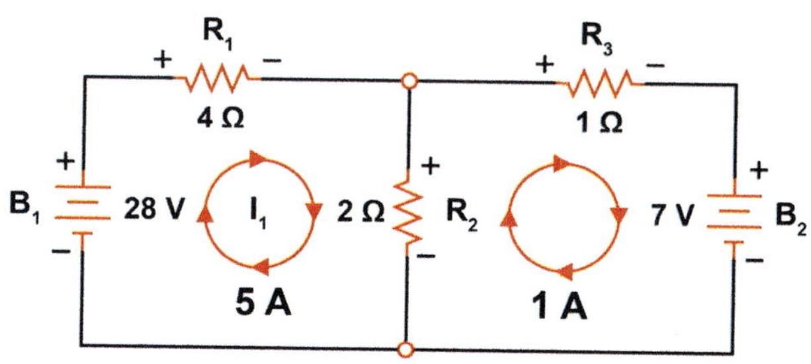

Figure 2.40. *Circuit with corrected mesh current direction for I_2.*

Source: https://www.allaboutcircuits.com/textbook/direct-current/chpt-10/mesh-current-method/.

We can now calculate the branch currents for our circuit using these mesh currents. Since only mesh current I_1 passes through B_1 and R_1, we can easily see that the current through those two circuit components is 5 A. Likewise, 1 A of current is passing through R_3 and into B_2.

The R_2, which has two mesh currents flowing through it, may then raise the question, "What about it?" Mesh currents I_1 and I_2 are 5 A and 1 A, respectively, flowing "down" and "up" through R_2.

We need to observe how mesh currents I_1 and I_2 interact in order to ascertain the true current flowing through R_2. They are opposing one another in this instance and moving in different directions. We can algebraically add them to arrive at a final value:

$$I_{R2} = I_1 - I_2 = 5 - 1 = 4 \text{ A} \quad (110)$$

A 4 A value, or "down," must be the current flowing through R_2. The branch currents are all displayed in Figure 2.41.

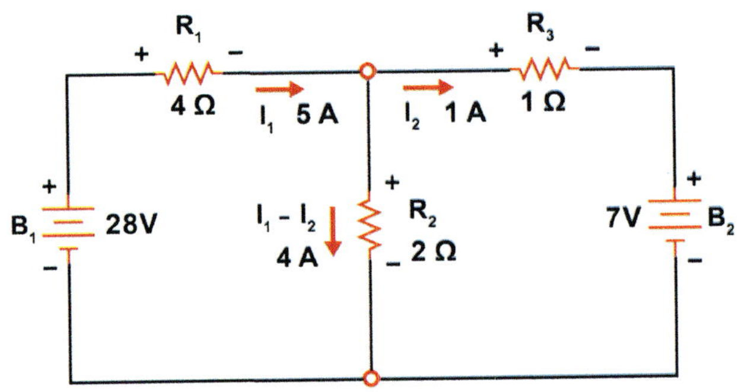

Figure 2.41. *Circuit with calculated branch currents.*

Source: https://www.allaboutcircuits.com/textbook/direct-current/chpt-10/mesh-current-method/.

- **Step 6: Calculate the Voltage Drops:** Now that we are aware of every branch current, we can use Ohm's law to determine the circuit's unknown voltage drops across the resistors.

$$V_{R1} = I_{R1}R_1 = I_1R_1 = 5{\cdot}4 = 20 \text{ V} \quad (111)$$

$$V_{R2} = I_{R2}R_2 = (I_1 - I_2)\ R_1 = (5-1){\cdot}2 = 8 \text{ V} \quad (112)$$

$$V_{R3} = I_{R3}R_3 = I_2R_3 = 1{\cdot}1 = 1 \text{ V} \quad (113)$$

We could check our results by returning to Kirchhoff's voltage law for our two loops:

$$V_{B1} - V_{R1} - V_{R2} = 28 - 20 - 8 = 0 \text{ V} \quad (114)$$

$$V_{R2} - V_{R3} - V_{B2} = 8 - 1 - 7 = 0 \text{ V} \quad (115)$$

2.13.2. Advantages of Mesh Current vs. Branch Current Methods

Mesh current analysis's main benefit is that it typically makes it possible to solve a big network with fewer simultaneous equations and unknown values. The branch current method in our example circuit requires the solution of three equations, whereas the mesh current method only requires the solution of two equations. This benefit may be more noticeable in networks with higher levels of complexity, like the circuit depicted in Figure 2.42.

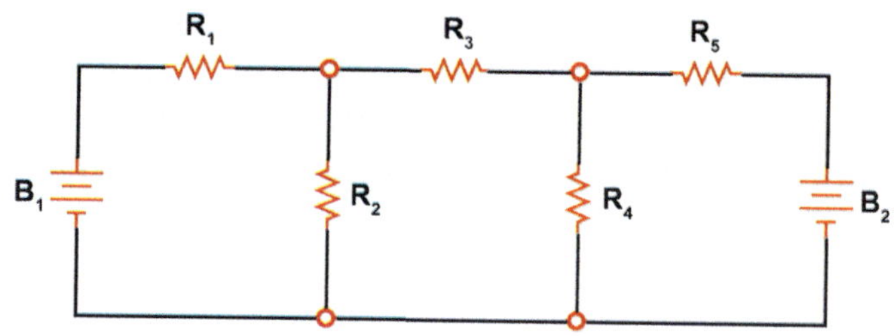

Figure 2.42. *Example circuit with increased complexity.*

Source: https://www.allaboutcircuits.com/textbook/direct-current/chpt-10/mesh-current-method/.

As shown in Figure 2.43, in order to solve this network using branch currents, we would need to create five variables (I_1 through I_5), one for each unique current in the circuit.

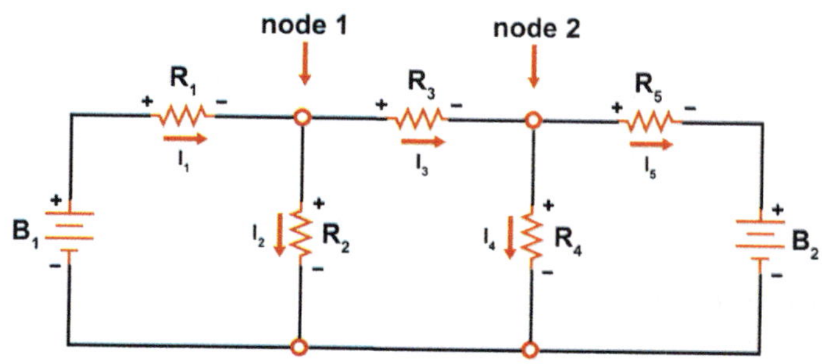

Figure 2.43. *Complex circuit setup for branch current analysis.*

Source: https://www.allaboutcircuits.com/textbook/direct-current/chpt-10/mesh-current-method/.

To solve this, we would need five equations: three KVL equations for the loops and two KCL equations at the nodes.

For the circuit of Figure 2.43, our five equations would be:

$I_1 - I_2 - I_3 = 0$ KCL at node 1 (116)

$I_3 - I_4 - I_5 = 0$ KCL at node 2 (117)

$V_{B1} - I_1R_1 - I_2R_2 = 0$ KVL of the left loop (118)

$I_2R_2 - I_3R_3 - I_4R_4 = 0$ KVL of the middle loop (119)

$I_4R_4 - I_5R_5 - V_{B2} = 0$ KVL of the right loop (120)

Overall, you might not mind using the branch current method of analysis for this circuit if you have nothing better to do with your time than solve for five unknown variables using five equations. That being said, the mesh current method—which requires only three unknowns and three equations to solve—is much simpler for those of us with more important things to do with our time. Figure 2.44 illustrates this.

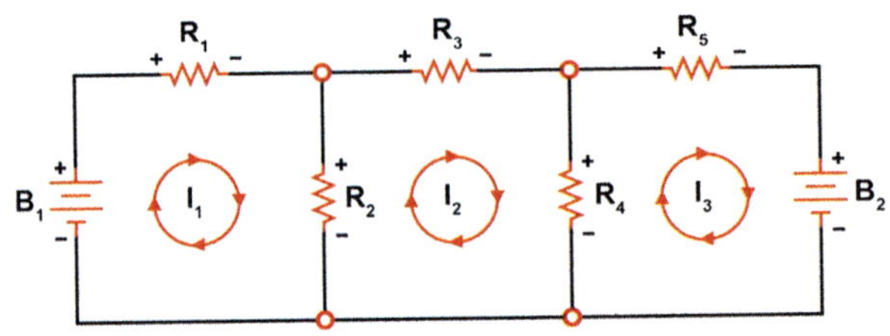

Figure 2.44. *Complex circuit setup for mesh current analysis.*

Source: https://www.allaboutcircuits.com/textbook/direct-current/chpt-10/mesh-current-method/.

Our three KVL equations using the mesh current method are:

$V_{B1} - I_1R_1 - (I_1 - I_2) R_2 = 0$ KVL of the left loop (121)

$(I_2 - I_1)R_2 - I_2R_3 - (I_2 - I_3) R_4 = 0$ KVL of the middle loop (122)

$(I_3 - I_2)R_4 - I3R_5 - V_{B2} = 0$ KVL of the right loop (123)

This approach has a distinct advantage because it requires fewer equations to solve, particularly when solving simultaneous equations manually without the aid of a calculator or computer.

ROLE MODEL

GUSTAV KIRCHHOFF

Gustav Robert Kirchhoff (12 March 1824 – 17 October 1887) was a German physicist and mathematician who contributed to the fundamental understanding of electrical circuits, spectroscopy, and the emission of black-body radiation by heated objects. He coined the term black-body radiation in 1860.

Source: https://upload.wikimedia.org/wikipedia/commons/thumb/f/fe/Gustav_Robert_Kirchhoff.jpg/220px-Gustav_Robert_Kirchhoff.jpg.

Several different sets of concepts are named "Kirchhoff's laws" after him, which include Kirchhoff's circuit laws, Kirchhoff's law of thermal radiation, and Kirchhoff's law of thermochemistry. The Bunsen–Kirchhoff Award for spectroscopy is named after Kirchhoff and his colleague, Robert Bunsen.

Life and Work:

Gustav Kirchhoff was born on 12 March 1824 in Königsberg, Prussia, the son of Friedrich Kirchhoff, a lawyer, and Johanna Henriette Wittke. His family were Lutherans in the Evangelical Church of Prussia. He graduated from the Albertus University of Königsberg in 1847 where he attended the mathematico-physical seminar directed by Carl Gustav Jacob Jacobi, Franz Ernst Neumann and Friedrich Julius Richelot. In the same year, he moved to Berlin, where he stayed until he received a professorship at Breslau. Later, in 1857, he married Clara Richelot, the daughter of his mathematics professor Richelot. The couple had five children. Clara died in 1869. He married Luise Brömmel in 1872.

Kirchhoff formulated his circuit laws, which are now ubiquitous in electrical engineering, in 1845, while he was still a student. He completed this study as a seminar exercise; it later became his doctoral dissertation. He was called to the University of Heidelberg in 1854, where he collaborated in spectroscopic work with Robert Bunsen. In 1857, he calculated that an electric signal in a resistanceless wire travels along the wire at the speed of light. He proposed his law of thermal radiation in 1859, and gave a proof in 1861. Together Kirchhoff and Bunsen invented the spectroscope, which Kirchhoff used to pioneer the identification of the elements in the Sun, showing in 1859 that the Sun contains sodium. He and Bunsen discovered caesium and rubidium in 1861. At Heidelberg he ran a mathematico-physical seminar, modeled on Franz Ernst Neumann's, with the mathematician Leo Koenigsberger. Among those who attended this seminar were Arthur Schuster and Sofia Kovalevskaya.

He contributed greatly to the field of spectroscopy by formalizing three laws that describe the spectral composition of light emitted by incandescent objects, building substantially on the discoveries of David Alter and Anders Jonas Ångström. In 1862, he was awarded the Rumford Medal for his researches on the fixed lines of the solar spectrum, and on the inversion of the bright lines in the spectra of artificial light. In 1875 Kirchhoff accepted the first chair dedicated specifically to theoretical physics at Berlin.

He also contributed to optics, carefully solving the wave equation to provide a solid foundation for Huygens' principle (and correct it in the process).

In 1864, he was elected as a member of the American Philosophical Society.

In 1884, he became foreign member of the Royal Netherlands Academy of Arts and Sciences.

Kirchhoff died in 1887, and was buried in the St Matthäus Kirchhof Cemetery in Schöneberg, Berlin (just a few meters from the graves of the Brothers Grimm). Leopold Kronecker is buried in the same cemetery.

Kirchhoff's Circuit Laws:

Kirchhoff's first law is that the algebraic sum of currents in a network of conductors meeting at a point (or node) is zero. The second law is that in a closed circuit, the directed sums of the voltages in the system is zero.

Kirchhoff's Three Laws of Spectroscopy:

1. A solid, liquid, or dense gas excited to emit light will radiate at all wavelengths and thus produce a continuous spectrum.
2. A low-density gas excited to emit light will do so at specific wavelengths, and this produces an emission spectrum.
3. If light composing a continuous spectrum passes through a cool, low-density gas, the result will be an absorption spectrum.

Kirchhoff did not know about the existence of energy levels in atoms. The existence of discrete spectral lines was known since Fraunhofer discovered them in 1814. And that the lines formed a discrete mathematical pattern was described by Johann Balmer in 1885. Joseph Larmor explained the splitting of the spectral lines in a magnetic field known as the Zeeman Effect by the oscillation of electrons. But these discrete spectral lines were not explained as electron transitions until the Bohr model of the atom in 1913, which helped lead to quantum mechanics.

- **Kirchhoff's Law of Thermal Radiation:** It was Kirchhoff's law of thermal radiation in which he proposed an unknown universal law for radiation that led Max Planck to the discovery of the quantum of action leading to quantum mechanics.
- **Kirchhoff's Law of Thermochemistry:** Kirchhoff showed in 1858 that, in thermochemistry, the variation of the heat of a chemical reaction is given by the difference in heat capacity between products and reactants:

$$\left(\frac{\partial \Delta H}{\partial T}\right)_p = \Delta C_p.$$

Integration of this equation permits the evaluation of the heat of reaction at one temperature from measurements at another temperature.

- **Kirchhoff's Theorem in Graph Theory:** Kirchhoff also worked in the mathematical field of graph theory, in which he proved Kirchhoff's matrix tree theorem.

CLASS ACTIVITY

A series circuit has a 120 V power source, a 10 Ω resistor, and a 15 Ω resistor. Students calculate the current flowing in the circuit and the voltage drop across each of the resistor.

SUMMARY

- A network comprises number of a circuit elements interconnected in a required manner and energized by one or more voltage or current sources. Network analysis involves the calculation of voltage and currents in the circuit. It may be required to calculate the current flowing through a particular circuit element or the potential difference between any two points in the network.
- Series circuit is an electrical circuit configuration where components are linked to each other end-to-end, creating a single path for the current flow. Thus, the current in the circuit will flow through every component successively.
- In a series circuit, the same amount of current flows through each component in the circuit. This is because there is only one path for the current flow. Since electric charge flows through conductors like marbles in a tube, the rate of flow (marble speed) at any point in the circuit (tube) at any specific point in time must be equal.
- A parallel circuit is where the electric current has multiple paths to flow through. It has two or more paths for current flow. Parallel circuits are one of the two types of circuits in which the devices have a common ending point making them appear to be in parallel.
- When two or more resistors are individually connected between two separate points (nodes) in a circuit, they are in parallel with each other. A parallel circuit provides more than one path for current.
- The voltage across any given branch of a parallel circuit is equal to the voltage across each of the other branches in parallel. As you know, each current path in a parallel circuit is called a branch.
- Kirchhoff's voltage law deals with voltages in a single closed path. Kirchhoff's current law applies to currents in multiple paths.
- When resistors are connected in parallel, the total resistance of the circuit decreases. The total resistance of a parallel circuit is always less than the value of the smallest resistor.
- A current source is a type of energy source that provides a constant current to a load even if the resistance of that load changes. A transistor can be used as a current source; therefore, current sources are important in electronic circuits.
- A parallel circuit acts as a current divider because the current entering the junction of parallel branches "divides" up into several individual branch currents.
- Total power in a parallel circuit is found by adding up the powers of all the individual resistors, the same as for series circuits.
- The node voltage method is a DC network analysis technique based on the structured application of Kirchhoff's current law. This method involves converting voltage sources to current sources and replacing resistances with equivalent conductances.

- The mesh current method is a network analysis technique where mesh (or loop) current directions are assigned arbitrarily, and then Kirchhoff's voltage law (KVL) and Ohm's law are applied systematically to solve for the unknown currents and voltages.

REVIEW QUESTIONS

1. What is a series circuit?
2. How to calculate voltage drop in a series circuit? Describe.
3. Describe the applications of series circuit.
4. What is a parallel circuit?
5. How are the resistors connected in a parallel circuit?
6. State Kirchhoff's current law in two ways.
7. Write the general current-divider formula.
8. What is the purpose of a shunt resistor in an analog meter movement?
9. What is the node voltage method?
10. What is the mesh current method?

REFERENCES

1. Alexander, C., & Sadiku, M. (2007). *Fundamentals of Electric Circuits* (p. 392). McGraw-Hill. ISBN 978-0-07-128441-7.
2. Nilsson, J., & Riedel, S. (2015). *Electric Circuits* (p. 85). Pearson Education Limited. ISBN 978-1-292-06054-5.
3. Oldham, K. T. S. (2008). *The doctrine of description: Gustav Kirchhoff, Classical Physics, and the "Purpose of all Science" in 19th-century Germany* (p. 52). Ph. D. dissertation, University of California, Berkeley. ProQuest Dissertations Publishing. (Docket 3331743).
4. Paul, C. R. (2001). *Fundamentals of Electric Circuit Analysis*. John Wiley & Sons. ISBN 0-471-37195-5.
5. Williams, T. (2005). *The Circuit Designer's Companion*. Butterworth-Heinemann. ISBN 0-7506-6370-7.

CHAPTER

3 AC Network Analysis

LEARNING OBJECTIVES

After studying this chapter, you will be able to:

- Discuss on AC fundamentals
- Describe the network theorems
- Explain the single-phase AC circuit
- Define the resonance circuit

KEY TERMS FROM THIS CHAPTER

Alternating current (AC)
Capacitive Circuit
Direct current (DC)
Frequency Selectivity
Inductive Circuit
Parallel LC Circuit
Series RC Circuit
Square waves
Triangular waves
Audio and radio signals
Current Change
Electric circuit theorems
Fully Resistive Circuit
linear network
Reciprocity theorem
Series Resonance
Superposition theorem
Voltage Change

INTRODUCTION

AC is an electric current that changes direction periodically and varies in magnitude continuously over time, unlike DC which flows in only one direction. AC is commonly used to power businesses and homes, and is the type of electricity used when plugging in appliances like kitchen appliances, televisions, fans, and lamps. The terms AC and DC are often used to represent alternating and direct current, respectively, particularly when referring to current or voltage. The typical waveform of AC in electric circuits is a sine wave, where the positive half-period aligns with the current direction. Different waveforms, such as triangular waves or square waves, are utilized in specific applications like guitar amplifiers. Examples of alternating current include audio and radio signals transmitted through electrical wires. This type of current carries information like sound, images, and video through modulation of an AC carrier signal. The frequencies of these currents are typically higher than those used in power transmission (Kuehni, & Lorraine, 1938).

3.1. AC FUNDAMENTALS

Essential principles of AC (Alternating Current) focus on electrical currents that change direction periodically, unlike DC (Direct Current) which flows in a single direction. The main feature of AC is its sinusoidal waveform, commonly identified by characteristics like amplitude (peak value), frequency (number of cycles per second, measured in Hertz), and phase (the waveform's position compared to a reference point). The voltage in an AC system can be expressed as $V(t)=V_m \sin(\omega t+\phi)$ where V_m is the maximum voltage, ω is the angular frequency ($\omega=2\pi f$, with f being the frequency), and ϕ is the phase angle.

AC systems are widely used because they offer numerous benefits in power generation, transmission, and distribution. High voltage AC transmission helps decrease energy losses when delivering electricity over long distances, and transformers make it simple to adjust AC voltages to suit different needs. In AC circuits, the presence of inductors and capacitors introduces reactance, along with resistance, to create impedance (Z), a complex quantity that combines resistive and reactive effects. The connection between voltage and current in these circuits is explained using complex numbers, allowing for phasor analysis to make calculations easier. AC circuits exhibit power through three distinct quantities: real power (P), representing work performed, reactive power (Q), which alternates between the source and load, and apparent power (S), the result of multiplying the RMS voltage and current, showcasing the overall power supplied. The power factor, which is the ratio of real power to apparent power, signifies the efficiency of power utilization. A power factor of 1 corresponds to purely resistive loads, while values below 1 point to the existence of reactance in the circuit (N. N. Bhargava & D. C. Kulshreshtha (1983)).

AC fundamentals also include understanding RMS (Root Mean Square) values, which provide a measure of the equivalent DC value that would deliver the same power to a load. For a sinusoidal voltage, the RMS value is $V_{RMS}= \frac{V_m}{\sqrt{2}}$.. This concept is crucial for accurately measuring and comparing AC and DC power levels.

Overall, AC fundamentals encompass the principles of sinusoidal waveforms, impedance, phasor analysis, and power relationships, forming the basis for analyzing and designing AC electrical systems.

3.1.1. Passive Components in AC Circuits

Resistance can be denoted as R, capacitance as C, and inductance as L. Regardless of whether resistors are utilized in DC or AC circuits, their resistance value remains constant regardless of the supply frequency. This is due to resistors being considered pure, with parasitic properties like zero inductance (L = 0) and infinite capacitance (C = ∞). In fully resistive circuits, voltage and current are always in phase, allowing for the calculation of power consumption at any moment by multiplying the voltage and current.

Capacitors and inductors have a unique type of AC resistance called reactance ($_{XL}$ and X_C), which was previously mentioned. Reactance impedes the flow of current, but unlike a resistor with a fixed resistance value, the amount of reactance in a capacitor or inductor is not constant. The reactance of these components depends on the frequency of the supply current and the DC value of the element itself.

In AC circuits, there is a well-known list of passive components along with equations that can be used to determine their impedance and reactance value for circuit current. It is important to note that in this discussion, we have assumed ideal components without any resistance. However, in practical applications, components typically have some resistance in addition to their other properties.

1. **Fully Resistive Circuit:** Resistors control the flow of current in a specific direction by impeding, regulating, or adjusting it. They can also limit the voltage within an electrical circuit based on this current flow. Resistors have a resistance measure known as impedance, denoted as R, which is measured in Ohms (Ω). Resistors come in either fixed values or variable values, such as potentiometers. Impedance and current value can be found using the following equations:

$$Z = \frac{V_R}{I_R} = R$$

$$Z = \angle 0^\circ = R + J0$$

$$I_S = \frac{V_S}{R} \quad (1)$$

2. **Fully Capacitive Circuit:** A capacitor is a component with the ability to store energy in the form of an electrical charge, similar to a small battery. The capacitance of a capacitor is measured in Farads (F), and in a DC circuit, it has infinite impedance (acting as an open circuit). Conversely, at very high frequencies, a capacitor has zero impedance (acting as a short circuit). Impedance and current value can be found using the following equations:

$$X_C = \frac{V_C}{I_C} = \frac{1}{2\pi f C}$$

$$Z = \angle -90^\circ = 0 - jX_C$$

$$I_S = \frac{V_S}{X_S} \quad (2)$$

3. **Fully Inductive Circuit:** An inductor is composed of a wire coil that generates a magnetic field either within the coil itself or in a central core when an electrical current passes through it. The inductance of an inductor is typically measured in Henries (H). In direct current (DC) circuits, an inductor acts as a short circuit with zero impedance. However, at high frequencies, an inductor behaves as an open circuit with infinite impedance. Impedance and current value can be found using the following equations:

$$X_L = \frac{V_L}{I_L} = 2\pi f L$$

$$Z = \angle 90^\circ = 0 + jX_C$$

$$I_S = \frac{V_S}{X_L} \quad (3)$$

3.1.2. Series AC Circuits

We can connect passive components together in series combinations in AC circuits to form RC, RL, and LC circuits, as explained below (Figure 3.1).

- **Series RC Circuit:** The circuit and the equation for the series RC circuit are:

$$Z = \sqrt{R^2 + X_C^2}$$

$$Z = \angle - \phi = R - jX_C$$

$$\phi(90 \to 0) = tan^{-1}(-\tfrac{X_C}{R})$$

$$I_S = \frac{V_S}{\sqrt{R^2+X_C^2}}$$

$$V_S = \sqrt{V_R^2 + V_C^2} \qquad (4)$$

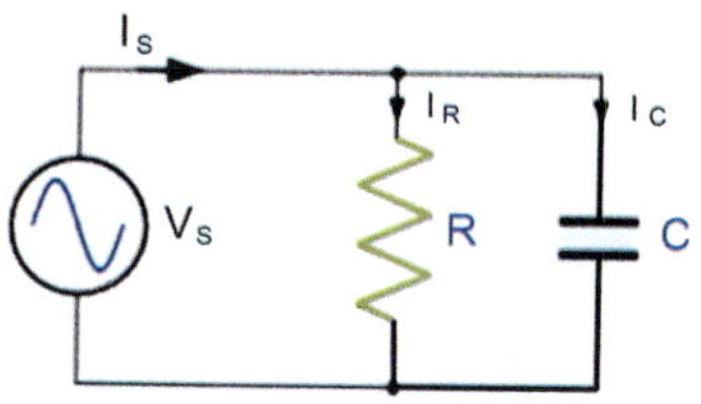

Figure 3.1. *Series RC circuit.*

Source: https://www.linquip.com/blog/what-is-ac-circuit/.

- **Series RL Circuit:** The circuit diagram and the equation for the series RL circuit are (Figure 3.2):

$$Z = \sqrt{R^2 + X_C^2}$$

$$Z = \angle - \phi = R - jX_C$$

$$\phi(90 \to 0) = tan^{-1}(-\tfrac{X_C}{R})$$

$$I_S = \frac{V_S}{\sqrt{R^2+X_C^2}}$$

$$V_S = \sqrt{V_R^2 + V_C^2} \qquad (5)$$

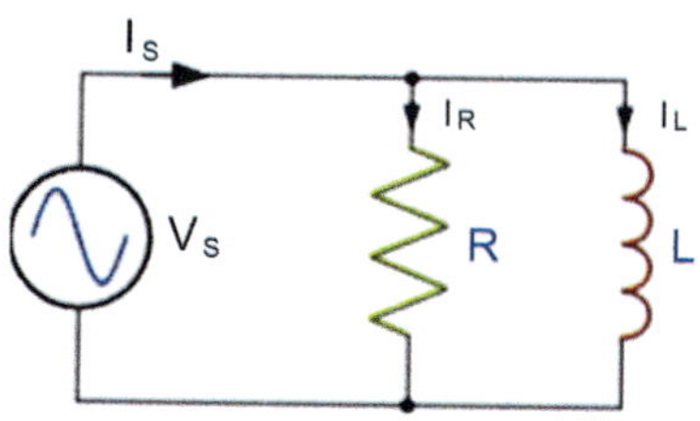

Figure 3.2. *Series RL circuit.*

Source: https://www.linquip.com/blog/what-is-ac-circuit/.

- **Series LC Circuit**

The circuit diagram and the equation for the series LC circuit are (Figure 3.3):

$$Z = \sqrt{(X_C - X_L)^2}$$

$$\therefore Z = X_C - X_L \; or \; X_L - X_C$$

$$Z = \angle(\phi_1 + \phi_2) = 0 + jX_L - jX_C$$

$$f_r = \frac{1}{2\pi\sqrt{LC}}$$

$$I_S = I_L = I_C \qquad (6)$$

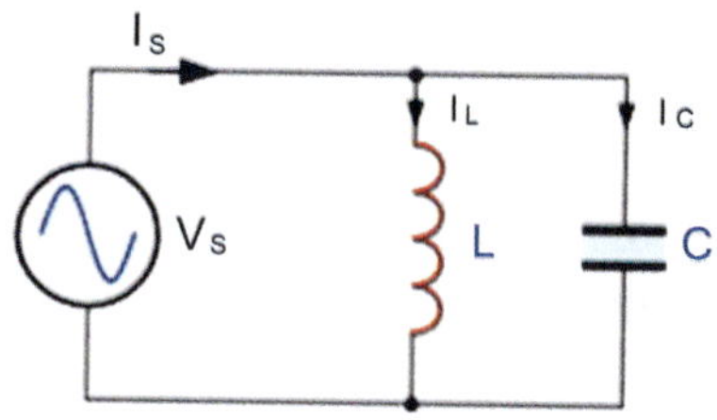

Figure 3.3. *Series LC circuit.*

Source: https://www.linquip.com/blog/what-is-ac-circuit/.

3.1.3. Parallel AC Circuits

We can connect passive components together in series combinations in AC circuits to form RC, RL and LC circuits, as explained below.

- **Parallel RC Circuit:** The circuit diagram and the equation for the parallel RC circuit are (Figure 3.4):

$$I_R = \frac{V_s}{R}, I_C = \frac{V_S}{X_C}$$

$$I_S = \sqrt{I_R^2 + I_C^2}$$

$$\phi = tan^{-1}(\tfrac{I_C}{I_R})$$

$$Y = \tfrac{1}{Z} = \sqrt{G^2 + B_c^2}$$

$$V_S = V_C = V_R \qquad (7)$$

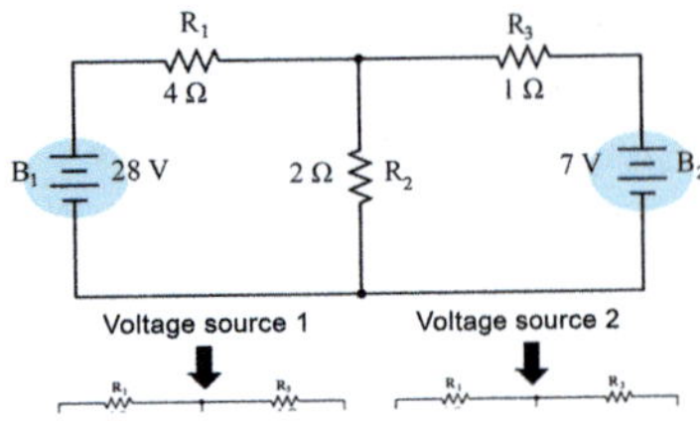

Figure 3.4. *Parallel RLC circuit.*

Source: https://www.linquip.com/blog/what-is-ac-circuit/.

- **Parallel RL Circuit:** The circuit diagram and the equation for the parallel RL circuit are (Figure 3.5):

$$I_L = \frac{V_s}{R}, I_C = \frac{V_S}{X_L}$$

$$I_S = \sqrt{I_R^2 + I_L^2}$$

$$\phi = tan^{-1}(\frac{I_L}{I_R})$$

$$Y = \frac{1}{Z} = \sqrt{G^2 + B_L^2}$$

$$V_S = V_L = V_R \quad (8)$$

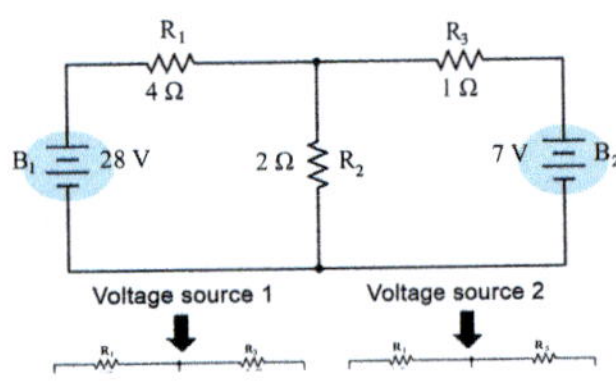

Figure 3.5. *Parallel RL circuit.*

Source: https://www.linquip.com/blog/what-is-ac-circuit/.

- **Parallel LC Circuit:** The circuit diagram and the equation for the parallel LC circuit are (Figure 3.6):

$$B_L = \frac{1}{X_L}, B_C = \frac{1}{X_C}$$

$$Y = \frac{1}{Z} = B_L + B_C$$

$$f_R = \frac{1}{2\pi\sqrt{LC}}$$

$$V_S = V_L = V_C \quad (9)$$

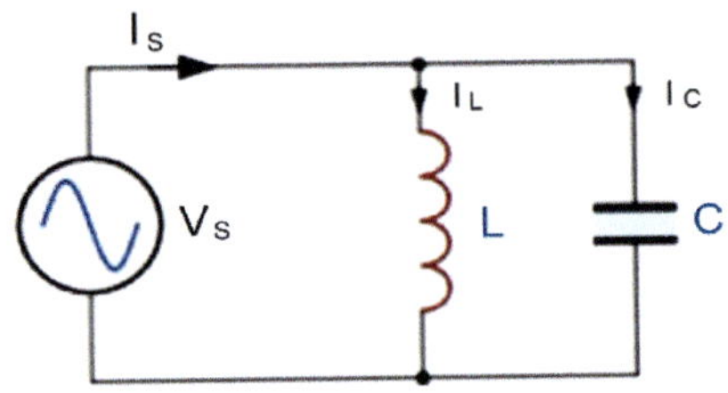

Figure 3.6. *Parallel LC circuit.*

Source: https://www.linquip.com/blog/what-is-ac-circuit/.

3.2. NETWORK THEOREMS

Electric circuit theorems are useful tools for determining voltage and current values in complex circuits with multiple loops. These theorems utilize fundamental principles and mathematical equations to analyze the basic elements of electrical parameters, such as voltage, current, and resistance. The fundamental theorems in circuit analysis consist of foundational principles like Superposition, Tellegen's, Norton's, Maximum Power Transfer, and Thevenin's theorems. Another set of network theorems commonly utilized in circuit analysis are Compensation, Substitution, Reciprocity, Millman's, and Miller's theorems.

All the network theorems are briefly discussed below:

1. **Super Position Theorem:** The Superposition theorem allows for the analysis of currents and voltages in a circuit with multiple sources by considering each source individually. It states that in a linear network with various voltage or current sources and resistances, the current in any branch is the sum of the currents caused by each source acting alone (Figure 3.7).

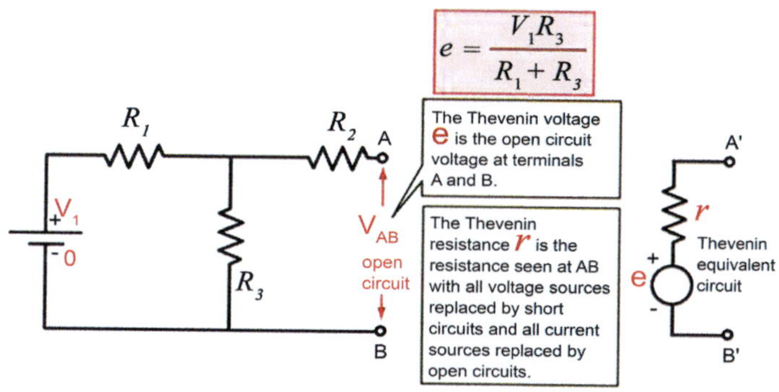

Figure 3.7. *Super position theorem.*

Source: https://www.elprocus.com/basics-of-network-theorems-in-electrical-engineering/.

The Superposition theorem is specifically utilized in linear networks, including both AC and DC circuits, to create Thevenin and Norton equivalent circuits. The circuit shown above is divided into two separate circuits based on the theorem mentioned. This division simplifies the overall circuit design, making it easier to analyze. By simplifying each individual circuit and then combining them again, one can easily determine specific parameters such as voltage drops across each resistor, node voltages, currents, and more.

2. **Thevenin's Theorem:** A linear network consisting of a number of voltage sources and resistances can be replaced by an equivalent network having a single voltage source called Thevenin's voltage (V_{th}) and a single resistance called (R_{th}) (Figure 3.8).

Vth = Voc

Rth = Voc/Isc

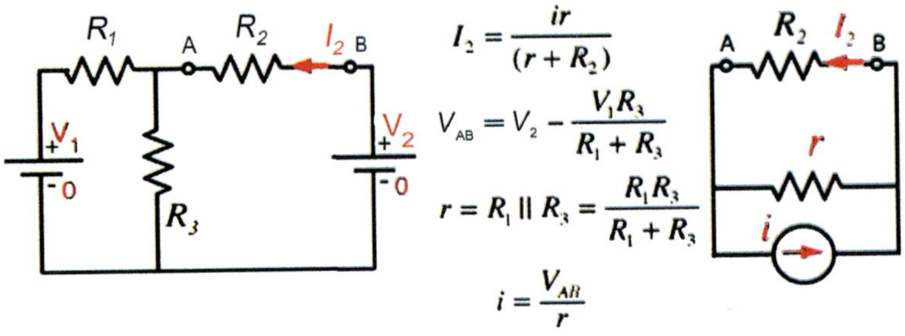

Figure 3.8. *Thevenin's theorem.*

Source: https://www.elprocus.com/basics-of-network-theorems-in-electrical-engineering/.

The above figure explains how this theorem is applicable for circuit analysis. Thevinens voltage is calculated by the given formula between the terminals A and B by breaking the loop at the terminals A and B. Also, Thevinens resistance or equivalent resistance is calculated by shorting voltage sources and open circuiting current sources as shown in the figure.

This theorem can be applied to both linear and bilateral networks. It is mainly used for measuring the resistance with a Wheatstone bridge.

3. **Norton's Theorem:** This theorem asserts that a linear circuit with multiple energy sources and resistances can be substituted by a single fixed current generator connected in parallel with a single resistor (Figure 3.9).

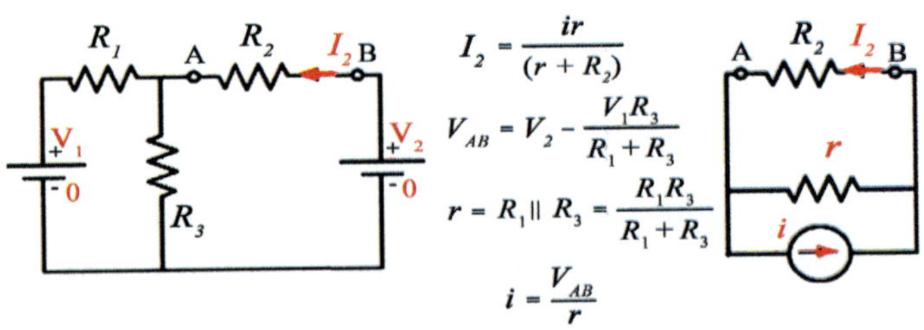

Figure 3.9. *Norton's theorem.*

Source: https://www.elprocus.com/basics-of-network-theorems-in-electrical-engineering/.

This concept is similar to Thevenin's theorem, where we calculate Thevenin's equivalent voltage and resistance values. However, in this case, we determine the equivalent current values. The process of finding these values is shown as given in the example within the Figure 3.9.

4. **Maximum Power Transfer Theorem:** This theorem describes the optimal conditions for maximum power transfer to a load within different circuit scenarios. It states that the power transferred from a source to a load is at its maximum when the resistance of the load matches the internal resistance of the source in the network. In AC circuits, the load impedance should be matched with the source impedance for optimal power transfer, regardless of the load's power factor (Figure 3.10).

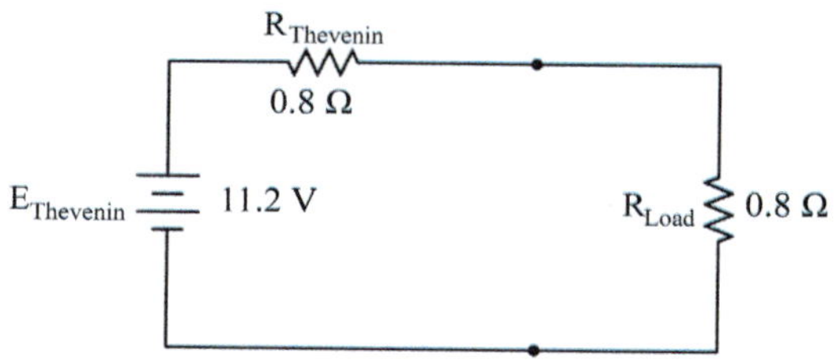

Figure 3.10. *Maximum power transfer theorem.*

Source: https://www.elprocus.com/basics-of-network-theorems-in-electrical-engineering/.

For instance, the above figure depicts a circuit diagram wherein a circuit is simplified up to a level of source with internal resistance using Thevenin's theorem. The power transfer will be maximum when this Thevinens resistance is equal to the load resistance. The Practical application of this theorem includes an audio system wherein the resistance of the speaker must be matched to the audio power amplifier to obtain maximum output.

5. **Reciprocity Theorem:** The reciprocity theorem allows for the identification of the second solution without the need for additional analysis after examining the circuit for one solution. It asserts that within a linear passive bilateral network, the excitation source and its response can be swapped (Figure 3.11).

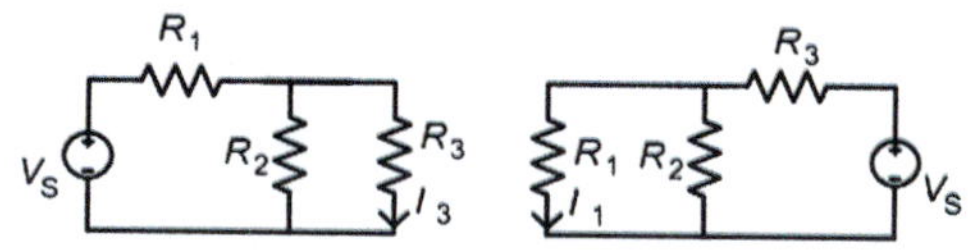

Figure 3.11. *Reciprocity theorem.*

Source: https://www.elprocus.com/basics-of-network-theorems-in-electrical-engineering/.

In the Figure 3.11 shown, the current in the R_3 branch is labeled as I_3 and is supplied by a single source V_s. If this source is moved to the R_3 branch and the original source location is shorted, then the current flowing from the original location (labeled as I_1) will be equal to I_3. This method allows us to find equivalent solutions for the circuit by analyzing it with a single solution.

6. **Compensation Theorem:** In a bilateral active network, adjusting the impedance from its original value to a different value carrying current I will induce changes in other branches equivalent to those caused by inserting a voltage source in the modified branch with a negative sign. This compensation theorem can be seen Figure 3.12, illustrating its application in circuit analysis (Figure 3.12).

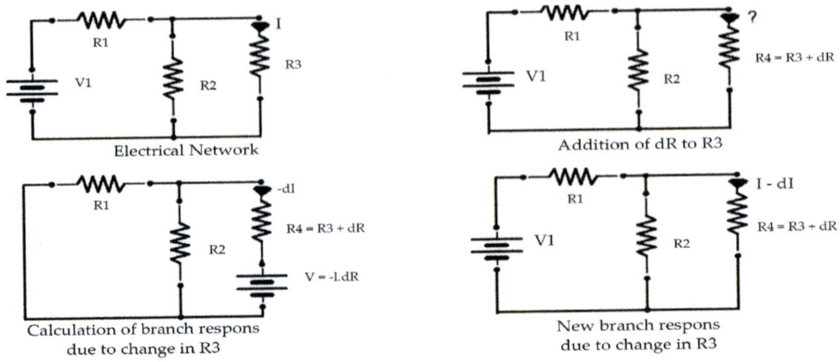

Figure 3.12. *Compensation theorem.*

Source: https://www.elprocus.com/basics-of-network-theorems-in-electrical-engineering/.

7. **Millman's Theorem:** According to Millman's theorem, multiple voltage sources with finite internal resistance can be simplified into a single voltage source with series equivalent impedance when operating in parallel. The equivalent voltage for these sources can be calculated using the formula provided in the figure above (Figure 3.13).

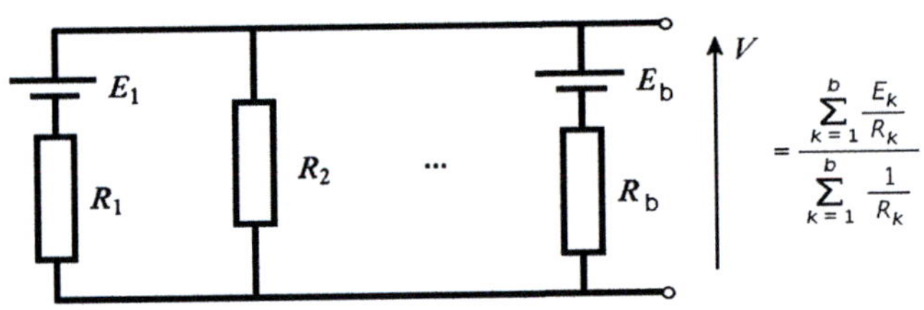

Figure 3.13. *Millman's theorem.*

Source: https://www.elprocus.com/basics-of-network-theorems-in-electrical-engineering/.

8. **Tellegen's Theorem:** This theorem applies to circuits containing linear or nonlinear, passive or active, and hysteric or non-hysteric networks. It asserts that the total instantaneous power in a circuit with n branches is always zero.

$$\sum_{k=1}^{n} P_k = V_k \times I_k = 0 \qquad (10)$$

9. **Substitution Theorem:** The substitution theorem asserts that replacing one branch in a network with another branch, as long as the new branch has the same terminal voltages and currents as the original branch, will not disrupt the overall currents and voltages in the network. This theorem is applicable to both linear and nonlinear circuits.

10. **Miller's Theorem:** This theorem explains that in a linear circuit, if there is a branch with impedance Z between two nodes with nodal voltages, that branch can be substituted by two branches connecting the nodes to the ground with

two impedances. Utilizing this theorem is not only beneficial for simplifying circuits, but also for creating new electronic circuit designs involving impedance changes (Figure 3.14).

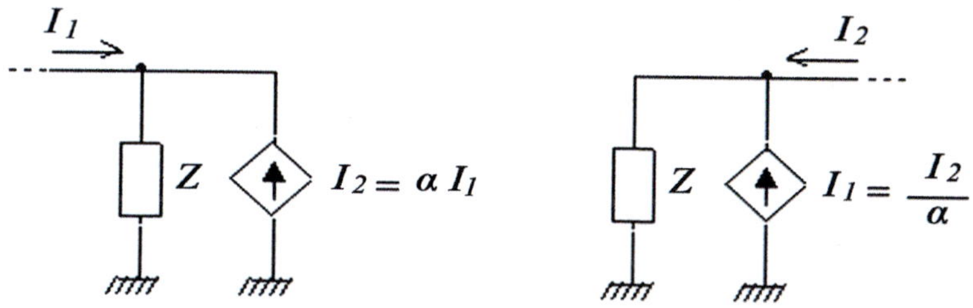

Figure 3.14. *Miller's theorem.*

Source: https://www.elprocus.com/basics-of-network-theorems-in-electrical-engineering/.

These are all basic network theorems used widely in the electrical or electronic circuit analysis. We hope that you might have got some basic ideas about all these theorems.

3.3. SINGLE-PHASE AC CIRCUIT

Alternating Current (AC) is an electric power type that oscillates in both magnitude and direction at consistent intervals. AC circuits can provide power in either a Single Phase or Three Phase system. In a single-phase system, circuit, or device, there is only one alternating voltage source present. In electrical engineering, single-phase voltage source power involves distributing electricity through a system where all supply voltages change simultaneously.

3.3.1.1. Single-Phase Power Supply

Single-phase voltage power is a two-wire AC power system comprising a single phase wire and a single neutral wire. This type of power, also known as residential voltage, is commonly used in homes and for small machines due to its lower electricity demand. Single-phase power connections utilize both the neutral and phase supply cables for average power transmission. In contrast, three-phase power is used for heavy loads and can operate large machinery in factories (Figure 3.15).

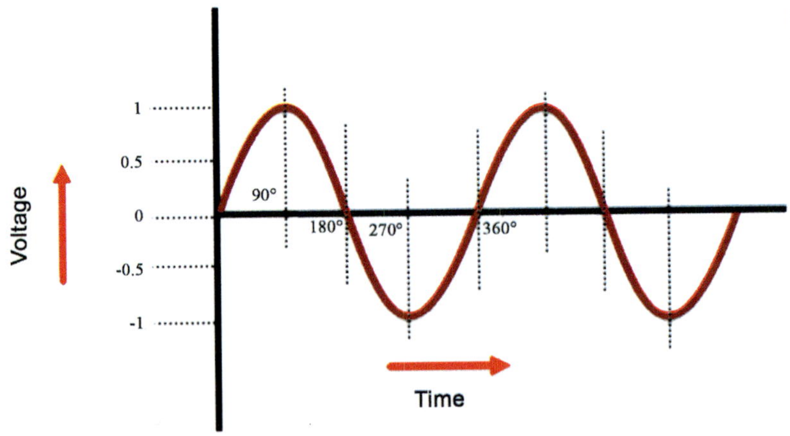

Figure 3.15. *Sine wave.*

Source: https://www.bartleby.com/subject/engineering/electrical-engineering/concepts/single-phase-ac-circuit.

In a single-phase power connection, the split-phase voltage begins at 230 Volts Amperes with a frequency of approximately 50 Hertz. Due to the fluctuation of the high voltage source across the single-phase connection, there is a lack of consistent power delivery to the load.

3.3.1.2. AC Circuit

Switching circuits are powered by a Switch Source, which can be either current or voltage. Alternating Voltage or Current occurs when the value of voltage or current changes and reverses direction periodically. Nowadays, most home and industrial systems and appliances run on alternating line current. Devices that use direct current are charged by converting alternating line current. This method was developed in the 1980s by Nikola Tesla as a solution to the limitations of Thomas Edison's DC generators (Figure 3.16).

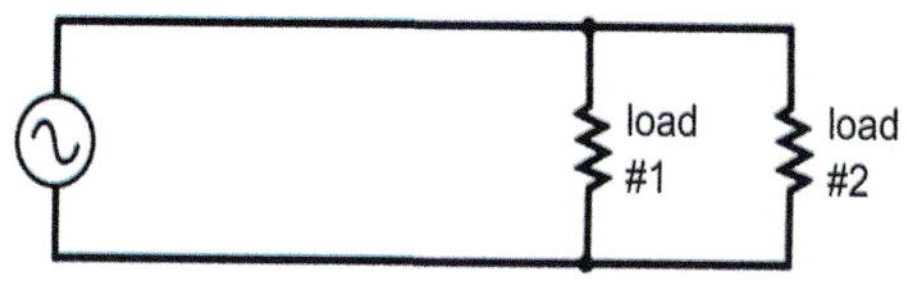

Figure 3.16. *Circuit with load.*

Source: https://www.bartleby.com/subject/engineering/electrical-engineering/concepts/single-phase-ac-circuit.

3.3.1.3. Various Terms Commonly Used in the AC Circuit

- Apparent power is the product of root mean square (rms) values of current and voltage. It is measured in volt-amperes or kilo-volt amperes (kVA).
- Reactive power is the product of apparent power and sine of the angle between voltage and current. This is also called wattless power and is expressed in reactive volt-amperes or kilo-volt amperes reactive (kVA R).
- True power is the product of the apparent power and the power factor. It is expressed in watts or kilo-watts (kW).
- Power factor is the ratio of true power to apparent power. It is a diensionless quantity. It is never be greater than 1 and is expressed either as fraction or as a percentage.

Q–Factor of Coil is the reciprocal of power factor. It is also known as the figure of merit of a coil or quality factor.

3.3.1.4. Phasor and Phasor Diagram in a Single-Phase AC Circuit

Vectors possess both magnitude and direction. The value of a sinusoidal variable can be depicted by a rotating line known as a Phasor. The Phasor is a straight line that rotates counterclockwise at a constant angular speed. The diagram below illustrates the waveform and mathematical representation of alternating quantities, which are sinusoidal in nature. Phasors are commonly used to represent these quantities (Figure 3.17).

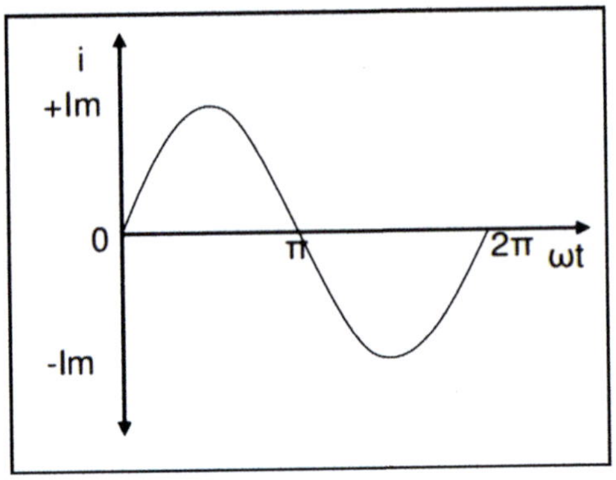

Figure 3.17. *Phasor diagram in a single-phase AC circuit.*

Source: https://www.bartleby.com/subject/engineering/electrical-engineering/concepts/single-phase-ac-circuit.

3.3.1.5. EMF Generation in a Single-Phase AC Circuit

In Single-phase EMF generation an alternating voltage can be generated

- By rotating the coil in the magnetic field
- By rotating the magnetic field inside the immovable coil

The amount of electricity generated depends on

- Number of turns on the coil
- Field strength
- Speed

3.3.1.6. Use of a Single-Phase AC Circuit

Single-phase power provides electricity for living rooms and household appliances, as these devices typically consume lower levels of energy. Examples of such appliances are fans, heaters, televisions, refrigerators, and lamps.

3.4. RESONANCE CIRCUIT

The phenomenon of resonance in electric circuits is essential for altering circuit behavior and transmitting electrical signals effectively. Resonance is critical in a wide range of applications, from radio frequency tuning to improving power transfer in electrical systems. It occurs at a constant frequency when impedance and reactance cancel each other out. In AC circuits, resonance happens when inductive reactance (X_L) and capacitive reactance (X_C) offset each other, resulting in zero net reactance. This leads to increased current flow and voltage amplitude, making the circuit more responsive to the applied AC frequency.

3.4.1. Key Components in Resonance

- **Inductive Reactance (X_L):** Inductive Reactance is the opposition that an inductor presents to an alternating current. It is directly proportional to the frequency of the AC signal. It is represented by (XL) and measured in ohms (Ω). It is mostly low for lower frequencies and high for higher frequencies. However, it is negligible for steady DC current. The inductive reactance formula is given as follows: $XL = 2\pi fL$ .As the frequency increases the inductive reactance also increases.
- **Capacitive Reactance (X_C):** A capacitor is a device used to store electrical energy. The capacitance of a capacitor determines the amount of charging a capacitor can achieve. The measure of the opposition to alternating current by the capacitor is called Capacitive Reactance. It is the resistance of a circuit element to changes in current or voltage. Its standard unit of measurement ohms (Ω). It is represented by the symbol Xc. It is inversely proportional to the frequency of the AC signal. As the frequency increases the capacitive reactance decreases. The capacitive reactance formula is given as follows: $XC = 1/2\pi fC$
- **Resonant Frequency (f_0):** The frequency at which the inductive and capacitive reactances are equal in magnitude but opposite in phase resulting in cancelation. It can be calculated using the formula:

$$f_0=1/2\pi\sqrt{(LC)}$$

Where L is the inductance and C is the capacitance.

- Resonant frequency in electronics is expressed when a circuit exhibits a maximum oscillatory response at a specific frequency. This is observed for a circuit that consists of an inductor and capacitor.

3.4.2. Effect of Resonance

Resonance, a phenomenon in which an object's frequency excites vibrations near the same frequency, is not only responsible for a singer breaking a glass with their voice but can also cause damage to high buildings, bridges, and electrical circuits. Earthquakes can also be triggered by resonance if their frequency matches that of a building, leading to potential structural damage. Before constructing any building or bridge, it is crucial to consider the effects of resonance to prevent potential disasters. In electrical circuits, resonance can increase current and voltage, resulting in a purely resistive impedance and maximum flow of electricity. If excessive voltage or current at the resonant frequency is applied, components can overheat and damage the circuit, leading to overcurrent in inductive and capacitive systems.

3.4.3. Characteristics of Resonance

- **Stability:** Maintaining the desired resonant frequency and response characteristics under varying loads and environmental conditions requires stability.
- **Current Change:** The current flowing through the circuit changes depending on the type of resonance.
- **Frequency Selectivity:** Resonance circuits are highly selective at resonant frequency the circuit exhibits maximum impedance (in a series resonance circuit) or minimum impedance (in a parallel resonance circuit), resulting in a sharp peak in the circuit's response. The sharp peak allows resonance circuits to filter out unwanted frequencies.
- **Voltage Change:** Individual components experience voltages higher than the source voltage at resonance.
- **Phase Shift:** In series resonance circuit voltage and current are in phase while in a parallel resonance circuit they are 180° out of phase.

3.4.4. Types of Resonance

There are many types of resonance:

- Series Resonance
- Parallel Resonance
- **Series Resonance:** The inductor (L), capacitor (C), and resistor (R) are connected in series in a series resonance circuit. At the resonant frequency (f_0), the inductive and capacitive reactance cancel each other out, resulting in a minimum impedance. This makes the circuit highly responsive to the applied frequency (Figures 3.18 and 3.19).

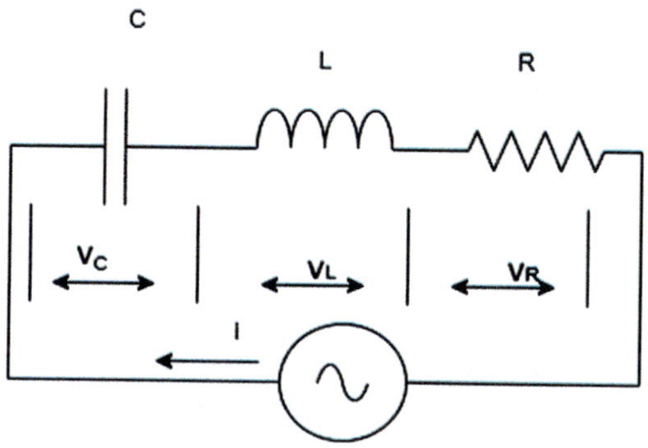

Figure 3.18. *Series resonance circuit.*

Source: https://www.geeksforgeeks.org/resonance-in-electric-circuits/.

In this circuit, the voltage across the inductor and capacitor is equal, i.e.,

VL=VC

At resonance frequency;

XL=XC

Where; XL is the inductive reactance and XC is the capacitive reactance. Voltage can be obtain by applying KVL to the series RLC circuit.

V=VR+VL+VC

If I is the current flowing through the circuit, then;

VR=IR

VL=IXL

VC=IXC

Therefore, the voltage equation can be written as;

V=IR+IXL+IXC

Also, the reactance XL and Xc are given by;

XL=jωL=j2πfL

XC=1/jωC=1/j2πfL

Therefore;

V=I[R+j(Ωl–1/ωC]

Hence, the above equation is in the form of V = IZ, where Z is called the impedance of the circuit i.e.,

Z=R+j(ωL–1/ωC)

Electrical Quantities and Parameters at Series Resonance

- **Resonance Frequency:** The supply frequency at which the inductive reactance and capacitive reactance become equal to each other is called the resonance frequency. Resonant frequency is expressed when a circuit exhibits a maximum oscillatory response at a specific frequency. This is observed for a circuit that consists of an inductor and capacitor. It is denoted by fr.

At series resonance;

XL=XC

ωL=1/ωC

Here, ω = ωr, angular resonance frequency.

ωr=1/√LC

The linear resonance frequency will be;

fr=1/2π√(LC)

- **Impedance:** Impedance is the total opposition that the circuit presents to the flow of alternating current. It is a combination of resistance and reactance due to the presence of resistive, capacitive, and inductive elements within the circuit. The impedance of a series RLC circuit is given by;

$Z=R+j(XL-Xc)$

At series resonance;

XL=XC

Therefore;

$Z=R$

- **Current:** Electric current is the movement of electric charge through the elements of a circuit in a continuous path. In a series circuit, all components, such as resistors, capacitors, and inductors, are connected in a line, allowing the same current to flow through each component. At series resonance, the inductive reactance (X_L) is equal to the capacitive reactance (X_C).

Then;

I=V/R

The current flowing from the source in the circuit is solely dependent on the resistance of the circuit, which acts as a bottleneck for the maximum current flow in the series RLC circuit. The diagram illustrates the connection between the current, impedance, and resonance frequency of the series resonance circuit (Figure 3.19).

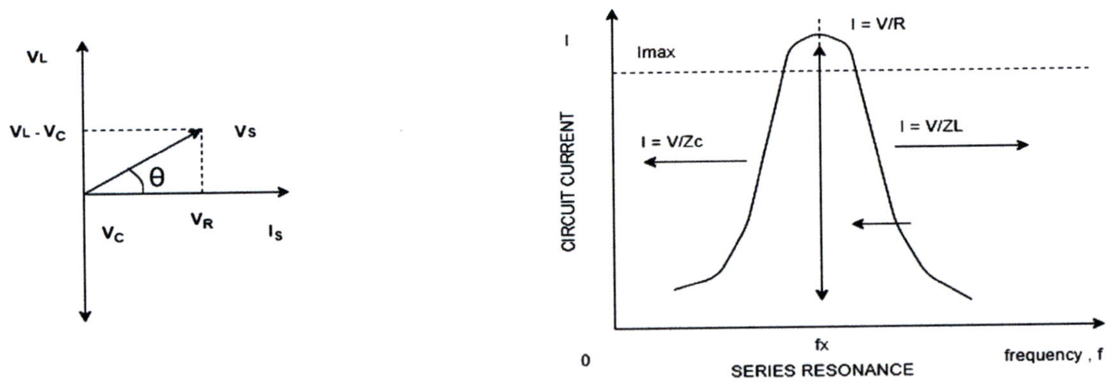

Figure 3.19. *Series resonance.*

Source: https://www.geeksforgeeks.org/resonance-in-electric-circuits/.

- **Quality Factor:** The quality factor of a series RLC circuit is defined as a ratio of energy stored in each cycle to the energy dissipated in each cycle, i.e.,

$$Q=1/R*\sqrt{L/C}$$

- **Parallel Resonance:** In a parallel resonance circuit, the inductor (L) and capacitor

(C) are connected alongside each other, with a resistor (R) usually connected in series with the inductor. When operating at the resonant frequency (f_0), the impedance of the inductor and capacitor balance each other, causing a significant spike in current flowing through the circuit (Figure 3.20).

The resonance occurs when the instantaneous values of currents IL and IC are equal and opposite to each other.

Applying KCL to this parallel RLC circuit we get;

I=IR+IL+IC

According to Ohm's law;

IR=V/R; IL=V/jXL; IC=V/(−jXC)

Therefore, the total circuit current is given by;

I=V/R+V/jXL−V/(jXC)

Hence, I=VY

Where;

Y=1/R+j(1/XC−1/XL)

Condition of Parallel Resonance

XL=XC

- Electrical Quantities at Parallel Resonance
  - **Resonance Frequency:** The resonance frequency of a parallel resonance circuit is the value of the supply frequency where the inductive reactance XL becomes equal to the capacitive reactance XC i.e.,

XL=XC

2πfL=1/2πfC

f2=1/4π2LC

fr=1/2π√LC

Also;

ωr=1/√LC

The resonant frequency of a parallel RLC circuit depends on the value of capacitance and inductance.

- **Impedance:** At resonance, the circuit's admittance reaches its lowest value, causing a disruption in the flow of electricity.

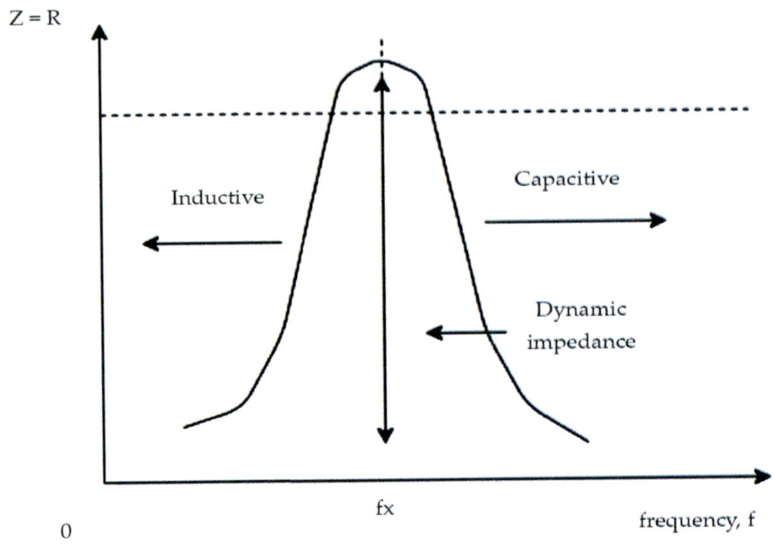

Figure 3.20. *Parallel resonance.*

Source: https://www.geeksforgeeks.org/resonance-in-electric-circuits/.

The impedance of the circuit is maximum at resonance and the circuit draws minimum current.

$$Z=R$$

- **Admittance:** Admittance measures the ability of current to flow through a device or circuit, and is the inverse of impedance, much like how conductance is the opposite of resistance. The SI unit for admittance is the siemens, denoted by the symbol S.

Admittance is given by:

$Y=1/R+j(1/XC-1/XL)$

At resonance: $XL=XC$

Putting it in the admittance equation we get

$Y=1/R$

At resonance the admittance Y being equal to the resistance reciprocal.

- **Voltage:** Voltage refers to the potential difference across each branch of the circuit. In Parallel circuit the voltage remains the same across each branch.

At parallel resonance;

XC=XL

Therefore;

I=VR

V=IR

At parallel resonance, the voltage across each element will be equal to the voltage across the resistance, representing the maximum voltage that appears across each element.

- **Current:** Current is the movement of electric charge through each separate branch of the circuit. In a parallel circuit, the total current is divided among the different branches according to their respective resistances or impedances.

The circuit current I at parallel resonant condition is given by;

I=IR

- **Quality Factor:** At parallel resonance;

fr=1/2π√LC

Q=R√C/L

- **Bandwidth***:* The difference in upper frequency and lower frequency denotes the bandwidth of the parallel resonance circuit. The power dissipation at the upper and lower frequencies is half of the full power dissipated at the resonance frequency fr.

The bandwidth of the parallel resonance circuit is expressed by the following formula.

BW=fupper-flower

BW=fr/Q

Table 3.1. Differentiate Between Series and Parallel Resonance

Series Resonance	Parallel Resonance
Series Resonance circuit is an acceptor circuit.	Parallel Resonance circuit is an rejecter circuit.
At resonance the impedance is a maximum.	At resonance the impedance is maximum nearly equal to infinity.
Current at resonance is maximum = V/R.	Current at resonance is minimum = V/(L/CR).
Power factor is unity.	Power factor is unity.
Series circuit magnifies voltage.	Parallel circuit magnifies current
The series resonance is widely used in tuning, oscillator circuits, voltage amplifiers, high frequency filters, etc.	The parallel resonance is used in current amplifiers, induction heating, filters, radio-frequency amplifiers, etc.

Source: https://www.geeksforgeeks.org/difference-between-series-resonance-and-parallel-resonance/

3.4.5. Application of Resonance

It has some vast practical applications in the electrical engineering field. Below are five applications of resonance in RLC circuits:

- Used for Voltage multiplier
- Used to filter a signal by blocking some frequencies and passing others.
- Used for Pulse discharge circuits
- Used in different types of oscillator circuit.
- These are used in the tuning circuits of analog radios.

Advantages of Resonance

- **Selective tuning:** Filtering out unwanted signals and focusing on the desired frequency band in communication systems.

- **Amplification:** Resonance circuits provide strong amplification to signals resulting in increased signal power. It is used in the applications communication systems.
- **Impedance Matching:** Resonance circuits are also used for impedance matching purposes ensuring maximum power transfer between the source and load. It is used in antennas and transmission lines.
- **Energy Efficiency:** Resonance circuits can store and transfer energy efficiently between the inductive and capacitive elements. The applications of energy efficiency are wireless power transfer systems and energy storage.

Disadvantages of Resonance

- **Bulkiness:** The inductive and capacitive components can be bulky and heavy limiting the downsize of resonant circuits.
- **Narrow Bandwidth:** Resonance circuits sometimes have a narrow bandwidth which means that they are effective for a limited range of frequencies. This limitation can restrict their capability in applications requiring broader frequency coverage.
- **Complex Design Requirements:** Designing resonance circuits with specific performance characteristics can be complex and require detailed analysis. Achieving desired resonance properties such as bandwidth and Q factor often necessitates careful consideration of circuit and component values.

CASE STUDY

FAILURE ANALYSIS OF ALTERNATING CURRENT

Failure analysis of alternating current (AC) systems is a critical aspect of ensuring the reliability and safety of electrical power systems. This case study focuses on a comprehensive failure analysis conducted on a 230 kV substation's AC auxiliary system that experienced a significant outage, affecting a large urban area. The substation in question was a crucial node in the power grid, supplying electricity to both residential and industrial sectors. The failure occurred during peak load conditions, resulting in a blackout and subsequent investigation to determine the root cause and prevent future occurrences.

The incident began with a sudden loss of power in the substation's auxiliary systems, which are essential for operating control systems, protection relays, and communication equipment. Initial observations indicated that the primary AC supply to the auxiliary systems had been interrupted. Backup systems, including batteries and diesel generators, failed to engage as designed, exacerbating the situation. An immediate response team was deployed to restore power and begin the diagnostic process.

The first step in the failure analysis involved a thorough inspection of the substation's infrastructure. Technicians examined physical components such as switchgear, transformers, and circuit breakers for any signs of damage or malfunction. Visual inspections revealed no apparent mechanical failures or damage due to external factors like weather or vandalism. Attention then shifted to the electrical and control systems. Detailed testing and diagnostics were conducted on the AC auxiliary supply circuits, including insulation resistance tests, continuity checks, and thermographic inspections to identify overheating components.

During these tests, it was discovered that a critical component, an Automatic Transfer Switch (ATS), had failed to operate correctly. The ATS is designed to switch the auxiliary power supply from the main source to a backup generator in the event of a primary supply failure. Detailed examination of the ATS revealed a malfunction in its control circuitry, preventing it from detecting the loss of primary power and initiating the switch to the backup generator. Further analysis identified that the control circuitry failure was due to a defective relay, which had not been replaced during routine maintenance cycles.

The failure of the ATS prompted a review of the maintenance and testing protocols for critical components within the substation. Maintenance records indicated that the relay in question had been flagged for replacement during a previous inspection but was not replaced due to inventory shortages. This oversight highlighted gaps in the maintenance scheduling and inventory management processes. Additionally, it was found that the testing procedures for the ATS did not adequately simulate real-world failure conditions, leading to a false sense of security regarding its operational readiness.

Beyond the immediate cause, the investigation also examined why the backup systems, such as the batteries and diesel generator, did not function as expected. The battery backup system was found to have degraded over time, with several cells showing reduced capacity due to aging and insufficient maintenance. The diesel generator, while mechanically sound, had not been tested under load conditions for an extended period, leading to issues with fuel delivery and startup during the actual event. This revealed a critical lapse in comprehensive testing and readiness verification of backup systems.

The failure analysis concluded with several key recommendations. First, a more rigorous and realistic testing protocol for the ATS and other critical components was established. This included periodic simulated failure scenarios to ensure that all systems would respond correctly under actual fault conditions. Second, an enhanced maintenance schedule was implemented, with stricter adherence to component replacement timelines and better inventory management to prevent shortages. Third, the backup systems were overhauled, including the replacement of aged batteries and a more frequent testing regime for the diesel generator to ensure its reliability.

In summary, this case study highlights the importance of thorough and realistic testing, strict adherence to maintenance schedules, and comprehensive readiness checks for backup systems in preventing failures in AC auxiliary systems. The incident underscored the interconnected nature of substation components and the cascading effects that a single point of failure can have on the overall power system. The lessons learned from this failure analysis have been instrumental in improving the reliability and resilience of the substation's AC auxiliary systems, ensuring better preparedness for future contingencies.

CLASS ACTIVITY

AC GENERATOR

1. Your teacher will explain and demonstrate a hand-operated, alternating current (AC) generator. During the demonstration, make the observations necessary to gain the information needed to answer these questions:

- When the AC generator is used to light a bulb, describe the brightness of the bulb when the generator is cranked slowly, and then rapidly. Write your observations in your log.
- When the AC generator is connected to a galvanometer, describe the action of the galvanometer needle when the generator is cranked slowly, and then rapidly.

2. It is easier to understand the creation of a current if you think of a set of invisible threads to signify the magnetic field of the permanent magnets. The very thin threads fill the space and connect the north pole of one magnet with the south pole of the other magnet. If the wire of the generator is imagined to be a very thin, sharp knife, the question you must ask is whether the knife (the wire) can "cut" the threads (the magnetic field lines). If the wire moves in such a way that it can cut the field lines, then a current is generated. If the wire moves in such a way that it does not cut the field lines, then no current is generated.

- Look at the diagrams of the magnetic fields shown. In which case, I, II, or III will a current be generated?

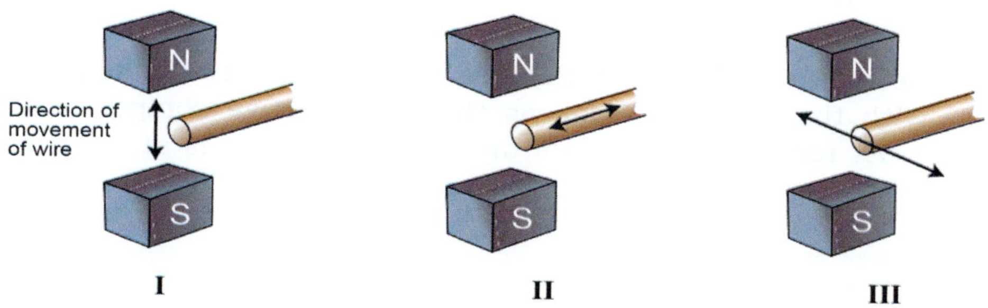

SUMMARY

- Alternating currents are also accompanied usually by alternating voltages. Besides, alternating current is also easily transformed from a higher voltage level to a lower voltage level.
- Alternating current can be produced or generated by using devices that are known as alternators. However, alternating current can also be produced by different methods where many circuits are used. One of the most common or simple ways of generating AC is by using a basic single coil AC generator, which consists of two-pole magnets and a single loop of wire having a rectangular shape.
- The average value is usually defined as the average of the instantaneous values of alternating current over a complete cycle. The positive half cycle of asymmetrical waves such as a sinusoidal voltage or current waveform will be equal to the negative half cycle.
- RMS value is defined as the square root of means of squares of instantaneous values. It can also be described as the amount of AC power that generates the same heating effect as an equivalent DC power.
- The phasor diagram is used to determine the phase relationships between two or more sine waves propagating with the same frequency.
- If an AC circuit consists of resistance only, the value of the impedance is the same as the resistance, and Ohm's law for an AC circuit, I = E/Z, is exactly the same as for a DC circuit.
- A resonant circuit is formed when a capacitor and inductor (coil) are in parallel or in series. The two circuit elements will block or pass a single specific frequency out of a divers mix. For this reason, resonant circuits make possible radio and TV transmission and reception and perform many other useful tasks.
- Resonance occurs in a parallel RLC circuit when the total circuit current is "in-phase" with the supply voltage as the two reactive components cancel each other out. At resonance the admittance of the circuit is at its minimum and is equal to the conductance of the circuit.

REVIEW QUESTIONS

1. How do you calculate the power dissipated in a fully resistive circuit?
2. Explain how to calculate the total current in a series AC circuit.
3. Describe how the power factor in a series RL circuit is calculated.
4. State the Superposition Theorem and its significance in AC circuit analysis.
5. What are the theorems in circuit analysis?
6. Discuss the phasor and phasor diagram in a single-phase AC circuit.

7. Describe how Thevenin's Theorem simplifies the analysis of AC circuits with multiple loads.
8. What is resonance in AC circuits?

REFERENCES

1. Bhargava, N. N., & Kulshreshtha, D. C. (1983). *Basic Electronics & Linear Circuits* (p. 90). Tata McGraw-Hill Education.
2. Hughes, T. P. (1993). *Networks of Power: Electrification in Western Society, 1880–1930* (p. 96). The Johns Hopkins University Press. ISBN 0-8018-2873-2.
3. Kuehni, H. P., & Lorraine, R. G. (1938). A new A-C network analyzer. *Transactions of the American Institute of Electrical Engineers*, 57, 67.
4. Robinson, A. J., & Snyder-Mackler, L. (2007). *Clinical Electrophysiology: Electrotherapy and Electrophysiologic Testing* (3rd ed., p. 10). Lippincott Williams & Wilkins. ISBN 978-0-7817-4484-3.
5. Suzuki, J. (2009). *Mathematics in Historical Context*. Mathematical Association of America. ISBN 978-0-88385-570-6. Available from Google Books.
6. Zipernowsky, K., Déri, M., & Bláthy, O. T. (1886). *Induction coil (U.S. Patent No. 352105).* U.S. Patent and Trademark Office. Archived (PDF) from the original on October 9, 2022.

CHAPTER 4

AC Power – Generation and Distribution

LEARNING OBJECTIVES

After studying this chapter, you will be able to:

- Understand three-phase system
- Explain power in AC circuits
- Perform three-phase power measurement
- Discuss balanced and unbalanced load
- Describe electrical power system
- Focus on residential wiring
- Identify circuit protective devices and safety precautions

KEY TERMS FROM THIS CHAPTER

- Conductors
- Electrical power
- Insulation
- Open Delta Connection
- Substation
- Toroidal Transformers
- Vibration
- Windings
- Currents
- High Leg Delta Connection
- Magnetic field
- Scott-T Connection
- Three-phase transformer
- Transformer
- Voltage

INTRODUCTION

The alternating current is so easy to generate and circulate, it is the most common form of electrical force. It is easy to move the substitute voltage up or down to match the required voltage level. To prevent power losses in the conductor, electrical power is transmitted at high voltages and low flows. To meet the needs of the buyer, it is then ventured down at the conveyance and shopper level. Larger electrical and electronic hardware components require AC power at 220–240 volts, or 110–120 volts for office and homegrown uses, and 415 volts for mechanical use.

Almost all electrical power is produced, transferred, and distributed in AC form. A distribution system often starts at a substation, from which a transmission network delivers the power. When consumers are situated close to the generating station, for example, the distribution system may initiate there. The use of primary and secondary distribution is also possible in larger or industrial areas.

4.1. THREE-PHASE SYSTEM

The term "three phase system" refers to a system in which the fault current is passed to the earth through one neutral wire and three wires through which the current flows. Stated differently, a three phase system is one that employs three wires for distribution, transmission, and generation. If the neutral and one of the three phases are removed from the system, the three-phase system can also be used as a single-phase system. In a three-phase system, the phase difference between the line currents is 120°, and their sum equals zero. There are four wires in the three-phase system, i. e. the one neutral conductor, the three conductors carrying the current. Half of the live wire's cross-section area is occupied by the neutral conductor. The neutral wire's current equals the total of the three wires' line currents, which means it's √3 times the current's zero phase sequence components.

Compared to a single-phase system, a three-phase system requires fewer conductors, which is one of its many benefits. Additionally, it provides the load with a constant supply. The three-phase system uses less energy and is more efficient. The generator produces a three-phase voltage with equal magnitude and frequency when the three-phase system is induced. It offers a continuous power supply, i.e., the power is still supplied by the other two phases of the system in the event that one of them is disrupted. One phase's current magnitude is equal to the total of the system's two other phases' current magnitudes. For the system to function properly, there must be a 120° phase difference between the three phases. If not, the system gets harmed.

4.1.1. Three-Phase Winding Connections

A three-phase transformer can be made by connecting the three windings of the transformer primary and secondary in a delta or wye configuration. Three-phase transformer connections come in four different varieties:

- Delta/delta
- Delta/wye
- Wye/delta
- Wye/wye

It is customary to list the connections for the H and X windings sequentially. The same three-phase primary line voltage can be converted into the same three-phase secondary line voltage using any type of connection.

Occasionally, though, the phase voltages of the various connections will differ from one another. One could, for instance, use a delta/delta transformer connection with three single-phase transformers set between 12,500 and 480 V, a wye/delta transformer connection with three single-phase transformers set between 7,225 and 480 V, a delta/wye connection with three single-phase transformers set between 12,500 and 277 V, or a wye/wye connection with three single-phase transformers set between 7,225 and 277 V. Everyone has benefits and drawbacks.

This section's transformer connection diagrams, which show the angular displacement between the primary and secondary windings, are for standard connections. The amount that one voltage is out of phase with another is measured by angular displacement. Three single-phase transformers connected in a different way than the standard connections depicted here will produce secondary line voltages from the transformers that may be the same value but will be out of phase with the secondary voltages from the standard connection. The secondary voltages cannot be paralleled when they are out of phase.

4.1.1.1. Delta/Delta Connection

When only one secondary voltage is needed or when three-phase equipment makes up the majority of the load, like in factories with large three-phase 480 V or 240 V motor loads and relatively small 120 V lighting and receptacle loads, the delta/delta connection, as illustrated in Figure 4.1, is utilized. The necessary primary and secondary voltage is directly correlated with the number of turns needed for the primary and secondary windings. Therefore, there is no benefit to transforming voltages with large differences using this transformer connection.

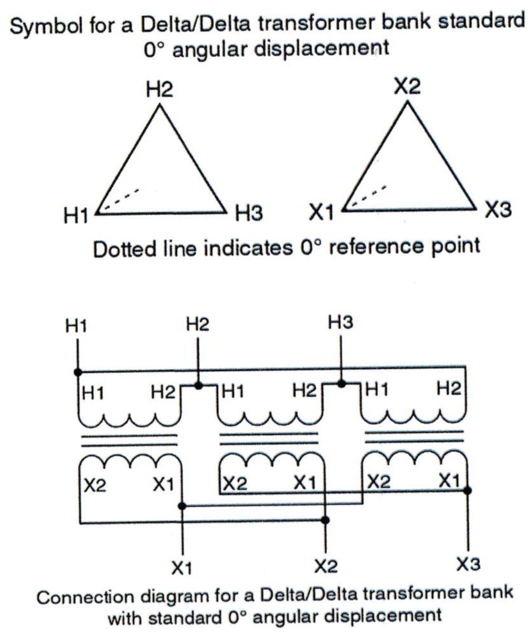

Figure 4.1. *A standard 0° angular displacement delta/delta connection.*

Source: By Ahmed Sheikh, https://eepower.com/technical-articles/pros-and-cons-three-phase-transformer-connections/#.

Advantages:

- The conductors of each single-phase transformer do not need to be as large as the line conductors supplying the three-phase load because the phase current is only 57.8% of the line current.
- The transformer effectively isolates electrical noise between the primary and secondary phases, and harmonic currents typically cancel each other out.
- The secondary voltage will fluctuate less with surges in load currents.
- In the event that one of the three single-phase transformers fails, three-phase voltage can still be supplied in the open delta, albeit at a reduced capacity of 58%.

Disadvantages:

- Since the secondary can only provide a single voltage, additional transformers and possibly receptacles are needed to supply the lighting.
- For high primary voltages, additional insulation is needed between the conductors of the primary winding in order to ensure that they are protected for the entire primary voltage.
- There isn't going to be a common ground on any secondary point. Voltages to the ground can therefore increase to extremely high values.

4.1.1.2. Delta/Wye Connection

When more than one voltage is needed on the secondary, such as in the case of three-phase, 208 V motor loads and 120 V lighting and outlet loads, the transformer connection type depicted in Figure 4.2 is utilized. Another common use for the delta/wye connection is as a step-up transformer connection.

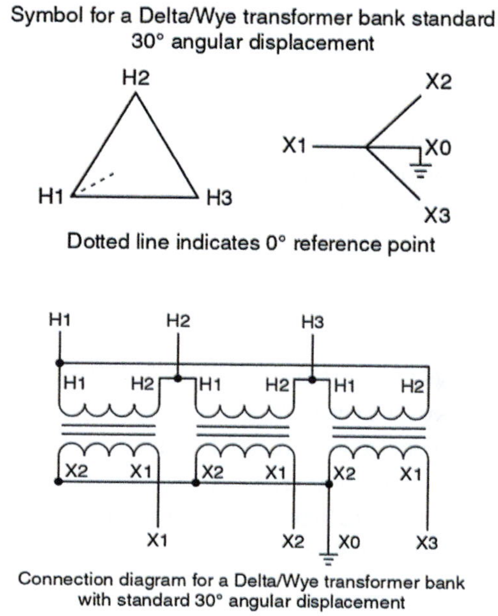

Figure 4.2. *A standard 30° angular displacement delta/wye connection.*

Source: By Ahmed Sheikh, https://eepower.com/technical-articles/pros-and-cons-three-phase-transformer-connections/#.

Advantages

- Because of the wye connection, the secondary line voltage of a single-phase transformer will be 1.73% higher when the primary and secondary windings of each transformer have the same number of turns applied to them. When using this transformer as a step-up transformer, this is advantageous.
- As a step-up transformer, the transformer requires less insulation since the secondary winding conductors won't need to be

shielded for the entire secondary line voltage.

- In a three-phase system with a 208 V line voltage, multiple voltages on the secondary could eliminate the need for additional transformers to supply 120 V loads.
- The system's voltage potential is limited to ground by a common point on the secondary, preventing it from rising above the secondary-phase voltage.

Disadvantages

- When utilizing this connection to step down high voltages, additional insulation may be needed for the primary windings since they must be insulated for the entire three-phase line voltage.
- Harmonic currents cannot be canceled out by the secondary wye connection.
- Since the secondary winding conductors will carry the entire three-phase line current, they need to be larger than the conductors of a delta system with an equivalent capacity.

4.1.1.3. Wye/Delta Connection

When only one secondary voltage is needed or when three-phase equipment makes up the majority of the load, such as in factories with sizable three-phase, 480 V or 240 V motor loads and relatively small 120 V lighting and outlet loads, the wye/delta transformer connection, as illustrated in Figure 4.3, is utilized. As a step-down transformer connection, the wye/delta connection is also frequently used.

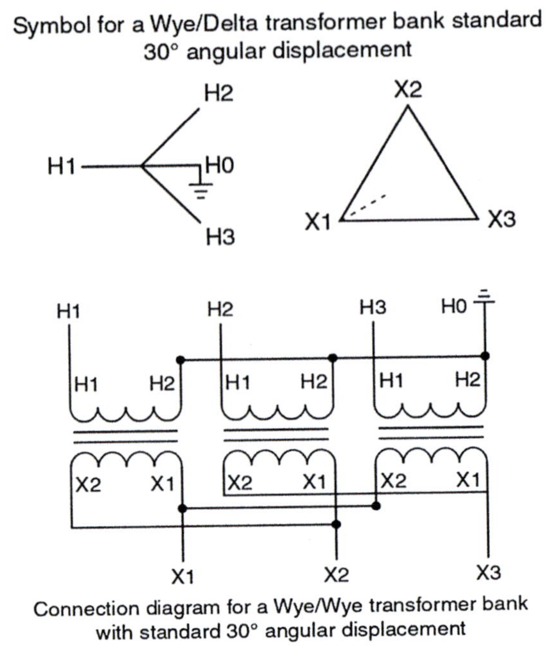

Figure 4.3. *A standard 30° angular displacement wye/delta connection.*

Source: By Ahmed Sheikh, https://eepower.com/technical-articles/pros-and-cons-three-phase-transformer-connections/#.

Advantages

- Because of the wye connection, the secondary line voltage of each single-phase transformer will be lower by a factor of 1.73 (57.8%) when the same number of turns are applied to the primary and secondary windings. When using this transformer as a step-down transformer, this is advantageous.
- Secondary harmonic currents cancel out in this transformer connection, and the transformer effectively isolates the primary and secondary noise.
- Stepping down from a high voltage may result in less insulation since the primary windings do not require

insulation for the entire three-phase line voltage.

- In the event that one of the three single-phase transformers fails, three-phase voltage can still be supplied in the open delta, albeit at a reduced capacity of 58%.

Disadvantages

- Since the secondary can only provide a single voltage, additional transformers and possibly receptacles are needed to supply the lighting.
- Similar to the delta/delta, there won't be a common ground at any point on the secondary. Voltages to the ground can therefore increase to extremely high values.
- The primary windings' conductors, which must carry the entire three-phase line current, are larger than those of a primary that is delta-connected and has the same capacity.
- In order to prevent voltage fluctuations caused by unbalanced loads, the common point of the wye primary windings should be connected to a system neutral.

4.1.1.4. Wye/Wye Connection

Seldom is the wye/wye transformer connection, as illustrated in Figure 4.4, utilized. Connected ground is frequently the point where the primary and secondary windings meet. Consequently, the secondary receives the noise from the primary and vice versa. In addition, harmonics from the loads and the magnetizing currents in the transformer core are inherent to the system. Communications may also be interfered with by the connection. Phase voltages that are unstable are not unusual.

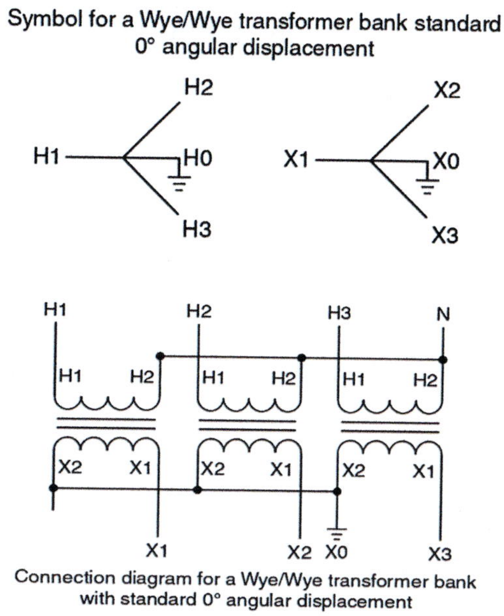

Figure 4.4. *A standard 0° angular displacement wye/wye connection.*

Source: By Ahmed Sheikh, https://eepower.com/technical-articles/pros-and-cons-three-phase-transformer-connections/#.

4.1.1.5. Other Types of Three-phase Transformers

These four permutations are not the only kinds of three-phase transformers available. The basic star and delta windings are modified to create the other types. The following are the other kinds.

- **Open Delta Connection:** In order to create an open delta connection, or V-V, two matching windings must be removed from the primary and secondary sides of a delta-delta circuit. When there is no load on the secondary terminals of this connection, three equal three-phase voltages are available. When the three-phase load is too small to install a complete three-phase transformer, this is typically done.

The three-phase load that can be carried for a given open delta transformer bank without going over the transformers' ratings is only 58% of the initial load. It could become a delta-delta connection if the load is increased and the open delta closes (Figure 4.5).

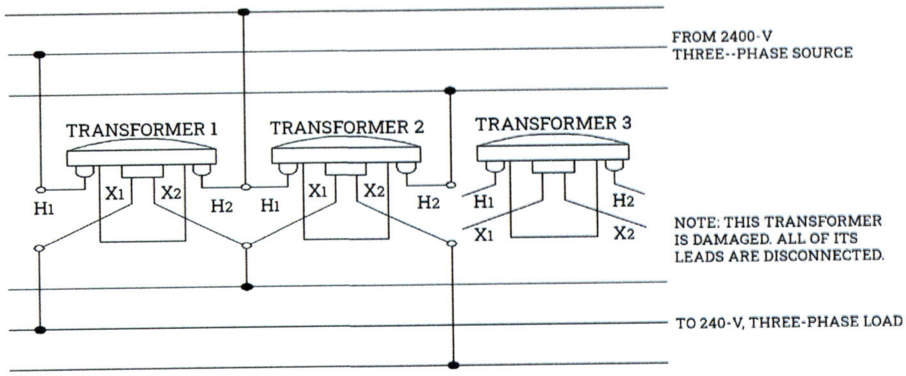

Figure 4.5. *Open delta connection.*

Source: https://www.iqsdirectory.com/articles/electric-transformer/three-phase-transformer.html.

- **Scott-T Connection:** Another name for this is a T-T connection. To achieve this, two transformers are used, one of which has center taps on the primary and secondary windings. The other transformer, referred to as the teaser transformer, has a 0.87 tap, while the main transformer has center taps. Because the teaser transformer only runs at 87% of its rated voltage, the transformer's full rating is not being used. Figure 4.6 shows how the winding is represented. A three-phase system is connected to a two-phase system using this kind of connection. The power supply on an electric furnace with a two-phase system is a typical use case for this conversion.

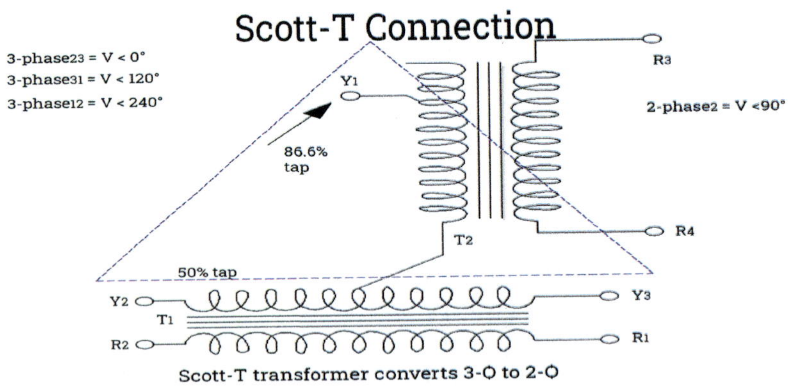

Figure 4.6. *Scott-T connection.*

Source: https://www.iqsdirectory.com/articles/electric-transformer/three-phase-transformer.html.

- **High Leg Delta Connection:** To perform a high leg delta, center tap the secondary side of the delta and connect the center tap to the ground. A three-phase, delta-connected supply and a single-phase supply are produced by this configuration. This is primarily utilized in residential and commercial distribution systems where customers can receive 120 V line to neutral for lights and appliances or 240V line-to-line for large equipment without the need for an additional transformer.

Red leg connections, wild leg connections, and orange leg connections are some other names for high leg delta connection transformers. High leg delta connections are typically utilized in North America, primarily in the United States. Referred to as a 208 V supply, the orange leg high leg delta connection is identified by an orange outer finish (Figure 4.7).

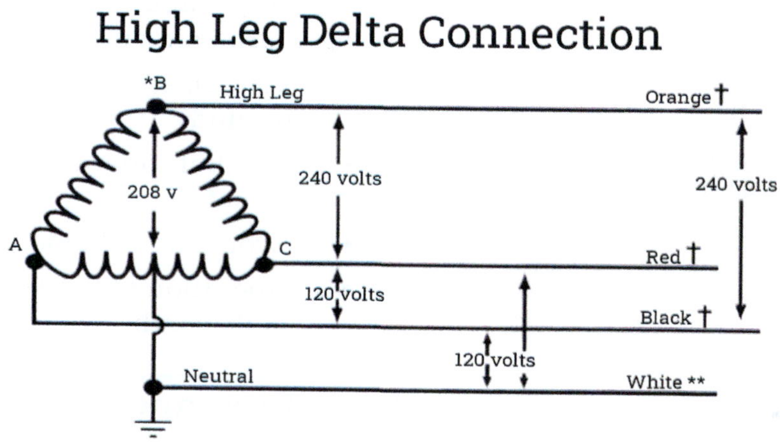

Figure 4.7. *High leg delta connection.*

Source: https://www.iqsdirectory.com/articles/electric-transformer/three-phase-transformer.html.

- **Toroidal Transformers:** Conventional transformers are inefficient with energy and noisy. Toroidal transformers defy convention, run extremely quietly, and generate very little heat. A toroidal transformer's core has a donut-shaped design, with insulation separating the primary and secondary windings to reduce noise-causing magnetic leakage.

A toroidal transformer's power is adjusted based on the magnetic fields that flow through its secondary coil. The secondary coil's winding count determines the voltage level; a winding-to-winding ratio of two to one doubles the voltage. Electrons can move freely in any direction because of the positive magnetic field produced by the primary coil.

The magnetic field peaks and then drops to zero voltage with an alternating current (AC) current signal from a toroidal transformer, completing the first half cycle. The electrons flow through the coil in the opposite direction as a negative magnetic field once the first cycle is finished. To finish a cycle, the voltage is brought back to zero by the switch pointing in the direction of the magnetic field.

Because of the magnitude of the magnetic field they produce, toroidal transformers have far lower noise levels than other transformers. It is possible to have extremely tight windings that are incapable of vibrating because of the donut-shaped core. Because there is no vibration, noise is completely eliminated, and a toroidal transformer is extremely secure (Figure 4.8).

Figure 4.8. *Toroidal transformer coir.*

Source: https://www.iqsdirectory.com/articles/electric-transformer/three-phase-transformer/toroidal-transformer-core.jpg.

Advantages of Toroidal Transformers

- Higher efficiency
- Low operating temperature
- Produces a hum
- Toroidal transformers are smaller because of their compact design.
- Light-weight
- Low stray magnetic field
- Low electromagnetic interference (EMI) emissions

4.1.2. Factors Affecting Choice of Connection

The factors affecting the choice of connection are listed below.

- Availability of neutral connection for grounding and load connection.
- Availability of a path for zero-sequence current and third harmonic current.
- Voltage stress and insulation to the ground.
- Parallel operation with another transformer.
- Economic consideration.
- Operation of the transformer under fault condition.

4.1.3. Pros and Cons of Transformer Connections

A thorough knowledge of the different three-phase transformer connections is essential for professionals in the power systems industry. Being able to choose the right connection type – whether it be delta/delta, delta/wye, wye/delta, or wye/wye – is crucial for efficient power distribution to meet specific electrical requirements. Understanding the pros and cons of these connections is important for ensuring that transformers are set up to work effectively, taking into account factors like conductor size, insulation needs, and harmonic currents. This knowledge is the key foundation for the reliable operation of electrical systems, playing a vital role in effective power distribution and management across various industrial and commercial environments.

4.2. POWER IN AC CIRCUITS

Instantaneous power in an electric circuit is the rate at which energy flows through a specific point in the circuit. Energy storage components like inductors and capacitors in alternating current circuits can cause periodic reversals in the direction of energy flow. The watt is its SI unit. Instantaneous active power is the portion of instantaneous power that, when averaged over an entire cycle of the AC waveform, results in a net transfer of energy in one direction. Active power, also known as real power, is the time average of this portion of instantaneous power. Instantaneous reactive power is the portion of instantaneous power that, because of stored energy, oscillates between the source and load in each cycle rather than resulting in a net transfer of energy; its amplitude is the absolute value of reactive power (Thomas Roland, Rosa Albert, & Gregory J. Toussaint, 2016).

4.2.1. Active, Reactive, Apparent, and Complex Power in Sinusoidal Steady-State

Both the current and the voltage in a basic circuit with alternating current (AC) that consists of a source and a linear time-invariant load are sinusoidal at the same frequency (Das, 2015). The two quantities simultaneously reverse their polarity if the load is solely resistive. Because of this, the instantaneous power, which is determined by multiplying the voltage by the current, is always positive, ensuring that the energy flow is always in the direction of the resistor. Only active power is transferred in this instance.

The voltage and current are 90° out of phase if the load is only reactive. The product of voltage and current is positive for two quarters of every cycle, but it is negative for the remaining two quarters, meaning that, on average, the load absorbs exactly the same amount of energy as it releases. Every half cycle has no net energy flow. In this instance, only reactive power moves: electrical power moves along the wires and returns by moving in reverse along the same wires, but there is no net energy transfer to the load. Even though the perfect load device doesn't use any energy at all, the current needed for this reactive power flow releases energy in the line resistance. Normal loads will receive both active and reactive power because practical loads have both resistance and inductance, also known as capacitance.

The product of the RMS voltage and current values yields apparent power. When building and running power systems, apparent power is taken into consideration because reactive power requires current that must come from the power source even though it is not used at the load. It is necessary to size generators, transformers, and conductors to carry the entire current, not just the current that performs useful work. Under certain operating conditions, an electrical grid's voltage levels can drop due to insufficient reactive power, which can cause the network to collapse (a blackout). Another effect is that unless two loads have the same phase difference between their voltage and current (the same power factor), adding their apparent powers will not yield the total power with accuracy.

Reactive power is traditionally handled as though capacitors produced it and inductors consumed it. When an inductor and a capacitor are connected in parallel, the currents that flow through them usually cancel out rather than increase. This is the basic method by which the power factor in electric power transmission is regulated; reactive power "generated" (or "consumed") by the load is partially compensated for by adding capacitors (or inductors) to a circuit. There are two types of circuits that supply reactive power: pure capacitive circuits, where the current waveform leads the voltage waveform by 90°, and pure inductive circuits, where the current waveform lags the voltage waveform by the same amount. Capacitive and inductive circuit elements consequently have a tendency to cancel each other out (Figure 4.9).

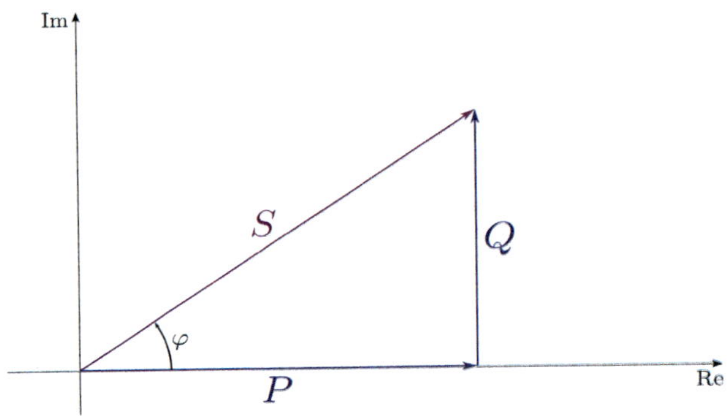

Figure 4.9. *The power triangle. The complex power is the vector sum of active and reactive power. The apparent power is the magnitude of the complex power.*

Source: By Eli Osherovich, CC BY-SA 3.0, https://commons.wikimedia.org/w/index.php?curid=15308452.

Active power, P

Reactive power, Q

Complex power, S

Apparent power, |S|

Phase of voltage relative to current, φ

When describing energy flow in a system, engineers use the following terms (giving each one a unique unit to distinguish it from the others):

- Active power, *P*, or real power: watt (W);

- Reactive power, Q: volt-ampere reactive (var);
- Complex power, S: volt-ampere (VA);
- Apparent power, $|S|$: the magnitude of complex power S: volt-ampere (VA);
- Phase of voltage relative to current, φ: the angle (in degrees) at which voltage and current differ; $\varphi = \arg(V) - \arg(I)$. Three vectors: current leading voltage (quadrant IV), current lagging voltage (quadrant I).

These are all denoted in the adjacent diagram (called a power triangle).

The diagram displays P as the active power, Q as the reactive power (positive in this case), S as the complex power, and the apparent power is indicated by the length of S. Since reactive power is non-productive, it is shown as the vector diagram's imaginary axis. Active power is the true axis because it is effective. The watt is the unit of power (symbol: W). Since apparent power is the product of RMS voltage and RMS current, it is frequently expressed in volt-amperes (VA). Volt-ampere reactive, or var, is the unit of measurement for reactive power. Reactive power is sometimes referred to as wattless power since it imparts no net energy to the load. It does, however, play a crucial role in electrical grids, and the 2003 Northeast blackout has been linked in large part to its absence. The key to comprehending power engineering is knowing how these three quantities relate to one another. The mathematical relationship between them can be expressed as $S = P + jQ$ (where j is the imaginary unit) or as vectors.

4.2.2. Power Factor

The power factor in a circuit is the ratio of apparent power to active power. Because energy stored in the load returns to the source, the system with the lower power factor will have higher circulating currents for two systems transmitting the same amount of active power. The transmission efficiency is decreased overall and higher losses are caused by these higher currents. For the same amount of active power, a circuit with a lower power factor will have higher apparent power and higher losses. When the voltage and current are in phase, the power factor is 1.0. When the current is 90° ahead of or behind the voltage, it is zero. A home with solar panels on the roof feeding power into the power grid during the day is an example of a load feeding energy into the source when the voltage and current are 180° out of phase and the power factor is zero. When expressing the sign of the phase angle of the current with respect to the voltage, power factors are typically expressed as leading or lagging. Current is regarded as either leading or lagging voltage since voltage is defined as the base to which current angle is compared. The cosine of the phase angle (φ) between the sinusoidal waveforms of voltage and current when the waveforms are purely sinusoidal is the power factor. Power factor is frequently shortened to cos ϕ on equipment nameplates and data sheets for this reason.

As an illustration, the voltage and current have a phase angle of 45.06° and an active power of 700 W. It is cos(45.6°) = 0.700 for the power factor. Thus, 700 W / cos (45.6°) = 1000 VA is the apparent power. Using an example, the idea of power dissipation in an AC circuit is clarified and demonstrated.

For example, when the power factor is 0.68, it indicates that only 68% (in magnitude) of the total current supplied is actually working at the load; the remaining current is not working. In power sector substations, power factor is crucial. A minimum power factor is required for the sub sectors to be connected to the national grid. If not, there are numerous losses. The necessary typically vary by at least 0.90 – 0.96. More power factor with less loss is preferable.

4.2.3. Reactive Power

The power applied to a load in a direct current circuit is proportional to the product of the current passing through it and the potential drop across it. Reactive power is the power that results from an inductor or capacitor. The reason for this is that components like capacitors and inductors are AC-only. Energy travels from the source to the load in a single direction. Both the voltage and the current in AC power fluctuate roughly sinusoidally. The voltage and current waveforms do not exactly align in circuits with inductance or capacitance. There are two parts to the power flow: reactive power, which is the part that cannot be used at the load because of the phase angle delay between the voltage and current, and the component that flows from the source to the load. It can be compared to current that arrives at the incorrect moment (either too late or too early). Reactive power is measured in volt-amperes reactive, or var, to differentiate it from active power. Although these units are left as var to indicate that they do not represent any actual work output, they can be simplified to watts.

Reactive power flow is the result of energy stored in the network's capacitive or inductive components. Over the network, reactive power flow has a significant impact on voltage levels. It takes careful control over voltage levels and reactive power flow to operate a power system within reasonable bounds. Reactive power supplied from transmission lines is reduced and provided locally through a process called reactive compensation, which lowers apparent power flow to a load. For instance, a shunt capacitor is placed close to the load in order to compensate for an inductive load. This enables the capacitor to supply all of the reactive power required by the load, negating the need for transmission lines to be used. By lowering the quantity of energy needed to be produced by the utility to accomplish the same amount of work, this practice saves energy. It also makes it possible to design transmission towers more optimally and create transmission lines with smaller or fewer bundled conductors for greater efficiency.

4.2.3.1. Capacitive vs. Inductive Loads

An offset between the waveforms of the current and voltage is caused by stored energy in the magnetic or electric field of a load device, such as a motor or capacitor. An apparatus that stores energy as an electric field is called a capacitor. An opposing voltage develops across the capacitor as a result of charge accumulation as current flows through it. This voltage rises to a maximum that the structure of the capacitor specifies. The voltage across a capacitor in an AC network varies continuously. The current leads the voltage in phase because the capacitor resists this shift. Reactive power is said to be sourced by capacitors, which results in a leading power factor.

Among the most prevalent load types in today's electric power system are induction machines. These devices store energy in the form of a magnetic field using inductors, which are big coils of wire. The inductor strongly resists this change in current and magnetic field when a voltage is first applied across the coil, which delays the current's maximum value. As a result, the current is out of phase with the voltage. It is said that inductors cause a lagging power factor because they "sink" reactive power. Reactive power can be sourced or sunk by induction generators, giving system operators some degree of control over the flow of reactive power and consequently voltage. These devices can be used to neutralize each other's effects because they have opposite effects on the phase angle between voltage and current. This is typically accomplished by using capacitor banks to offset the induction motor-induced lagging power factor.

4.2.3.2. Reactive Power Control

In most cases, reactive power flow requires transmission-connected generators. For instance, the Grid Code Requirements on the transmission system in the United Kingdom mandate that generators supply their rated power at the designated terminals between the limits of 0.85 power factor lagging and 0.90 power factor leading. In order to keep the reactive power balance equation intact and maintain a safe and economical voltage profile, the system operator will switch between actions.

Generator MVRs + System gain + Shunt capacitors = MVAR Demand + Reactive loss + Shunt reactors (4.1)

In the power balance equation above, reactive power is generated by the transmission network's capacitive nature and is primarily sourced from the "system gain." Early in the morning, before demand spikes, decisive switching actions can maximize system gain and contribute to system security throughout the day. There will be a need to use pre-fault reactive generators in order to balance the equation. Additionally, shunt reactors, shunt capacitors, static VAR compensators, and voltage control circuits will be employed as sources of reactive power.

4.2.4. Unbalanced Sinusoidal Polyphase Systems

Any system's active and reactive powers are clearly defined, but one of the most contentious issues in power engineering is how to define apparent power for unbalanced polyphase systems. Appearance power was first only a matter of merit. Steinmetz's *Theoretical Elements of Engineering* (1915) and Stanley's *Phenomena of Retardation in the Induction Coil* (1888) are credited with providing key definitions of the idea. Nevertheless, it became evident that the definitions of apparent power and power factor could not be applied to unbalanced polyphase systems with the advent of three phase power distribution. To address the problem, the AIEE and the National Electric Light Association convened a Special Joint Committee in 1920. They took into account two meanings.

$$S_A = |S_a| + |S_b| + |S_c|$$
$$\mathrm{pf}_A = \frac{P_a + P_b + P_c}{S_A}, \tag{4.2}$$

that is, the arithmetic sum of the phase apparent powers; and

$$S_V = |P_a + P_b + P_c + j(Q_a + Q_b + Q_c)|$$
$$\mathrm{pf}_V = \frac{P_a + P_b + P_c}{S_V}, \tag{4.3}$$

that is, the magnitude of total three-phase complex power.

The 1920 committee was unable to reach a consensus, and the subject remained the center of conversation. Another committee was established in 1930, but they were unable to find an answer. The conversations they had are captured in the longest and most contentious transcripts the AIEE has ever released [Emanuel, Alexander (1993)]. The discussion on this topic was not further resolved until the late 1990s.

In 1993, Alexander Emanuel proposed a new definition for an unbalanced linear load supplied with asymmetrical sinusoidal voltages, based on the theory of symmetrical components.

$$S = \sqrt{(|V_a^2| + |V_b^2| + |V_c^2|)\,(|I_a^2| + |I_b^2| + |I_c^2|)}$$
$$\text{pf} = \frac{P^+}{S}, \tag{4.4}$$

that is, the product of the squared sums of line voltages and line currents multiplied by their root. The positive sequence power is indicated by P^+:

$$P^+ = 3|V^+||I^+|\cos(\arg(V^+) - \arg(I^+)) \tag{4.5}$$

V^+ is the product of the squared sums of the line voltages and the squared sums of the line currents. The positive sequence power is indicated by I^+.

4.2.5. Multiple Frequency Systems

Any waveform can have its RMS value determined, and from there apparent power can be computed. At first glance, it would seem that calculating numerous product terms and averaging them all would be required for active power. Nevertheless, a more thorough examination of one of these product terms yields an extremely intriguing outcome.

$$\begin{aligned} &A\cos(\omega_1 t + k_1)\cos(\omega_2 t + k_2) \\ &= \frac{A}{2}\cos[(\omega_1 t + k_1) + (\omega_2 t + k_2)] + \frac{A}{2}\cos[(\omega_1 t + k_1) - (\omega_2 t + k_2)] \\ &= \frac{A}{2}\cos[(\omega_1 + \omega_2)t + k_1 + k_2] + \frac{A}{2}\cos[(\omega_1 - \omega_2)t + k_1 - k_2] \end{aligned} \tag{4.6}$$

Nonetheless, as long as ω is nonzero, the time average of a function of the form $\cos(\omega t + k)$ is zero. Thus, the only terms in the product where the frequency of voltage and current match have an average that is not zero. Stated differently, the calculation of active (average) power can be achieved by treating each frequency independently and summing the results. Furthermore, this demonstrates that harmonic currents are undesirable if the mains supply voltage is taken to be a single frequency, which it usually is. They won't affect the active power transferred, but they will raise the apparent power and RMS current (because non-zero terms will be added). Therefore, the power factor will decrease due to harmonic currents. The device's input can have a filter installed to lessen harmonic currents. Usually, this will be a network of capacitors and inducers or just a capacitor (which depends on parasitic resistance and inductance in the supply). In general, the harmonic currents would be further reduced and the power factor would be kept closer to unity with an active power factor correction circuit at the input.

4.3. THREE-PHASE POWER AND ITS MEASUREMENT

Wattmeters are used to measure power in AC circuits. The potential coil (PC) and the current coil (CC) are the two coils that make up a wattmeter. Connected across the load, the high-resistance potential coil conducts current in proportion to the potential difference across the load. The load is connected in series with the low- resistance current coil.

There are several methods available for conducting three phase power measurements:

- One wattmeter method
- Two wattmeter method
- Three wattmeter method.

4.3.1. Three Wattmeter Method

This method is used to measure the power in a three-phase, three-wire circuit. Figure 4.10 depicts the connection. Each wattmeter measures power in its own phase because the neutral wire is shared by the three phases; the total power is determined by adding the readings from all three wattmeters.

Total power of load circuit, $P_{3\text{-}\varphi} = W_1 + W_2 + W_3$ (4.7)

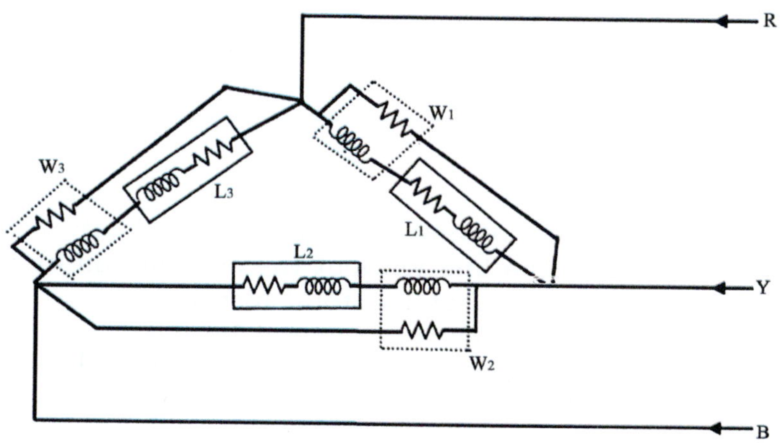

Delta connected load

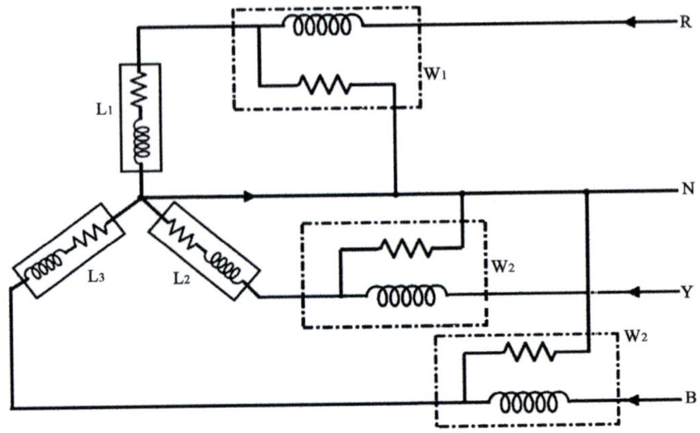

Star connected load

Figure 4.10. *In the case of delta connected circuits, power measurement by three wattmeter method is very difficult because phase coils of load are required to be broken for inserting the current coils of wattmeter.*

Source: https://www.yourelectricalguide.com/2017/05/power-measurement-in-three-phase.html#google_vignette.

4.3.2. One Wattmeter Method

Since the power in each phase of a balanced 3-wire, 3-phase load circuit is equal, the circuit's total power can be calculated by multiplying the power measured in any given phase. As a result, one wattmeter can be used to measure power in circuits with three phases and three wires. However, this approach has a drawback. A measurement error of this magnitude is produced by even a small imbalance in the loading.

4.3.3. Two Wattmeter Method

This method is typically used to measure the power in circuits with three-phase, three-wire loads. Two wattmeters' current coils are inserted into any two lines, and the potential coil is connected to the line without the current coil by means of its own current coil. It can be demonstrated that the total instantaneous power absorbed by the load is equal to the sum of the power measured by the two wattmeters, W_1 and W_2. However, in actual use, the inertia of the moving system causes wattmeters to read average power (Figure 4.11).

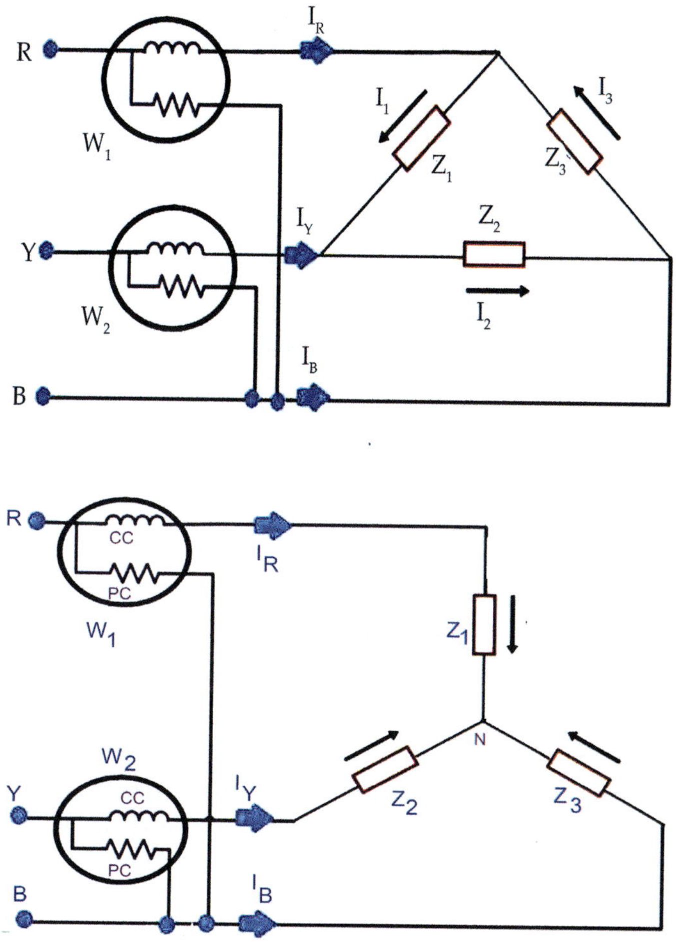

Figure 4.11. *The two wattmeter method of power measurement in three-phase circuits is suitable for every type of three-phase circuit weather circuit is balanced or unbalanced and star connected or delta connected.*

Source: https://www.yourelectricalguide.com/2017/05/power-measurement-in-three-phase.html#google_vignette.

4.3.4. Variation in Wattmeter Reading (on Lagging PF)

The power factor of the load affects both wattmeter readings when using the two wattmeter method. The following describes how various power factors affect wattmeter readings:

When $\varphi = 0^\circ$ i.e., $\cos \varphi = 1$

The readings from both wattmeters will be equal and positive.

Therefore; $P_{3\text{-}\varphi} = W_1 + W_2$

When $\varphi = 60^\circ$ i.e., $\cos\varphi = 0.5$

In this instance, one wattmeter displays a positive reading and the other a zero deflection.

Therefore; $P_{3\text{-}\varphi} = W_1$.

When $90^\circ > \varphi > 60^\circ$ i.e., $0 < \cos\varphi < 0.5$

Here, a positive reading will come from one wattmeter and a negative reading (down-scale reading) from another wattmeter.

To get a positive reading on W_2, the potential coil or current coil connection must be reversed. Following the reversal of the connection, all readings should be interpreted negatively.

Therefore; $P_{3\text{-}\varphi} = W_1 - W_2$

When $\varphi = 90^\circ$ i.e., $\cos\varphi = 0$

Both the wattmeters will give equal and opposite reading.

Therefore; $P_{3-\varphi} = W_1 + W_2 = 0$

4.3.5. Effect of Leading PF on Wattmeter Readings

Leading power factor causes the readings on two wattmeters to be swapped. The wattmeter with the higher reading is now the W_2. Regarding the lagging power factor, there are no changes to the other discussions.

4.4. BALANCED AND UNBALANCED LOAD

Three live wires and one path return make up the three-phase system. A lot of power is transmitted using the three-phase system. There are primarily two categories for the three phase system. There exist two types of three-phase systems: one balanced and the other unbalanced.

A system that has equal load distribution throughout its three phases is said to be in balance. Each of the three phases has the same voltage magnitude and is spaced out by a 120° angle.

4.4.1. Analysis of Balanced Three Phase Circuit

Solving balanced three-phase circuits based on each phase is always preferable. The line voltage is taken into account when the three-phase supply voltage is provided without regard to the line or phase value.

Below are the steps to solve the balanced three-phase circuits.

Step 1 – First of all draw the circuit diagram.

Step 2 – Determine $X_{LP} = X_L/\text{phase} = 2\pi f_L$.

Step 3 – Determine $X_{CP} = X_C/\text{phase} = 1/2\pi f_C$.

Step 4 – Determine $X_P = X/\ \text{phase} = X_L - X_C$

Step 5 – Determine $Z_P = Z/\text{phase} = \sqrt{R^2{}_P + X^2{}_P}$

Step 6 – Determine $\cos\phi = R_P/Z_P$; the power factor is lagging when $X_{LP} > X_{CP}$ and it is leading when $X_{CP} > X_{LP}$.

Step 7 – Determine the V phase.

For star connection $V_P = V_L/\sqrt{3}$ and for delta connection $V_P = V_L$

Step 8 – Determine $I_P = V_P/Z_P$.

Step 9 – Now, determine the line current I_L.

For star connection $I_L = I_P$ and for delta connection $I_L = \sqrt{3}\ I_P$

Step 10 – Determine the Active, Reactive and Apparent power.

4.4.2. Analysis of Unbalanced Three Phase Circuit

The load is connected either as a star or delta connection, and the analysis of the three-phase unbalanced system is a little challenging.

- **Interconnection of Three Phase System:** An AC generator operating in three phases has three windings. There are two terminals on each winding (start and finish). Each phase supplies as an independent load through a pair of wires if a distinct load is connected across each phase winding, as indicated in the Figure 4.12. As a result, to connect the load to a generator, six wires are needed. This will increase the system's complexity and expense.

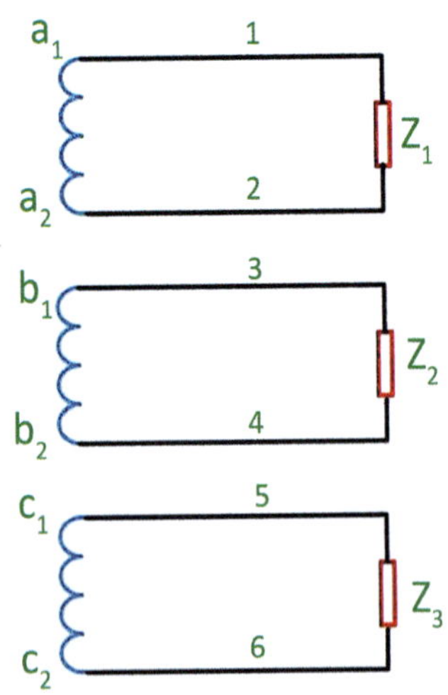

Figure 4.12. *Three windings of a three-phase AC generator.*

Source: https://circuitglobe.com/circuit-analysis-of-3-phase-system-balanced-condition.html.

Consequently, the three-phase windings of an AC generator are connected in order to minimize the number of line conductors. There are two methods available for connecting the windings in a three-phase system.

- **Connection of Three Phase Loads in Three Phase System:** Due to the star and delta connections used in the three-phase supply. In a similar manner, the three-phase loads are also connected using a delta or star connection. Figure 4.13 depicts the three-phase load connected in the star.

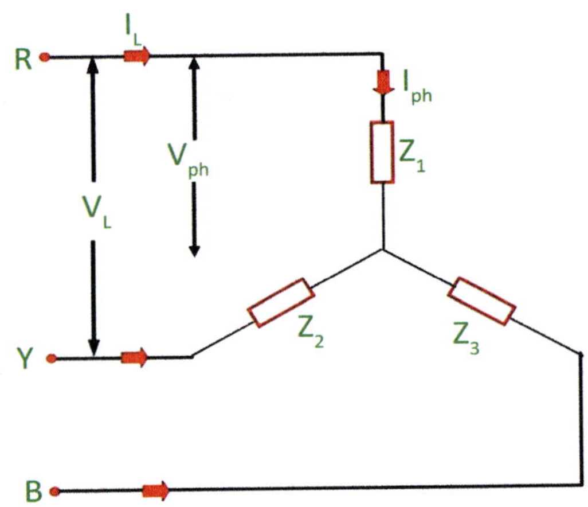

Figure 4.13. *The three-phase load connected in the star.*

Source: https://circuitglobe.com/circuit-analysis-of-3-phase-system-balanced-condition.html.

The following Figure 4.14 illustrates the delta connection for three-phase loads.

As mentioned above, the three-phase loads can be balanced or unbalanced. A three-phase load is considered balanced if its magnitude and phase angle are equal for all three loads (Z_1, Z_2, and Z_3). All of the phase currents and all of the phase voltages under such connections have the same magnitude.

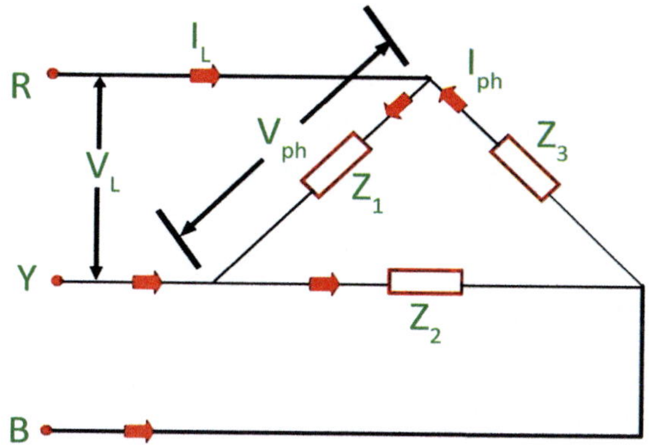

Figure 4.14. *The delta connection of three-phase loads.*

Source: https://circuitglobe.com/circuit-analysis-of-3-phase-system-balanced-condition.html.

4.5. ELECTRICAL POWER SYSTEM

A system of electrical components used to supply, transfer, and consume electricity is called an electric power system. The electrical grid, which supplies power to homes and businesses over a large area, is an example of a power system. The three main components of the electrical grid are the distribution system, which distributes power to surrounding residences and businesses, the transmission system, which moves power from the generating centers to the load centers, and the generators that provide the power.

Homes, businesses, hospitals, and industries all have smaller power systems. This entire system is represented by a single line diagram. Most of these systems are dependent on three-phase AC power, which is the industry standard for large-scale power distribution and transmission in the contemporary world. Automobiles, electric rail systems, ocean liners, submarines, and aircraft are examples of specialized power systems that don't always require three-phase AC power.

4.5.1. Structure of Power System

The following sub-systems can be separated out of the complex enterprise known as the power system. The power system's subsystems are described in detail below.

4.5.1.1. Generating Substation

The fuel used in the generating station (coal, water, nuclear energy, etc. is transformed into electrical power. The electrical power is step-upped for long-distance transmission and generated between 11 and 25 kV. There are three primary categories for the generating substation's power plant: i.e., nuclear power plants, thermal power plants, and hydropower plants.

The two primary parts of the generating station are the transformer and the generator. The mechanical energy is transformed into electrical energy by the generator. The combustion of coal, gas, and nuclear fuel, as well as gas turbines and occasionally internal combustion engines, provide the mechanical energy. From one level to another, the transformer transfers power with exceptionally high efficiency. With the exception of transformer losses, the power transfer from the secondary is roughly equal to that from the primary. Long-distance power transmission is made possible by the step-up transformer, which lowers line losses.

4.5.1.2. Transmission Substation

The overhead lines that move generated electrical energy from the generating to the distribution substations are carried by the transmission substation. It only provides the vast majority of its power to very large consumers or bulk power substations.

These two tasks are primarily carried out by the transmission lines.

- From generating stations to bulk receiving stations, it moves the energy.
- It links the two or more power plants together. Through the transmission lines, the nearby substations are also connected.

Operating at a voltage greater than 66 kv, the transmission voltage is standardized line-to-line at 69 kv, 115 kV, 138 kV, 161 kV, 230 kV, 345 kV, 500 kV, and 765 kV. Extra high voltage (EHV) is the term typically used to describe transmission lines above 230 KV.

Substations known as receiving substations, primary substations, or high voltage substations are where the high voltage line terminates. Stepping down the voltage to a suitable value for the subsequent portion of the flow toward the load occurs in a high voltage substation. It is possible to serve the very large industrial consumers straight to the transmission system.

4.5.1.3. Sub-Transmission Substation

The sub-transmission system is the section of the transmission system that links the distribution substations to the high voltage substations via the step-down transformer. There are 90 to 138 KV sub-transmission voltage levels available. A few sizable industries are directly served by the sub-transmission system. The substations contain the reactor and capacitor needed to keep the transmission line voltage constant.

The sub-transmission system functions similarly to a distribution system. Here's how it's different from a distribution system.

- The voltage level of a sub-transmission system is higher than that of a distribution system.
- It supplies only bigger loads.
- In contrast to a distribution system that supplies certain loads, it only supplies a small number of substations.

4.5.1.4. Distribution Substation

A distribution system is the part of an electrical power system that connects all of the consumers in a given area to the bulk power sources. Transmission lines link the generating substations to the bulk power plants. They supply power to a few substations, which are typically located in handy locations close to the load centers.

Power is distributed by the substations to residential, commercial, and small-scale users. Large power blocks are needed by the consumers, and these are typically provided by transmission systems or sub-transmission systems.

4.5.2. Generation of Electrical Power

There are several basic techniques for converting non-electric energy sources into electrical energy. Photovoltaic systems or revolving electric generators are the two ways utility-scale generation is accomplished. Batteries supply a tiny percentage of the electricity that utilities distribute. The triboelectric effect, piezoelectric effect, thermoelectric effect, and betavoltaics are additional methods of producing electricity that are employed in specialized applications.

- **Generators:** Typically, wind turbines are used in conjunction with other power-generating methods to generate electricity. Kinetic energy is converted into electrical power by electric generators. This method, which is based on Faraday's law, is the most widely used to generate electricity. By rotating a magnet inside of closed loops of conducting material, such as copper wire, it can be observed experimentally electromagnetic induction, which uses mechanical energy to force a generator to rotate, is the method used in almost all commercial electrical generation.
- **Electrochemistry:** The direct conversion of chemical energy into electrical energy, as in a battery, is known as electrochemistry. Electrochemical power generation plays a significant role in mobile and portable applications. Batteries provide the majority of electrochemical power nowadays. While secondary cells, such as rechargeable batteries, are used for storage systems instead of primary generation systems, primary cells, like the typical zinc-carbon batteries, function as direct power sources. Fuel cells are open electrochemical devices that can be used to generate electricity from synthetic or natural fuels. Osmotic power may occur in locations where fresh and saltwater combine.
- **Photovoltaic Effect:** The conversion of light into electrical energy, as in solar cells, is known as the photovoltaic effect. Solar radiation is directly converted to DC electricity by photovoltaic panels. If necessary, power inverters can then transform that into AC electricity. Even though sunlight is abundant and free, producing electricity through solar power still typically costs more than producing power through large-scale mechanical generation because of the cost of the panels. The cost of low-efficiency silicon solar cells has been declining, and multijunction cells with conversion efficiencies of nearly 30% are now offered for sale. In experimental systems, efficiency levels of over 40% have been shown.

Up until recently, photovoltaics were mostly utilized as a backup source of electricity for individual residences and businesses, or in remote locations without access to a commercial power grid. Recent developments in photovoltaic technology and manufacturing efficiency, along with environmental concerns-driven subsidies, have significantly accelerated the deployment of solar panels. The growth in installed capacity is approximately 20% annually, with the United States, China, India, Japan, Germany, and Japan leading the way.

4.5.3. Transmission of Electrical Power

The process of moving massive amounts of electricity generated at power plants—like large-scale solar facilities—over great distances in preparation for eventual consumer use is known as electrical power transmission. The North American transmission grid, a vast network of electric power lines and related facilities in the United States, Canada, and Mexico, receives electricity from power plants in

North America. Transmission typically happens at high voltage (69 kV or above) due to the characteristics of electricity and the significant amount of power involved. Typically, power is delivered to a substation close to a population center. High-voltage electricity is transformed into lower voltages at the substation so that it can be used by consumers. Low-voltage electric distribution lines are then used to deliver the finished product to end users.

In the event that there were no suitable transmission facilities already in place for newly built solar energy power plants, new transmission lines and related infrastructure would be needed. Various environmental impacts would result from the construction, operation, and decommissioning of high-voltage transmission lines and related facilities. The nature and extent of the effects resulting from the construction, operation, and decommissioning of transmission lines are contingent upon the size and type of the line, its length, and several other site-specific variables.

High-voltage electric transmission lines and their related infrastructure are primarily made up of:

- **Transmission Towers:** The part of the power transmission system that is most visible is the transmission tower. Their purpose is to maintain the high-voltage conductors, or power lines, isolated from both the environment and one another. There are many different types of tower designs that use a monopole or an open lattice work, but they are usually metal structures that are very tall (a 500 kv tower could be 150 feet tall with crossarms as wide as 100 feet).
- **Conductors (Power Lines):** The power lines that transport electricity to and from customers along the grid are known as conductors. Every electrical circuit typically consists of a tower with multiple conductors strung together. Although twisted metal strands make up the majority of conductor construction, more recent conductors may include ceramic fibers within an aluminum matrix for increased strength and reduced weight.
- **Substations:** Substations transform the extremely high voltages used in electric transmission into lower voltages for end users. Substations come in a variety of shapes and sizes, sometimes spanning several acres; they are usually gravel-surfaced and devoid of vegetation. They are typically accessible via a permanent access road and are fenced. Substations often have a range of buildings, wires, gates, lights, and other elements that give them an industrial look.
- **Rights of Way (ROWs):** Land set aside for the transmission line and related facilities, required to ease maintenance and reduce the risk of fires and other accidents, is included in the right of way for a transmission corridor. It offers a safety gap between the nearby buildings and vegetation and the high-voltage lines. For reasons of safety and/or accessibility, some vegetation clearance might be required. Typically, native vegetation or plants chosen for their advantageous growth patterns—slow growth and low mature heights—make up a ROW. Access roads, on the other hand, can occasionally make up a portion of the ROW and offer more practical

access for vehicles used for maintenance and inspection. The width of a ROW varies from 50 feet to roughly 175 feet or more for 500 KV lines, depending on the voltage rating of the line.

- **Access Roads:** It is typically necessary to have paved or gravel access routes to transmission line structures for both line construction and maintenance. Building an access road may need clearing vegetation and/or rerouting land. During a transmission line project's construction and decommissioning phases, more temporary roads might also be required.

4.6. RESIDENTIAL WIRING

Every electrical appliance in your house or building runs on AC power. The electricity that powers your home's main circuit comes from sources outside the building. The power voltages in your neighborhood are converted by transformers into the 120 V electricity needed for your home. Watts are the result of multiplying voltage by current, as previously discussed in a post. This is crucial because a single circuit breaker has a power limit and is meant to trip when the power becomes too great.

Three wires make up a continuous electrical circuit: a hot wire, a neutral wire, and a ground wire. The hot wire will allow electricity to enter a receptacle in your house or into the device you want to charge. The circuit is then completed when the electricity returns through the neutral wire. The purpose of the ground wire is to eliminate any excess power by sending a signal back to the main power source through the breaker.

4.6.1. Color Coding Wires

AC wires are color-coded for safety in the United States. The colors designate each wire's purpose in a 110 V home electrical circuit. The hot wire, carrying electricity, will always be the black wire. It is never acceptable to use the black wire for grounding or neutrality. Another wire that is regarded as hot is a red wire. Always take caution when moving forward.

The white wire serves as the neutral wire. The current in your electrical circuit is returned by this wire. The grounding wires are the bare green or green wires with yellow tape on them. Only other green grounding wires should be connected to by green grounding wires at all times. When a grounding wire comes into contact with any conductive part of the frame, it shields users and appliances from shock (Figure 4.15).

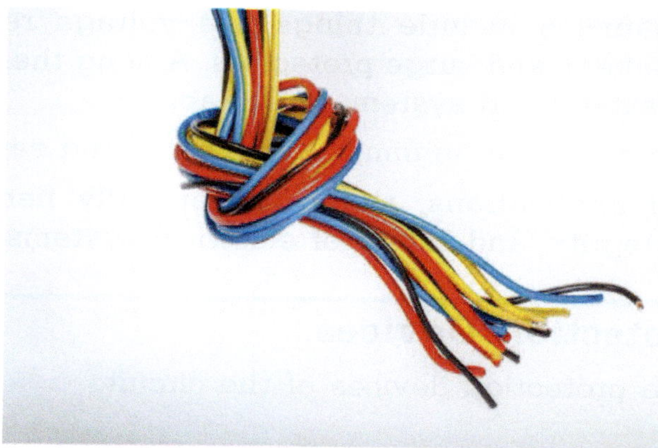

Figure 4.15. *Types of electrical wires.*

Source: https://erieit.edu/learn-basics-of-home-electrical-wiring/.

4.7. CIRCUIT PROTECTIVE DEVICES AND SAFETY PRECAUTIONS

Protection devices shield electrical systems and equipment from a range of hazards, which is crucial for preserving security and dependability. Devices come in a variety of forms, including RBCC, switch breaker, and fuse and each one is designed to address specific operational requirements and threats.

Although these protective devices offer numerous benefits, such as equipment protection and safety, they are not without limitations. When a fuse blows, it has to be replaced immediately, which causes downtime. However, circuit breakers could eventually break mechanically.

Despite certain drawbacks, protection devices are crucial for maintaining the longevity and integrity of electrical systems.

- Protection devices are crucial parts of electrical systems because they reduce the risk of overcurrent, short circuits, and ground faults, among other hazards.
- They provide voltage fluctuation protection, allowing equipment to continue operating safely.
- Protection mechanisms include things like voltage regulators, GFCIs, fuse boxes, circuit breakers, and surge protectors. Among their duties are equipment protection, user safety, and system dependability.
- They prevent electrocution, minimize downtime, and safeguard equipment.
- For a variety of applications, they are generally necessary to ensure the dependability, integrity, and safety of electrical systems.

4.7.1. Nature of Protection Devices

Here are the some of the protection devices of the circuit.

- Fuse
- Circuit Breaker
- PolySwitch
- RCCB
- Metal Oxide Varistor

- Inrush Current Limiter
- Gas Discharge Tube
- Lighting Arrester

4.7.1.1. Fuse

Fuse serves a specific purpose among all other electrical circuit protection devices. Through its metal strip, which liquefies the current when the flow is high, it guards against overcurrent. These days, different categories—like response time, breaking capacity, current ratings, and specific voltage—are helpful in different applications. As such, it is among the essential items that serve as helpful defense mechanisms.

Figure 4.16. A miniature time-delay 250 V fuse that will interrupt a 0.3 A current at after 100 s, or a 15 A current in 0.1 s. 32 mm (1 1/4") long.

Source: By DeltaFalcon - Own work, CC BY-SA 3.0, https://commons.wikimedia.org/w/index.php?curid=8094316

4.7.1.2. PolySwitch

The resettable and passive electronic component are used to guard against overcurrent errors in electronic circuits. This device is also referred to as a multi- and poly-fuse. It can be used in many different fields, including computers and aerospace, to name a few, and replacing it is not an easy feat.

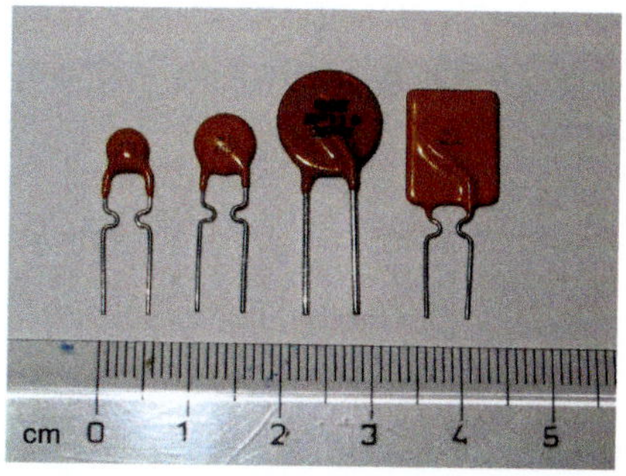

Figure 4.17. Resettable fuses - PolySwitch devices.

Source: By Shaddack - Own work, Public Domain, https://commons.wikimedia.org/w/index.php?curid=402736

4.7.1.3. RCCB

The purpose of the residual current circuit breaker (RCCB), a safety device, is to detect problems with your power supplies and, in ten to fifteen milliseconds, turn them off to prevent electric shocks. These devices can also be reset, just like the testable apparatus. However, another name for it is a residual current device (RCD).

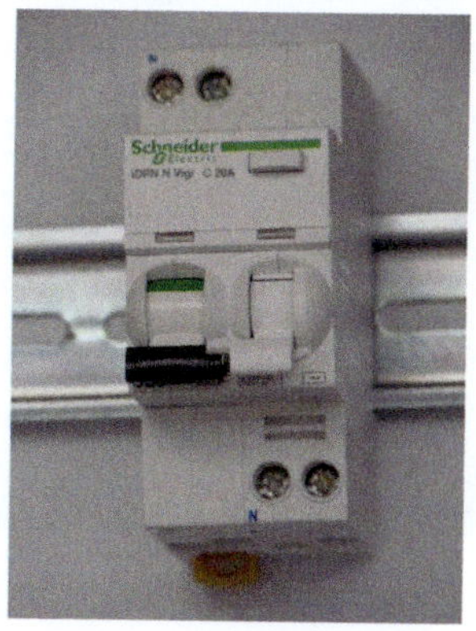

Figure 4.18. An example of a rail-mounted RCBO.

Source: By Dmitry G - Own work, CC BY-SA 3.0, https://commons.wikimedia.org/w/index.php?curid=28907017

4.7.1.4. Inrush Current Limiter

To be more precise, this kind of electrical component trips circuit breakers and blows fuses by stopping the flow of inrush current, preventing frequent damage to the device. In particular, these thermistors are typically larger than width-type thermistors and are specifically designed for power applications.

4.7.1.5. Metal Oxide Varistor

A voltage-dependent resistor, also known as a varistor, is an electrical component whose resistance varies depending on the operating voltage. The variable resistor is the source of the term "varistor." The MOV (metal oxide varistor) is the pinnacle of voltage-dependent resistors for recurrent applications.

Figure 4.19. Metal-oxide varistor manufactured by Siemens & Halske AG.

Source: By Michael Schmid, CC BY-SA 3.0, https://commons.wikimedia.org/w/index.php?curid=386147

4.7.1.6. Gas Discharge Tube

A collection of electrodes in gas enclosed in a temperature-resistant cover and sequestered is called a gas discharge tube or gas-filled tube. Although the cover is usually made of glass, military tubes frequently service glass that has wrinkles, and power tubes frequently hire ceramics.

Figure 4.20. A compact fluorescent bulb is a household application of a gas-filled tube.

Source: By CC BY 3.0, https://commons.wikimedia.org/w/index.php?curid=8689967

4.7.1.7. Lighting Arrester

This particular protective device is very useful in controlling the flow of current and stopping unwanted flow in the circuit. All of these protection mechanisms are therefore the circuit's best defense.

4.7.1.8. Circuit Breaker

The main purpose of the circuit breaker is to specifically stop the flow of current when a fault arises. One of its main advantages is that it can operate autonomously, requiring no outside assistance to continue with operations. There are numerous classifications for circuit breakers as well, each with a specific function. It should be noted that semiconductors are used in its practice for security.

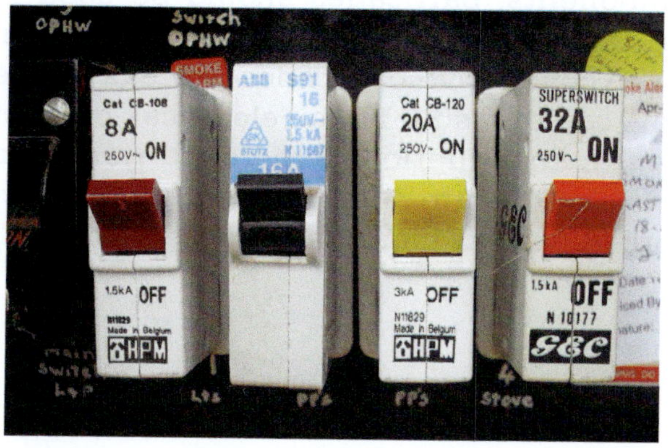

Figure 4.21. Four one-pole miniature circuit breakers.

Source: By Bidgee - Own work, CC BY-SA 3.0, https://commons.wikimedia.org/w/index.php?curid=7632868

4.7.2. Importance of Protection Devices

- Protection devices act as a system's safety guard, preventing equipment damage and ensuring user safety and system functionality.
- They are effective in identifying a wide range of hazards, such as sudden electrical fluctuations, short circuits, inadequate grounding, and overvoltage.
- By seeing these risks early on, they help to prevent accidents and equipment failures.
- Protective devices reduce the risk of injury and extend the life of equipment. It also increases operational efficiency by averting major disasters.

4.7.3. Use of Protective Equipment

When working in areas with potential electrical hazards, employees must be provided with and use electrical protective equipment suitable for the tasks they are performing and the parts of their bodies they need to protect. Protective equipment needs to be kept safe and dependable and undergoes regular testing or inspections in accordance with 29 CFR 1910.137, Electrical Protective Devices. If the insulating properties of protective gear are compromised while being used, the insulating material needs to be covered with leather or other suitable materials. Anywhere there is a risk of burns from contact with exposed energized parts or head injuries from electrical shock, nonconductive head protection must be worn. When there is a chance of eye or facial injuries from electric arcs, flashes, or flying objects brought on by electrical hazards, protective eyewear must be worn.

4.7.4. General Protective Equipment and Tools

When working close to exposed energized conductors or circuit parts, employees must use insulated tools and handling equipment if they could come into contact with the conductors or parts. Wherever tools and equipment are vulnerable to damage, the insulating material needs to be safeguarded. When the fuse terminals are energized, fuses must be installed or removed using fuse handling equipment that is insulated for the circuit voltage. When used close to exposed, energized parts, all ropes and handlines must be nonconductive. While working close to exposed, energized parts that could be touched or where hazardous electric heating or arcing could occur, workers must be protected from shock, burns, or other electrical-related injuries with protective shields, protective barriers, or insulating material. Normal enclosed live parts need to be guarded to keep unqualified people away from the live parts when they are exposed for maintenance or repair.

4.7.5. Alerting Techniques

To alert and safeguard workers against the risk of electrical shock, burns, or malfunctioning parts of electric equipment, alerting techniques must be used. Where appropriate, safety symbols, accident prevention tags, or safety signs must be used to alert staff members to potential electrical risks. When it's necessary to restrict or prevent employee access to work areas where they could be exposed to uninsulated energized conductors or circuit components, barricades should be used in conjunction with safety signs. It is forbidden to use conductive barricades in areas where they could result in an electrical contact hazard. In situations where barriers and signs are insufficient to warn or protect workers, an attendant should be stationed.

CASE STUDY

SURGE PROTECTION FAILURE

Problem

A technology company is experiencing frequent damage to their electronic equipment. The equipment is being damaged by voltage surges on the power grid, resulting in costly repairs and lost productivity.

Solution:

The electrical engineer identified that the surge protection devices in place were not sufficient to protect the equipment.

The engineer designed and implemented a new surge protection system that included a combination of surge arresters, surge suppressors, and voltage regulators to effectively protect the equipment from voltage surges.

The engineer also provided training to the maintenance team on how to properly maintain the new surge protection system.

CLASS ACTIVITY

ALTERNATIVE AND RENEWABLE ENERGY SOURCES

Objectives

The student will understand alternative and renewable electric power generation energy sources.

Materials

Internet for research

Background

Alternative energy comes from conventional resources but produces fewer emissions than conventional energy. Sources of alternative energy include: natural gas cogeneration, fuel cells, use of waste energy, electricity and natural gas used for transportation. Energy used from these resources fall into three categories: electricity generation, heat production and transportation fuels. Non-hydroelectric renewable energy refers to electricity supplied from the following renewable sources of power: solar, geothermal, biomass, landfill gas and wind. Installation of these renewable energy resources is growing in the United States. Air emissions associated with generating electricity from solar, geothermal and wind technologies are negligible because no fuels are combusted in these processes.

Procedure

1. Have students research various fuel sources of alternative energy: natural gas for cogeneration, fuel cells, burning garbage and landfill gas.
2. Have students review information on renewable energy fuel sources such as geothermal, solar and wind in section four.
3. Have students work individually or in groups to develop a creative delivery method of teaching the class about their chosen alternative or renewable fuel source. Delivery methods may be:
   - Edible replica of the fuel source (uncooked spaghetti wind turbine, gelatin and cookie layer landfill, peanut butter sandwich fuel cell, wafer cookie solar PV array, etc.). Replicas must list ingredients and what component they represent.
   - Write a Haiku for each alternative and renewable fuel source.
   - Chose a song and rewrite the lyrics to reflect information about an alternative or renewable fuel source.
   - Develop a board game with informational cards and decision-making questions.
4. Have students write a brief report about their chosen alternative or renewable fuel source to accompany their project.

To Know and Do More

Have students research the latest technologies that can advance the use of alternative or renewable fuel sources. Are there recent innovative developments?

SUMMARY

- The system which has three phases, i.e., the current will pass through the three wires, and there will be one neutral wire for passing the fault current to the earth is known as the three phase system.
- The three windings of the transformer primary and the transformer secondary can be connected in a delta or wye to create a three-phase transformer. There are four types of three-phase transformer connections, namely, delta/delta, delta/wye, wye/delta, and wye/wye.
- In an electric circuit, instantaneous power is the time rate of flow of energy past a given point of the circuit. In alternating current circuits, energy storage elements such as inductors and capacitors may result in periodic reversals of the direction of energy flow. Its SI unit is the watt.
- The ratio of active power to apparent power in a circuit is called the power factor. For two systems transmitting the same amount of active power, the system with the lower power factor will have higher circulating currents due to energy that returns to the source from energy storage in the load.
- In a direct current circuit, the power flowing to the load is proportional to the product of the current through the load and the potential drop across the load. The power that happens because of a capacitor or inductor is called reactive power.
- In ac circuits, power is measured with the help of wattmeter. A wattmeter is an instrument, which consists of two coils called the potential coil (PC) and the current coil (CC). The potential coil having high resistance is connected across the load and carries the current proportional to the potential difference across the load. The current coil having low resistance is connected in series with the load.
- The three phase system is divided mainly into two types. One is a balanced three-phase system and another one is an unbalanced three-phase system. The balance system is one in which the load are equally distributed in all the three phases of the system. The magnitude of voltage remains same in all the three phases and it is separated by an angle of 120°. In the unbalance system the magnitude of voltage in all the three phases becomes different.
- An electric power system is a network of electrical components deployed to supply, transfer, and use electric power. An example of a power system is the electrical grid that provides power to homes and industries within an extended area.
- Electrical power transmission is the process by which large amounts of electricity produced at power plants, such as industrial-scale solar facilities, is transported over long distances for eventual use by consumers.
- A continuous electrical circuit consists of three wires; a hot wire, a neutral wire, and a ground wire. The electricity will flow through the hot wire into the device you wish to power, or to a receptacle in your home.

- Protection devices are essential for maintaining security and reliability by protecting electrical systems and equipment from a variety of hazards. There are many different kinds of devices like fuse, switch breaker, RBCC, etc. and each one is made to handle particular threats and operational needs.

REVIEW QUESTIONS

1. What is the three-phase system?
2. What is the three-phase winding star connection?
3. What are the problems associated with the Y-Y connection of a three-phase transformer?
4. What are the advantages of the delta-delta connection of a three-phase transformer?
5. What is the tertiary winding in the three-phase transformer?
6. What is Scott connection of the transformer?
7. What are the advantages of a three-phase transformer over a three-phase transformer bank?
8. How many types of Three-Phase Transformer Connections are there?
9. What is the total power in an AC circuit?
10. What is the unit of 3-phase power?
11. How is electrical power generated?

REFERENCES

1. Amanulla, B., Chakrabarti, S., & Singh, S. N. (2012). Reconfiguration of power distribution systems considering reliability and power loss. *IEEE Transactions on Power Delivery, 27*(2), 918–926. https://doi.org/10.1109/TPWRD.2011.2179950.
2. Bernardon, D. P., Garcia, V. J., Ferreira, A. S. Q., & Canha, L. N. (2010). Multicriteria distribution network reconfiguration considering subtransmission analysis. *IEEE Transactions on Power Delivery, 25*(4), 2684–2691. https://doi.org/10.1109/TPWRD.2010.2041013.
3. Das, J. C. (2015). *Power System Harmonics and Passive Filter Design* (p. 2). Wiley, IEEE Press. ISBN 978-1-118-86162-2.
4. Emanuel, A. (1993). On the definition of power factor and apparent power in unbalanced polyphase circuits with sinusoidal voltage and currents. *IEEE Transactions on Power Delivery, 8*(3), 841–852. https://doi.org/10.1109/61.252612.
5. Fink, D. G., & Beatty, H. W. (Ed.). (1978). *Standard Handbook for Electrical Engineers* (11th ed., p. 18–17). McGraw Hill. ISBN 0-07-020974-X.
6. Franchi, C. M. (2019). *Electrical Machine Drives.* CRC Press. https://doi.org/10.1201/b22314.
7. Garud, R., Kumaraswamy, A., & Langlois, R. (2009). *Managing in the Modular Age: Architectures, Networks, and Organizations* (p. 249). John Wiley & Sons. ISBN 9780631233169.

8. Hughes, T. P. (1993). *Networks of Power: Electrification in Western Society, 1880–1930* (pp. 120–121). JHU Press. ISBN 9780801846144.
9. Sallam, A. A., & Malik, O. P. (2011). *Electric Distribution Systems* (p. 21). IEEE Computer Society Press. ISBN 9780470276822.
10. Short, T. A. (2014). *Electric Power Distribution Handbook* (pp. 1–33). CRC Press. ISBN 978-1-4665-9865-2.
11. Thomas, R. E., Rosa, A. J., & Toussaint, G. J. (2016). *The Analysis and Design of Linear Circuits* (8th ed., pp. 812–813). Wiley. ISBN 978-1-119-23538-5.
12. Tomoiagă, B., Chindriş, M., Sumper, A., Sudria-Andreu, A., & Villafafila-Robles, R. (2013). Pareto optimal reconfiguration of power distribution systems using a genetic algorithm based on NSGA-II. *Energies, 6*(3), 1439–1455. https://doi.org/10.3390/en6031439.
13. Yu, Q., Wang, X., Cheng, Y., & Tian, L. (2019). *Analysis and Mathematical Models of Canned Electrical Machine Drives*. Springer Singapore. https://doi.org/10.1007/978-981-13-2745-2.

CHAPTER 5

Electronic Instrumentation and Measurement

LEARNING OBJECTIVES

After studying this chapter, you will be able to:

- Examine the concepts of measurement in electronics
- Determine the static and dynamic characteristics of electrical measuring instruments
- Evaluate the units of electrical measurement
- Access the measurement of voltage, current, resistance and power

KEY TERMS FROM THIS CHAPTER

Digital multimeters
Electronic circuit
Hysteresis
Clamp Meter
Angular Frequency
Dynamic response
Instrumental Errors
Electrical energy
Electronic filters
Photovoltaic (PV) cell
Current transformer (CT)
Digital multimeter (DMM)
Electromotive force
Measurement errors

INTRODUCTION

To measure anything is to compare it to a set or standard value. Some of the physical quantities we utilize on a daily basis, such as length, mass, time, etc., may be measured. These values are measured using various scales and measuring equipment to create different instruments or for different scientific applications and research projects.

The term "electrical measurement" would then refer to the simple act of measuring electrical attributes or quantities. Electrical measuring tools are used to measure electrical parameters, which include terms like current, electricity, voltage, frequency, energy, and power. Every electrical instrument is chosen depending on its intended use and the type of quantity to be measured at the specific location. Different applications require different standards of measurement. A smaller scale measurement device can be used to measure things like household appliances and tiny electrical circuits that consume little electricity. However, locations with large electrical consumption, such as power plants and factories, require measuring devices that can measure parameters up to extremely high values.

5.1. CONCEPTS OF MEASUREMENT IN ELECTRONICS

Assigning numbers to physical quantities and events is the process of measurement. The sciences, engineering, building, and other technical domains, as well as practically every day activity, all depend on measurement. The components, circumstances, restrictions, and theoretical underpinnings of measurement have therefore been thoroughly examined.

As an alternative to using instruments, which can range in complexity from basic rules for measuring length to extremely complex systems designed to detect and measure quantities entirely beyond the capabilities of the senses, like radio waves from a distant star or the magnetic moment of a subatomic particle, measurements can be made using only human senses, in which case they are frequently referred to as estimates.

The defining of the quantity to be measured is the first step in the measurement process, which also always includes a comparison with a known quantity of the same kind. The amount or item to be measured is "transduced" into an analogous measuring signal if it cannot be directly compared. Since there is always some contact between the item and the observer or observing device during a measurement, accuracy is limited because of this energy exchange, which is minor in common applications but can become significant in specific types of measurement.

5.1.1. Measurement Theory

The study of measurement theory focuses on how numbers are assigned to objects and phenomena. It also addresses the types of items that may be measured, the relationships between various measurements, and the issue of measurement inaccuracy. Three fundamental issues need to be addressed by any general theory of measurement: uniqueness, which refers to the extent to which the selected representation is the only one that could be possible for the object or phenomenon under consideration, representation, which justifies the assignment of numbers, and error.

A number of axiom systems—basic principles and presumptions—have been developed to serve as the foundation for measurement theory. The following are some of the most significant categories of axioms: conjointness, axioms of difference, axioms of order, axioms of extension, and axioms of geometry. The order that numbers impose on objects is guaranteed by axioms of order to match the order determined by real observation or measurement. The representation of qualities like mass, length, and time duration is dealt with by axioms of extension. These attributes can be concatenated, or joined, for many objects that display the same feature. Interval measurement is governed by axioms of difference. The principles of conjointness suggest that attributes like loudness, intelligence, or hunger, which cannot be directly measured, can be inferred by observing how their constituent elements interact with one another. The principles of geometry dictate the representation of attributes with multiple dimensions using pairs, triples, or even n-tuples of numbers.

One of the main issues in measurement theory is the error problem. It was formerly thought that measuring errors may be completely eradicated with further advancements in science and technology. The majority of scientists no longer hold this idea, and nearly all physical measurements reported today include some indication of the likely degree of error or the limitation of precision. Errors of observation, including sampling errors, personal, systematic, random, and instrumental errors, as well as direct and indirect errors (where one inaccurate measurement is used to compute other measurements) are some of the different types of errors that need to be considered.

The history of measurement theory begins in the fourth century BC, when Euclid's Elements contained a theory of magnitudes created by the Greek mathematicians Eudoxus of Cnidus and Thaeatetus. The English mathematician Thomas Simpson conducted the first systematic study on observational error in 1757, but Joseph-Louis Lagrange and Pierre-Simon Laplace, two French astronomers from the 18th century, produced the essential work on error theory. The British utilitarian moralist Jeremy Bentham made an attempt to develop a theory for the measurement of value in the 18th century, which was also the first attempt to apply measurement theory to the social sciences. The foundation of modern axiomatic theories of measurement can be traced back to the contributions of Hermann von Helmholtz and Otto Hölder, two German scientists. The application of measurement theory to fields like psychology and economics has been greatly influenced by the work of Oskar Morgenstern and John von Neumann.

There has been little success in establishing standard measurement sequences or methodologies for social theories because most of them are speculative in character. The lack of generally accepted theoretical frameworks and quantifiable measures, sampling errors, issues with the measurer's intrusion on the object being measured, and the subjective nature of the data obtained from human subjects are a few of the issues with social measurement. The social science of economics has most likely been successful in implementing measurement theories because a large number of economic variables, such as quantity and price, are simply and reliably measured. Demography has also effectively used measurement techniques, especially with regard to mortality tables.

5.1.2. Types of Measurement Errors

Measurement errors are discrepancies that occur during the measurement process. They refer to the differences between the actual values of quantities being measured and the values obtained from the measurement. These errors can originate from different sources and have a substantial impact on the accuracy and reliability of the measurement results. Measurement errors can be classified into two main types: systematic errors and random errors.

Systematic errors are inaccuracies that occur consistently and repeatedly in measurements due to flaws in the measurement system. These errors can result from faulty instruments, calibration issues, or predictable environmental factors. Identifying and correcting systematic errors can be achieved through proper maintenance and calibration of the measurement instruments. For instance, if a scale always reads 0.5 kilograms more due to a calibration error, this discrepancy is a form of systematic error. It is essential to address systematic errors as they can introduce biases in results and impact the overall accuracy of measurements.

On the other hand, random errors are erratic and result from a variety of uncontrollable sources, such as fluctuating measurement processes, environmental factors, and observer variability. These mistakes are typically characterized by their randomness and lack of a clear pattern. For example, small discrepancies in readings caused by minute adjustments to the observer's angle or vibrations in the surrounding environment can be regarded as random mistakes when measuring an object's length numerous times. Random mistakes are difficult to totally eliminate, but their effects can be reduced by making several measurements and analyzing the results using statistical techniques.

The contrast between absolute and relative mistakes is another crucial component of measuring errors. The difference between the true value and the measured value, stated in the same units as the measurement, is known as the absolute error. Conversely, relative error is the ratio of the absolute error to the real value and is commonly represented as a percentage. Interpreting the precision and accuracy of the measurements requires an understanding of both kinds of faults.

Furthermore, human factors such as inaccurate data recording, misinterpreted instrument readings, or subjective judgments can also affect measurement mistakes. It's critical to create standardized protocols, provide thorough training for measurement process staff, and, when practical, use automated data recording technologies in order to reduce these errors.

In scientific research and industrial applications, accurate identification and control of measurement errors are essential to maintain the reliability and validity of data. Techniques such as proper experimental design, consistent calibration of equipment, conducting multiple measurements, and employing statistical analysis are utilized to address and minimize these errors. By effectively recognizing and correcting for measurement errors, researchers and professionals can enhance the accuracy of their data, resulting in more precise findings and informed decision-making. The three main types of measurement errors can be classified as follows:

1. **Gross Errors:** Gross errors are caused by inexperienced observers making mistakes while taking measurements. These errors can vary between different observers and can also result from using the wrong instrument. To reduce gross errors, we can take two steps.
   - Choose the best suitable instrument, based on the range of values to be measured.
   - Note down the readings carefully
2. **Systematic Errors:** Systematic

error occurs when an instrument consistently produces a uniform deviation during operation, usually due to the characteristics of the materials used in the instrument. The systematic errors can be classified into the following three types.

i. **Instrumental Errors:** This type of errors occur due to short-comings of instruments and loading effects.

ii. **Environmental Errors:** This type of errors occur due to the changes in environment such as change in temperature, pressure & etc.

iii. **Observational Errors:** This type of errors occur due to observer while taking the meter readings. Parallax errors belong to this type of errors.

3. **Random Errors:** Random errors are caused by unknown sources during the measurement process and cannot be eliminated or minimized. However, to obtain more accurate measurements without random errors, two steps can be taken.

   i. **Step1** – Take more number of readings by different observers.

   ii. **Step2** – Do statistical analysis on the readings obtained in Step1.

Following are the parameters that are used in statistical analysis.

- Mean
- Median
- Variance
- Deviation
- Standard Deviation

Now, let us discuss about these statistical parameters.

- **Mean:** Let $x_1, x_2, x_3, \ldots, x_N$ are the N readings of a particular measurement. The mean or average value of these readings can be calculated by using the following formula.

$$m=(x_1+x_2+x_3+\ldots+x_N)/N \quad (1)$$

Where; m is the mean or average value.

If the number of readings of a particular measurement are more, then the mean or average value will be approximately equal to true value

- **Median:** When there are a large number of readings for a specific measurement, it becomes challenging to determine the mean value. In this case, calculating the median value can provide an estimate of the mean value. To find the median value, first arrange the readings in ascending order. Use the following formula to calculate the median if there is an odd number of readings.

$$M=x_{((N+1)/2)} \quad (2)$$

We can calculate the median value by using the following formula, when the number of readings is an even number.

$$M=\frac{x_{(N/2)}+x_{([N/2]+1)}}{2} \quad (3)$$

- **Deviation from Mean:** The difference between the reading of a particular measurement and the mean value is known as *deviation from mean*. In short, it is called *deviation*. Mathematically, it can be represented as

$d_i = x_i - m$

Where;

d_i is the deviation of ith reading from mean.

x_i is the value of ith reading.

m is the mean or average value.

- **Standard Deviation:** The root mean square of deviation is called standard deviation. Mathematically, it can be represented as

$$\sigma = \sqrt{\frac{d_1^2 + d_2^2 + d_3^2 + \dots + d_N^2}{N}} \quad (4)$$

The above formula is valid if the number of readings, N is greater than or equal to 20. We can use the following formula for standard deviation, when the number of readings, N is less than 20.

$$\sigma = \sqrt{\frac{d_1^2 + d_2^2 + d_3^2 + \dots + d_N^2}{N-1}} \quad (5)$$

Where;

σ is the standard deviation

$d_1, d_2, d_3, \dots, d_N$ are the deviations of first, second, third, ..., Nth readings from mean, respectively.

Note – If the value of standard deviation is small, then there will be more accuracy in the reading values of measurement.

- **Variance:** The square of standard deviation is called variance. Mathematically, it can be represented as;

$V = \sigma^2$

Where; V is the variance; and σ is the standard deviation

The mean square of deviation is also called variance. Mathematically, it can be represented as

$$V = \frac{d_1^2 + d_2^2 + d_3^2 + \dots + d_N^2}{N} \quad (6)$$

The above formula is valid if the number of readings, N is greater than or equal to 20. We can use the following formula for variance when the number of readings, N is less than 20.

$$V = \frac{d_1^2 + d_2^2 + d_3^2 + \dots + d_N^2}{N-1} \quad (7)$$

Where; V is the variance

$d_1, d_2, d_3, \dots, d_N$ are the deviations of first, second, third, ..., N^{th} readings from mean, respectively.

So, with the help of statistical parameters, we can analyze the readings of a particular measurement. In this way, we will get more accurate measurement values.

5.1.3. Electrical Measuring Instruments

Electrical measuring instruments are tools used to quantify the strength of an electric current for various purposes. These instruments typically measure current, voltage, resistance, and power, with the results being expressed in units such as amps, volts, ohms, and watts, respectively. These portable devices enable quick and precise measurements to be taken, allowing for the collection of data that can be compared to existing information. By studying these characteristics and traits, one can better understand the electrical object under examination.

In simple language, using tools to measure power, intensity, resistance, and voltage helps ensure that machines and installations are working correctly. These tools also help identify how much electric energy a service is consuming.

5.1.3.1. Electrical Capacities the Measuring Instruments Measure

Depending on what needs to be determined about an electrical circuit, different magnitudes can be measured:

- Voltage is the power differential between two electrical current-conducting sites in an element. The volt is used as its measurement unit.
- The amount of current flowing through an electrical circuit is known as its intensity. The amp is used as its measurement unit.
- The value of a conductor in relation to the electron flow is called resistance. Ohms are used as its measurement unit.
- Capacitance is an electrical circuit element's ability to store charge. The amp is also its measurement unit.
- The combination of voltage (volts) and current (amps) produces electrical power. The watt is the name of its measurement unit.

5.1.3.2. Classification of Electrical Measuring Instruments

Measuring instruments are classified according to their function:

- Analog: based on electromagnetic systems; when a current or voltage passes through a conductor, a magnetic field is generated that moves a needle within a magnitude scale.
- Digital: these are the most common because they make the same measurement as analog versions but in a faster, more accurate, and more reliable way.
- Registers: these can be either analog or digital and are responsible for recording the values being measured in a graph where the value of the magnitude is shown over a period of time.

5.1.4. Main Electrical Measuring Instruments

Basic electrical measuring instruments are:

- **Ammeter:** An ammeter is a device that measures current and needs to be connected to the circuit in series. Since the ammeter receives all of the circuit's current, it must have very little resistance.
- **Voltmeter:** A voltmeter is a device that measures potential differences. It needs to be connected in parallel to the circuit element that needs the potential difference. A voltmeter needs to have a very high resistance in order to prevent a significant current from passing through it.
- **Ohmmeter:** An ohmmeter is a resistance measurement tool. The circuit being tested does not supply the energy required for the ohmmeter circuit to function, in contrast to the voltmeter or ammeter. This energy is provided to the ohmmeter by a separate voltage source, such as a battery.
- **Multimeter***:* Many shunts and series multipliers are used with a moving-coil meter in instruments to

produce a range of measurements on a single scale graduated to read voltage and current. Measurements of resistance are also possible if a battery is included.

These devices go by the names multimeters, universal instruments, or multirange instruments. A common example would be an "Avometer." With the use of distinct terminals or a selector switch, one can choose a certain range. It is possible to do only one measurement at once. When a rectifier is included into the instrument, these devices can frequently be used in both a.c. and d.c. circuits. The Fluke Digital Multimeter is a top performer in the industry, known for its precision, durability, reliability, and safety features. These digital multimeters are widely used for measuring DC and AC currents, voltages, resistance, continuity, temperature, and other variables.

- **Oscilloscope**: The oscilloscope is basically a graph-displaying device – it draws a graph of an electrical signal. In most applications, the graph shows how signals change over time. From the graph, it is possible to:
  - determine the time and voltage values of a signal
  - calculate the frequency of an oscillating signal
  - see the 'moving parts' of a circuit represented by the signal
  - tell if a malfunctioning component is distorting the signal
  - find out how much of a signal is d.c. or a.c.
  - tell how much of the signal is noise and whether the noise is changing with time

  Oscilloscopes are used by everyone from television repair technicians to physicists. They are indispensable for anyone designing or repairing electronic equipment.

- **Wattmeter**: A wattmeter is an instrument for the measurement of power in an electrical circuit.

  The instrument has two coils:
  - a current coil, which is connected in series with the load, like an ammeter
  - a voltage coil, which is connected in parallel with the load, like a voltmeter.
- **Megger:** A megger is an essential instrument for measuring high resistance and insulation resistance. It includes a hand-driven dc generator and a direct reading true ohmmeter, also known as a megohmmeter. Unlike an ordinary ohmmeter, a megger is specifically designed to measure the resistance of multimillions of ohms, such as conductor insulation. This portable instrument is commonly used to measure high resistance levels, typically one megaohm and above, in electrical cables and equipment. It helps check the insulation resistance of electric circuits in relation to the earth and other components.
- **Continuity Tester:** A continuity tester is the measurement of the resistance of a cable to discover if the cable is continuous, i.e., that it has no breaks or high-resistance joints.
- **Tachometer:** A tachometer is a device that measures the rotational speed of an engine shaft, commonly expressed in revolutions per minute. These days, a tachometer

is required to measure the speed at which several objects rotate. Knowing the length of time and the speed of a moving object is useful since it allows one to compute the distance and schedule maintenance to stop wear and tear.

- **Stroboscope:** A stroboscope is a tool used to observe a rotating object at fixed intervals, achieved through a rotating or vibrating shutter or a specially designed lamp that flashes regularly. When the time between views matches the object's revolution time and the view is brief, the object will appear motionless.
- **Fluxmeter:** A fluxmeter outperforms a ballistic galvanometer in certain magnetic measurements due to its unique design. It is essentially a modified version of a ballistic galvanometer with adjustments such as reducing suspension torque and increasing electromagnetic damping. In addition to being more portable, a fluxmeter does not require a rapid change in flux over a short period of time like a ballistic galvanometer. The deflection of a fluxmeter is consistent for a given flux change regardless of the duration of the change.
- **LCR meter:** The LCR (or RLC) meter, or bridge, is an electronic instrument specifically designed for the measurement of impedances having arbitrary phase angles (LCR stands for inductance, capacitance, resistance)
- **Non-contact Voltage Tester:** A non-contact voltage tester or detector is an electrical tester that helps to detect the presence of voltage. Voltage presence is useful information to have when troubleshooting or working on a failed asset. The first tool you'll reach for is a non-contact voltage detector.
- **Clamp Meter:** A clamp meter is a versatile electrical testing tool that combines a voltmeter with a clamp-type current meter. From its origins as a single-purpose test tool for electricians, modern clamp meters have evolved to offer a wide range of measurement functions, increased accuracy, and special features. These devices have transitioned from analog to digital technology, offering the functionality of a digital multimeter (DMM) along with the added benefit of a built-in current transformer.
- **Power Quality Analyzer:** Power quality analyzers are user-friendly devices that can accurately detect a wide range of power quality issues, including high-speed transients, voltage regulation violations, high-frequency harmonics, signaling, and more. These analyzers have advanced features that allow them to automatically identify the type of current transformer (CT) connected and its range to ensure proper connection.
- **Irradiance Meter:** The solar irradiance meter is a key component in estimating solar irradiation which is necessary and essential to designing sustainable energy systems such as photovoltaic (PV) systems.
- **Battery Analyzer:** The battery analyzer is an ideal test tool for maintenance, troubleshooting and performance testing of individual stationary batteries and battery banks used in critical battery backup applications.

5.2. STATIC AND DYNAMIC CHARACTERISTICS OF ELECTRICAL MEASURING INSTRUMENTS

It is crucial to comprehend the performance characteristics of a measurement system when selecting the appropriate instrument. Key characteristics include accuracy, precision, resolution, sensitivity, and others, which help users choose the best instrument for their specific measuring needs.

There are two basic performance characteristics of measuring instrument:

- Static characteristics: value of the measured variable change slowly.
- Dynamic characteristics: value of the measured variable change very fast.

5.2.1. Static Characteristics of Electrical Measuring Instruments

Certain applications require the measurement of quantities that remain constant or change gradually over time. In these cases, it is possible to establish a set of standards to evaluate the accuracy of the measurement without disrupting dynamic descriptions that rely on differential equations. These standards are known as static characteristics, and the primary ones include:

- **Accuracy:** Accuracy is defined as how closely an instrument reading aligns with the true value of the quantity being measured. It is crucial for electrical measuring instruments to be accurate as it ensures conformity to truth. Accuracy can be measured in terms of inaccuracy or limits of errors, and can be expressed in various ways.
  - **Point Accuracy:** This accuracy measurement pertains to just one specific point on the instrument's scale. It does not provide any details on the accuracy at different points on the scale, nor does it offer insight into the overall accuracy of the instrument.

- **Accuracy as Percentage of Scale Range:** When an instrument has uniform scale, it's accuracy may be expressed in terms of scale range.
- **Accuracy as Percentage of True Value:** The best way to conceive the idea of accuracy is to specify it in terms of the true value of the quantity being measured within +0.5% or –0.5% of true value.

- **Precision:** Precision is a measure of the degree of agreement within a group of measurements; that is, given a fixed value of quantity, it is a measure of the reproducibility of the measurements. The word "precise" refers to something that is precisely defined. Consider the following scenario to illustrate the distinction between the terms accuracy and precision: we have an ammeter with a high degree of precision due to its distinct, easily readable scale and knife-edge pointer that is arranged in mirrors to eliminate parallax. It is another crucial static feature of electrical measurement tools.

We can measure readings with this ammeter to 1/100 of an ampere with precision. The readings are consistent and clearly defined each time we take a measurement. However, the ammeter is not accurate due to a faulty zero adjustment, therefore the readings do not align with the truth (Figure 5.1).

- **Stability:** A measuring system's capability to consistently meet performance standards over an extended period is known as long-term stability. Zero stability refers to the instrument's ability to return to a zero reading once the input quantity is reset to zero, even with consistent external conditions.

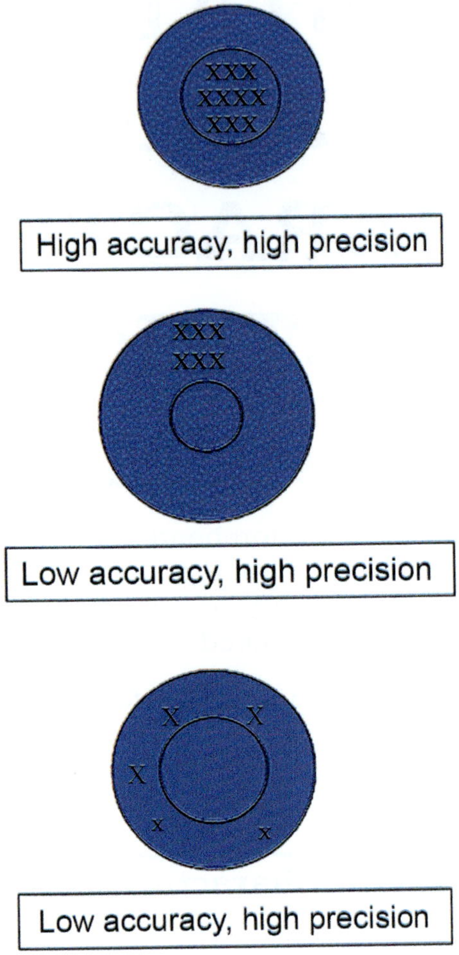

Figure 5.1. *Accuracy and precision.*

Source: http://ecoursesonline.iasri.res.in/mod/resource/view.php?id=3311.

- **Resolution:** If the input to an instrument slowly increases from a starting point that is not zero, it will be noticed that the output of the instrument remains constant until a certain minimum increase in the input occurs. This minimum increase is known as the resolution of the instrument, which is the smallest incremental change in the input that the measuring system can detect. This is considered one of the most important static characteristics of electrical measuring instruments. Resolving power, or discrimination power, refers to the system's capability to detect small changes in the input. The resolution of an instrument is greatly influenced by how finely its output scale is divided. When the input to an instrument is gradually increased from zero, there will be a minimum input value below which no changes in the output can be seen. This minimum input value is known as the threshold of the instrument.

- **Threshold:** When slowly increasing the instrument input from zero, there will be a minimum value where no output change can be noticed. This value is known as the threshold of the instrument. When specifying the threshold, the first noticeable output change is typically described as any measurable change that can be detected.
- **Drift:** Drift is a gradual change in the output signal of a transducer or measuring system that is not caused by any alterations in the input. It usually results from shifts in the internal operating conditions of the system's components. Drift can manifest as zero drift or sensitivity drift. Zero drift refers to a deviation in the instrument output over time from its initial reading, while all other measurement factors remain constant. This could stem from variations in component values due to environmental changes or aging. Zero drift is commonly measured in units such as volts per °C for a voltmeter affected by temperature fluctuations, known as the zero drift coefficient.
- **Repeatability:** It is a feature of electrical measuring tools with high precision. When the same input is applied repeatedly over a brief period of time, with the same measurement settings, same instrument and observer, same location, and same conditions of usage maintained throughout, it characterizes the proximity of output values. Drift and internal noise have an impact on it.

The percentage of the true value is used to express it. Transducers for measuring are used continuously in process control operations; consistency in product quality dictates that the repeatability of the transducer's performance is more significant than its accuracy.

- **Reproducibility:** Consistency in measuring the same input value is achieved through close monitoring at various times, conditions, and with different instruments. Output signals are examined for reliability over extended periods and in different settings to ensure perfect reproducibility, allowing for interchangeability of instruments and transducers.
- **Dead Zone:** The dead zone is the maximum input quantity change where there is no output from the instrument. This can be due to static friction, backlash, or hysteresis. Dead zone is also referred to as dead band or dead space. Primary transducers utilizing elastic mechanical elements will exhibit

hysteresis, creep, and elastic after-effects to a certain degree.

- **Backlash:** The maximum distance or angle through which any part of the mechanical system may be moved in one direction without applying appreciable force or motion to the next part in a mechanical sequence.
- **Hysteresis:** Whether a system is mechanical, electrical, or any other type, hysteresis is a phenomena that shows various output effects while loading and unloading. The term "hysteresis" refers to the variation in an instrument's readings depending on whether the input signal's fixed value is approaching or veering away from changing input values. That is, when the measurement is made of the identical value using the symmetry approach, the upscale and downscale deflections do not coincide. Hysteresis is the term used to describe the lack of synchrony between the loading and unloading curves.
- **Linearity:** It establishes the proportionality between the output signal and the input quantity. The calibration characteristic is linear and is represented by a straight line that passes through the origin if the sensitivity remains constant for all values of the measurement system, ranging from zero to the full scale value. A linear scale may be used if the instrument is an indicating or recording device. A measurement system's calibration curve's linearity is determined by how closely it resembles a straight line. An instrument may nonetheless be extremely accurate even if its calibration curve for the intended input is not a straight line. However, linear response is the most ideal in many situations.

It establishes the proportionality between the output signal and the input quantity. When the measuring system's sensitivity remains consistent across the whole range of values, the calibration characteristic is linear and resembles a straight line that passes through the origin. A linear scale may be used if the instrument is an indicating or recording device. When there is a zero error, the characteristic takes on the form of the equation y=mx+c, where m stands for slope, c for intercept, and y for output and x for input.

A measurement system's calibration curve's linearity is determined by how closely it resembles a straight line. An instrument may nonetheless be extremely accurate even if its calibration curve for the intended input is not a straight line (Figure 5.2).

In many applications, however, linear response is most desirable.

Linearity is defined as;

Linearity=Maximum deviation of o/p from idealized straight line/Actual readings

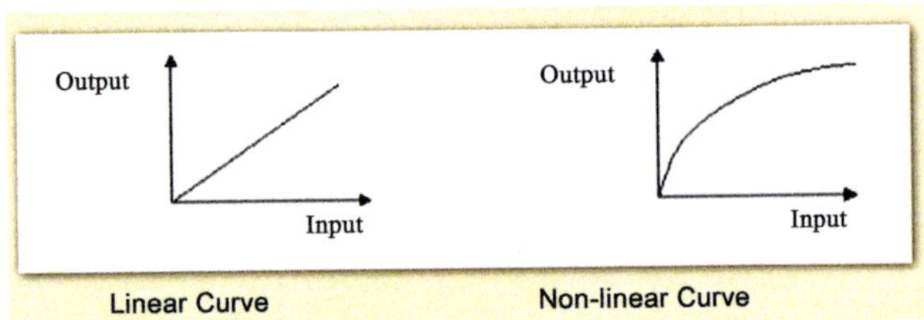

Figure 5.2. *Linearity.*

Source: https://blogger.googleusercontent.com/img/b/R29vZ2xl/AVvXsEiVVaed9e-2L56Kbfmox66LdpNI3f-ICzVLuNF6ycsf1K7TAIX_qH8pIFxDg1arlrqJhFlGUrKMmtdiLQRu Qgn20ULceWRMoIhz4LXXParXQTxM_44FlOcEmW59pD1YZy6WjWsPo0rBs9c/s400/Linearity1.png.

- **Range or Span:** The phrases span and range are used to describe the lower and APA calibration points, respectively. The span is just the difference between the full scale and lower scale value. Typically, the range of indicating instruments is from zero to the full scale value. Nevertheless, the identical instruments are biased to read voltages only, say, between 5 and 25 volts. A bias is used to prevent these instruments' zero from being indicated. The scale span in this instance is 25–5, or 20 V, while the scale range is defined as 5 V to 25 V.
- **Bias***:* Bias describes a constant error which exits over the full range of measurement of an instrument. The error is normally removable by calibration.
- **Tolerance:** This word, which indicates the maximum error that can be predicted in a given value, is closely related to accuracy. Although it isn't exactly a static feature of measuring devices, it is brought up here since some instruments' accuracy is occasionally expressed as a tolerance figure. When used appropriately, tolerance refers to the greatest deviation a produced part can have from a certain value. Resistors and other electrical circuit components have tolerances of about 5%.

5.2.2. Dynamic Characteristics of Electrical Measuring Instruments

Dynamic measurement systems consist of inputs that are constantly changing, leading to varying outputs. The system's response to these changing conditions is known as its dynamic response, with the dynamic characteristics of electrical measuring instruments outlined below:

- **Dynamic Error:** It is the difference of true value of the quantity changing with the time the value indicated by the instrument provided static error is zero. Total dynamic error is the phase difference between input and output of the measurement system.
- **Fidelity:** Fidelity refers to the system's capability to replicate output exactly as the input, without any time lag or phase difference considered in the definition. Ideally, a system should exhibit 100% fidelity, meaning the output mirrors the input without any distortion. Fidelity requirements vary depending on the application.
- **Bandwidth:** It is the frequency range at which its dynamic sensitivity is acceptable. It is necessary for measuring systems' dynamic sensitivity to be no more than 2% of their static sensitivity.

The aforementioned requirement is loosened for other physical systems, such as electrical filters and electronic amplifiers, with the result that their bandwidth specifications can reach frequencies where the dynamic sensitivity is 70.7% of zero or the mid-frequency.

- **Speed of Response:** It speaks to its capacity to react to abrupt shifts in the input signal's amplitude. For a step input function, it is often expressed as the amount of time it takes the system to approach steady state conditions. In order to describe the reaction of the various systems encountered in practice, terminology like time constant, measurement lag, settling time, and dead time dynamic range are used. As a result, the speed of response is evaluated based on knowledge of the system performance under transient situations. This is one of the crucial dynamic features of electrical measurement tools.

- **Time Constant:** It is related to how a first order system behaves and is known as the time it takes for the system to reach 0.632 of its final output signal amplitude. A system with a smaller time constant will reach its final output amplitude sooner than a system with a larger time constant, giving it a faster response time.
- **Measurement Lag:** Instrument response delay is the time it takes for an instrument to react to a change in the measurement. This delay is typically minimal, but it can be more substantial when fast measurements are needed. There are two types of measurement delays: retardation, where the instrument reacts immediately after a change in the measurement, and time delay, where the system reacts after a delay following the input application.
- **Settling Time:** Settling time refers to how long it takes for an instrument or measurement system to reach its final steady state position after receiving an input. For portable instruments, settling time is how long it takes for the pointer to come to rest within a range of –0.3% to +0.3% of its final scale length. Panel type instruments have a settling time of –1% to +1% of its final scale length for the pointer to come to rest. A shorter settling time indicates a faster response speed. Settling time can vary depending on system parameters and operating conditions, making it an important dynamic characteristic of electrical measuring instruments.

5.3. UNITS OF ELECTRICAL MEASUREMENT

The most common units for voltage, current, and resistance are the Volt [V], Ampere [A], and Ohm [Ω], respectively. These units are based on the International System of Units (SI), with other electrical units being derived from SI base units. In some cases, multiples or fractions of these standard units are used in electrical or electronic circuits and systems to account for very large or very small measurements.

The following Table 5.1 gives a list of some of the standard electrical units of measure used in electrical formulas and component values.

Table 5.1. Standard Electrical Units of Measure

Electrical Parameter	**Measuring Unit**	**Symbol**	**Description**
Voltage	Volt	V or E	Unit of Electrical Potential $V = I \times R$
Current	Ampere	I or i	Unit of Electrical Current $I = V \div R$
Resistance	Ohm	R or Ω	Unit of DC Resistance $R = V \div I$
Conductance	Siemen	G or ⊠	Reciprocal of Resistance $G = 1 \div R$
Capacitance	Farad	C	Unit of Capacitance $C = Q \div V$
Charge	Coulomb	Q	Unit of Electrical Charge $Q = C \times V$
Inductance	Henry	L or H	Unit of Inductance VL = -L(di/dt)
Power	Watts	W	Unit of Power P = V × I or I2 × R

Impedance	Ohm	Z	Unit of AC Resistance Z2 = R2 + X2
Frequency	Hertz	Hz	Unit of Frequency $f = 1 \div T$

Source: https://www.electronics-tutorials.ws/dccircuits/dcp_3.html.

5.3.1. Multiples and Sub-Multiples

Electrical and electronic engineering involves a wide range of values between the maximum and minimum standard units. For instance, resistance can be less than 0.01Ω or greater than 1,000,000Ω. To simplify notation, we use multiples and submultiples of the standard unit to avoid excessive zeros. Table 5.2 provides the names and abbreviations for these units.

Table 5.2. The Names and Abbreviations of Electrical Units

Prefix	Symbol	Multiplier	Power of Ten
Tera	T	1,000,000,000,000	10^{12}
Giga	G	1,000,000,000	10^{9}
Mega	M	1,000,000	10^{6}
kilo	k	1,000	10^{3}
none	none	1	10^{0}
centi	c	1/100	10^{-2}
milli	m	1/1,000	10^{-3}
micro	µ	1/1,000,000	10^{-6}
nano	n	1/1,000,000,000	10^{-9}
pico	p	1/1,000,000,000,000	10^{-12}

Source: https://www.electronics-tutorials.ws/dccircuits/dcp_3.html.

So, to display the units or multiples of units for either Resistance, Current or Voltage we would use as an example:

- 1 kV = 1 kilo-volt – which is equal to 1,000 Volts.
- 1 mA = 1 milli-amp – which is equal to 1/1000 of an Ampere.
- 47 kΩ = 47 kilo-ohms – which is equal to 47 thousand Ohms.
- 100 uF = 100 micro-farads – which is equal to 100/1,000,000 of a Farad.
- 1 kW = 1 kilo-watt – which is equal to 1,000 Watts.

- 1 MHz = 1 mega-hertz – which is equal to 1 million Hertz.

One must multiply or divide by the difference between the two values in order to convert from one prefix to another. Convert 1 MHz to kHz, for instance.

As 1 MHz equals 1,000,000 hertz and 1 kHz is equal to 1,000 hertz, as we know from the explanation above, 1 MHz is a thousand times larger than 1 kHz. Then, since 1MHz is equal to 1000 kHz, we must multiply megahertz by 1,000 in order to convert it to kilohertz.

Similarly, we would need to divide by 1,000 in order to convert kilohertz to megahertz. Depending on whether you need to multiply, moving the decimal point left or right would be a lot faster and easier way to do it.

As well as the "Standard" electrical units of measure shown above, other units are also used in electrical engineering to denote other values and quantities such as:

- Wh – The Watt-Hour: The energy usage of a circuit during a specific timeframe, typically measured in watt-hours (Wh), kilowatt-hours (kWh), or megawatt-hours (MWh). For example, a light bulb may use 100 watts of power for one hour.
- dB – The Decibel: The decibel, denoted as dB, is a logarithmic unit that represents gain in voltage, current, or power. It is one tenth of a Bel (B) and is often used to express the input-output ratio in amplifier, audio circuits, or loudspeaker systems.

For example, the dB ratio of an input voltage (V_{IN}) to an output voltage (V_{OUT}) is expressed as $20\log_{10}$ (Vout/Vin). The value in dB can be either positive (20dB) representing gain or negative (–20dB) representing loss with unity, i.e., input = output expressed as 0dB.

- θ – **Phase Angle:** the difference in degrees between the voltage waveform and the current waveform with the same periodic time is known as the phase angle. Depending on the circuit element, it can be either "leading" or "lagging" and indicates the time difference or shift. A waveform's phase angle is commonly expressed in radians or degrees.
- ω – **Angular Frequency:** The Angular Frequency, symbol ω, is another quantity that is mostly employed in a.c. circuits to describe the Phasor Relationship between two or more waveforms. This rotating unit has an angular frequency of $2\pi f$ and is measured in rads/s, or radians per second. One cycle is 360°, or 2π, in whole; so, 180°, or π rad, is the half of a rotation.
- τ – **Time Constant:** When a step response input is applied to an impedance circuit or linear first-order system, the Time Constant is the amount of time it takes for the output to reach 63.7% of its maximum or minimum output value. It is a gauge for response speed.

5.4. MEASUREMENT OF VOLTAGE, CURRENT, RESISTANCE AND POWER

Understanding current and voltage is crucial for any electrical circuit. These two characteristics are essential for creating electronic devices like smartphones, televisions, and refrigerators. That's why we're going back to the basics of electrical engineering to show you how to measure current and voltage.

5.4.1. The Basics – Parallel and Series Connection

In electrical engineering, there are two types of circuits: The parallel and series connections. You can see these two types of circuits here (Figure 5.3):

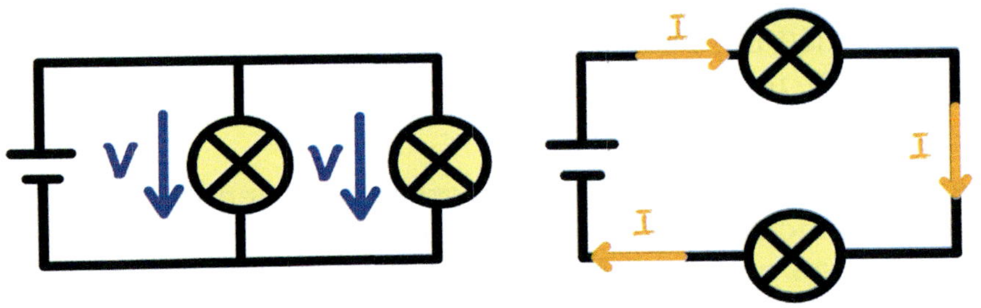

Figure 5.3. *Parallel (left) and series (right) connection.*

Source: https://www.dewetron.com/wp-content/uploads/2022/07/CurrentVoltage-1024x371.jpg.

Naturally, your next question is going to be, "What's the difference between series and parallel circuits, and why it matters for measuring voltage and current?"

Two parts (two bulbs in our example) are linked to the power supply in succession in a series circuit. The property of a series connection is that, regardless of the measurement point, the same current flows through it all. Consider this to be similar to a water pipe: The same volume of water must exit the pipe at both ends if it enters at one end. Additionally, the water flow in the pipe is uniform throughout.

In the series circuit, the electrical voltage varies throughout, though. A voltage drop occurs when a portion of the voltage is lost at the first lamp, and another portion is lost at the second lamp. There is less voltage left over for the last component the more components you connect in series.

In contrast, every component in a parallel circuit has the same voltage drop. The components are positioned next to one another rather than immediately behind one another, which is the reason behind this. However, this also implies that the circuit's total current is not constant. Once more, the water pipe analogy is useful in this situation.

When the water pipe splits into two smaller pipes, only a portion of the initial total flow goes through each pipe. However, we have already covered the fundamentals, so let's dive into how current and voltage measurements are conducted.

5.4.2. The Current Measurement

Ammeters and voltmeters are devices used to measure current and voltage. Multimeters or power analyzers are devices that have the ability to measure not just voltage but also current, and typically other quantities as well.

The components are connected to the ammeter in series if you wish to measure the current. Why in a sequence? Because only then—as previously mentioned—does the same current pass through both the component and the ammeter. A different current would be detected if the ammeter and the component were connected in parallel.

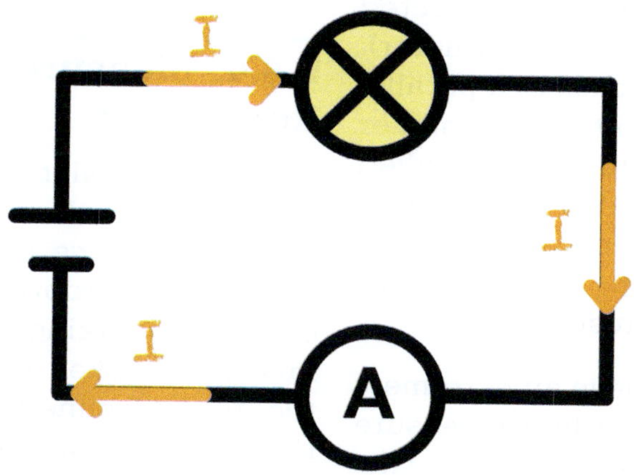

Figure 5.4. *Ammeter in an electric circuit.*

Source: https://www.dewetron.com/wp-content/uploads/2022/06/Current-1024x773.jpg.

Additionally, in order to ensure accurate measurements with minimal impact on the circuit, an ammeter must have a very low internal resistance. This ensures that the device does not interfere with the current flow in the circuit. If the internal resistance were high, according to Ohm's law, less current would pass through the circuit, affecting the measurement accuracy. Therefore, using an ammeter in parallel is not recommended, as it can lead to an overload of current flowing through the device. In such cases, the internal overcurrent protection will be triggered, causing the fuse to blow and rendering the ammeter inoperative until the fuse is replaced (Figure 5.4).

Unfortunately, there are situations where opening a circuit to install an ammeter is not feasible. In these cases, indirect current measurement is a more practical option. Instead of directly measuring the current, this approach involves observing the associated effects of the current flow. By analyzing these effects, the electric current can be determined. One example of an indirect current measurement tool is the current clamp, which detects the magnetic field created by the current to calculate the current flow in the conductor.

5.4.3. The Voltage Measurement

The voltage measurement and the current measurement operate in exactly the opposite ways. The component where the voltage drop is to be measured is connected in parallel to a voltage meter (voltmeter). Parallel because both branches of a parallel circuit have the same voltage. Consider what may occur if the meter was installed improperly, such as in series connection. In this instance, the meter itself would experience a voltage drop, leading to an inaccurate measurement result.

A voltmeter needs to have an extremely high internal resistance in order to measure voltage accurately. It is ideal for this resistance to be significantly higher than the resistance of the component you wish to measure the voltage drop across. This is required because, in the absence of it, the circuit's current flow and, consequently, the voltage drop across the component, would alter. Thus, in this scenario, the circuit current would be directly impacted by the voltmeter (Figure 5.5).

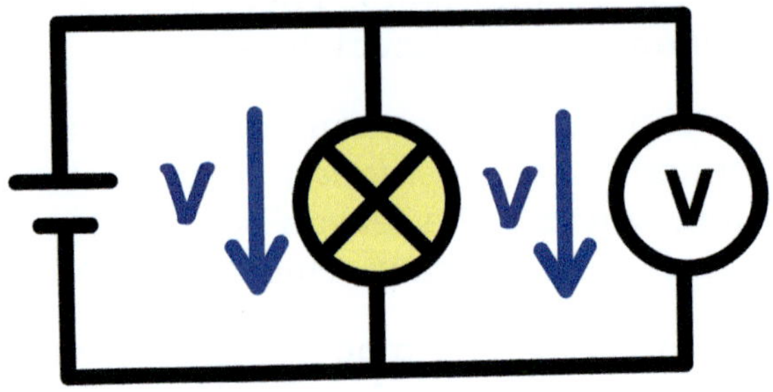

Figure 5.5. *Voltmeter in an electric circuit.*

Source: https://www.dewetron.com/wp-content/uploads/2022/06/Voltage-1024x541.jpg.

Indirect measurement of current is possible and also very common, but there is no possibility for indirect voltage measurement. However, it is possible to measure voltage without contact by means of electrometers and similar measuring devices. However, one uses these methods rarely in the field of electrical engineering.

5.4.4. Measurement of Resistance

Electrical resistance is crucial to the operation of electronic devices' circuitry. If the resistance in their circuitry deviates from the appropriate level, these devices can stop working. Electricity, however, is invisible. It takes a specialized measurement tool to determine whether a circuit has the right amount of resistance.

A device like an analog or digital multimeter is used to measure resistance. Both kinds of instruments have a wide range of applications because they are capable of measuring not just resistance but also current, voltage, and other variables.

Resistance measurement, however, does not entail determining the resistance value of the circuit. Rather, the circuit's applied voltage and current are measured in order to calculate resistance. The circuit (or resistance) under measurement exhibits a voltage (or more accurately, a voltage drop) when a current is supplied to it. Ohm's Law can be used to compute resistance by monitoring the current and voltage.

Therefore, if the measured values for current and voltage are known, it is possible to determine the resistance value of a circuit. Ohm's Law is the measurement principle used by both analog and digital multimeters to measure resistance.

5.4.4.1. Resistance Measurement with an Analog Tester

When using an analog multimeter to measure resistance, turn off the circuit being measured. Connect the black test lead to the COM input terminal and the red test lead to the positive input terminal marked with a "+." Put the instrument in Ω mode and adjust the range button according to the expected resistance of the circuit.

Utilizing the 0 Ω adjustment knob, short the black and red test pins and set the needle to 0 Ω. Next, connect the black and red test pins to both ends of the circuit being measured, and then read the value that the meter indicates.

Remember that when the instrument is in resistance mode, putting a voltage on the test leads could cause harm to the tester. Additionally, the analog multimeter's battery may be low if you are unable to do 0 Ω correction. If this happens to you, make sure the battery voltage is correct (Figure 5.6).

Always conduct zero adjustment when measuring resistance. (Mechanical and electrical zero adjustment) Situations where voltage is being applied is hazardous, therefore separation is critical.

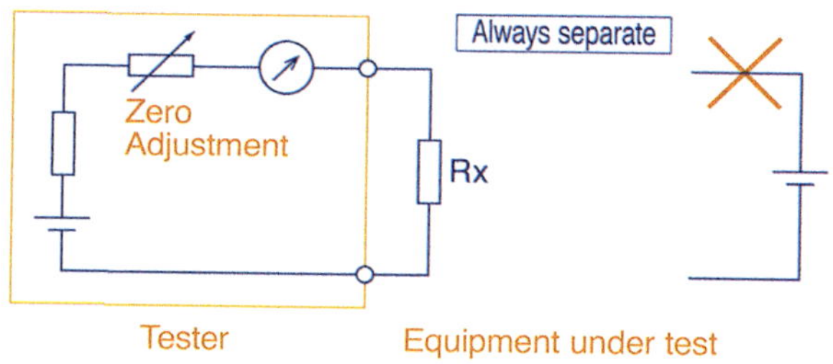

Figure 5.6. *Resistance measurement circuit of an analog meter.*

Source: https://www.hioki.com/themes/hioki/images/global/learning/methods/resistance-measurement-methods-img-05.png.

5.4.4.2. Resistance Measurement with a Digital Multimeter

Generally, resistance can be measured using a digital multimeter in a similar way to an analog multimeter. It is a straightforward process where the only noticeable distinction is the digital display instead of an analog needle. Despite this difference, the fundamental procedure is quite similar. However, digital multimeters offer two measurement methods.

- Two-terminal measurement; and
- Four-terminal measurement.

Typically, the two-terminal measurement method is commonly used when testing resistance with a digital multimeter. This technique involves applying a consistent current and recording the resistance value using the voltmeter on the device. Similar to analog multimeters, this method does have a drawback of including the resistance of the wiring in the final resistance value (Figure 5.7).

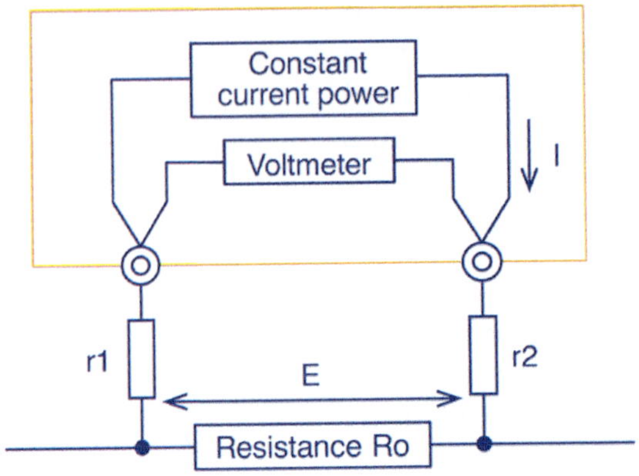

Figure 5.7. *Two-terminal measurement method.*

Source: https://www.hioki.com/themes/hioki/images/global/learning/methods/resistance-measurement-methods-img-06.png.

In order to reduce the impact of the extra resistance, the test leads are connected before taking a measurement to calibrate the resistance value to zero. Despite this method, the effects cannot be entirely eliminated. Thus, four-terminal measurement was developed to tackle this issue by utilizing four test leads and separate circuits for the voltmeter and ammeter (Figure 5.8).

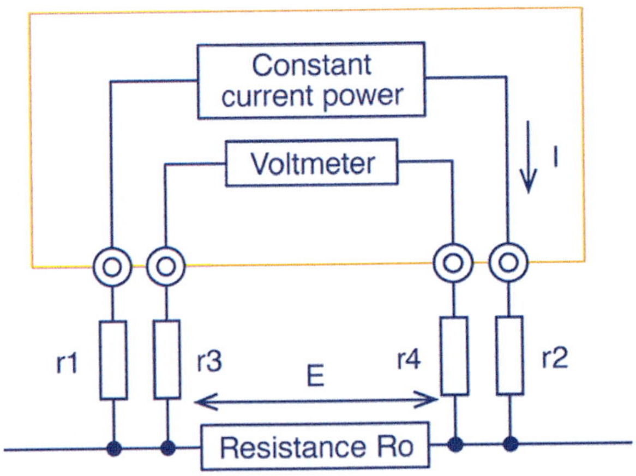

Figure 5.8. *Four-terminal measurement method.*

Source: https://www.hioki.com/themes/hioki/images/global/learning/methods/resistance-measurement-methods-img-07.png.

Different types of resistance, such as wire, relay, and connector resistance, along with the internal resistance of batteries, exist. It is crucial to select the appropriate instrument for the specific measurement required. Make sure to carefully choose an instrument that fits your needs when making a purchase (Figure 5.9 and 5.10).

Figure 5.9. *Clip type lead (clip-type lead for 4-terminal measurement).*

Source: https://www.hioki.com/themes/hioki/images/global/learning/methods/resistance-measurement-methods-img-08.jpg.

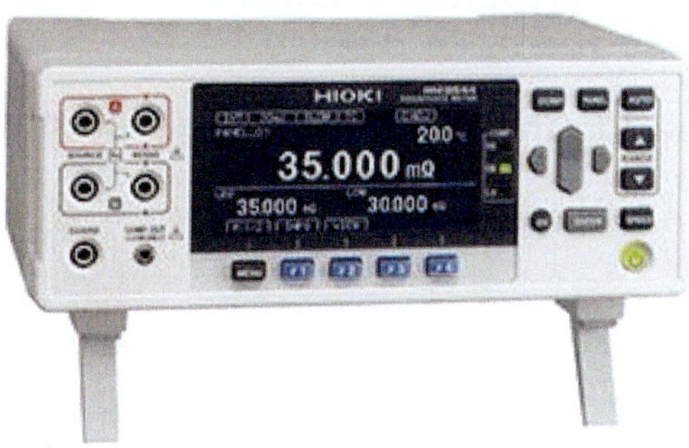

Figure 5.10. *Resistance meters.*

Source: https://www.hioki.com/themes/hioki/images/global/learning/methods/resistance-measurement-methods-img-09.jpg.

5.4.4.3. Sources of Error When Measuring Resistance

Test leads' wiring resistance is not the only factor that affects resistance value measurement results. The following factors also come into play (Figure 5.11):

- Electromotive force
- Thermal noise
- Leakage current
- Dielectric absorption
- Friction noise
- External noise
- Temperature, humidity, and wind

Resistance meters offer the capability to mitigate the impacts of temperature and other variables. This is achieved by comparing the readings from a temperature sensor connected to the meter with the reference temperature, and adjusting the resistance values accordingly. If there is inconsistency in the measured resistance values, it is important to identify the factors causing this issue and implement measures to resolve them.

$$Rt = Rt0 \times \{ 1 + \alpha t0 \times (t - t0) \}$$

Rt : Actually tested resistance [Ω]

Rt0 : Compensated resistance [Ω]

t0 : Reference temperature [°C]

t : Current ambient temperature [°C]

$\alpha t0$: Temperature coefficient at t0

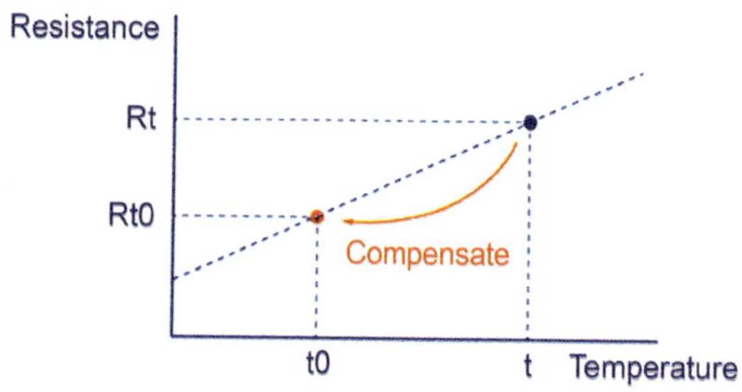

Figure 5.11. *Capturing the temperature-converted resistance value.*

Source: https://www.hioki.com/themes/hioki/images/global/learning/methods/resistance-measurement-methods-img-10.png.

Utilize a temperature-compensated resistance meter to automatically obtain the temperature-adjusted resistance value. For the most precise measurement of low resistance, opt for a four-terminal resistance meter.

Rather than measuring resistance directly, analog and digital multimeters compute resistance based on voltage and current using Ohm's Law. The primary method of use is the same for both kinds of instruments. For increased precision, digital instruments come with features like four-terminal measuring.

Numerous external factors can have an impact on resistance values. It's important to determine the source and take action if measurement results show instability.

5.4.5. Measure Electrical Power

Measuring electrical power involves determining the power in a circuit, typically expressed in watts. When it comes to centrifugal pump operations, both electrical and mechanical-electrical methods are utilized to measure electrical power (Figure 5.12).

Measurement of electrical power for:

- Direct current: voltage (V) and current (I) measured using voltmeters and ammeters
- Single-phase alternating current: effective power (P_{eff}) measured using a wattmeter
- Three-phase current: two power values (P_{eff1} and P_{eff2}) measured using two wattmeters (dual wattmeter method), with total electrical power calculated as;

$P_{eff} = P_{eff1} + P_{eff2}$. *See Fig. 5.12 Power measurement*

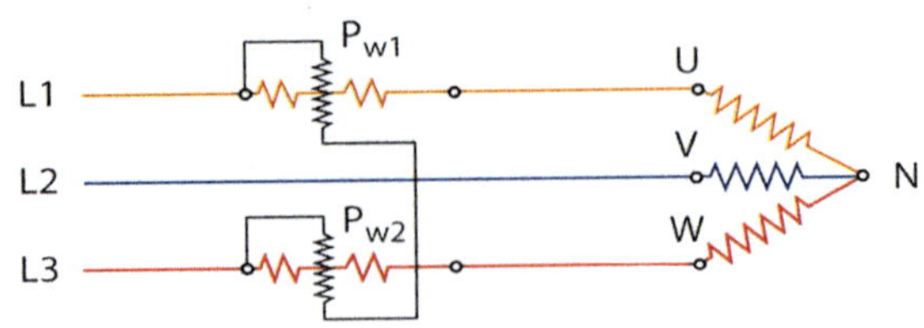

Figure 5.12. *Power measurement: Dual-wattmeter method (Aron configuration; consumer, e.g., motor).*

Source: https://acszigalen.cloudimg.io/v7/https%3A%2F%2Flive-resources-e2e-sales.ksb.com%2Fresource%2Fblob%2F1116908%2F23ce21318f61c4e62cc79a9f46bc26ac%2Fpic-power-measurement-data.jpg?ci_url_encoded=1&optipress=3.

It is frequently possible to combine these measurement capabilities into a single, electronic measuring device with an easily readable display.

See Figure 5.13 Power measurement.

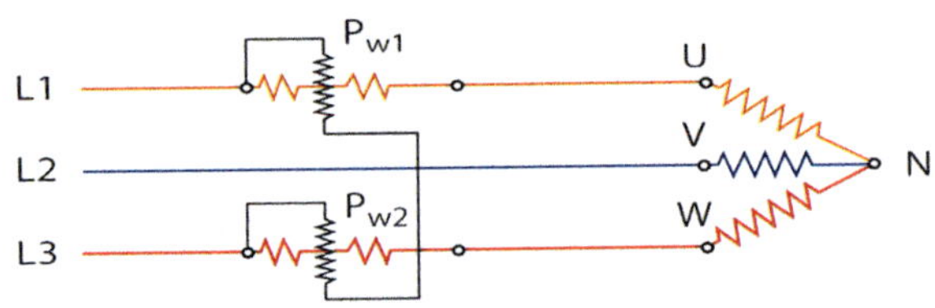

Figure 5.13. *Power measurement: simplified measurement with accessible or artificial neutral.*

Source: https://acszigalen.cloudimg.io/v7/https%3A%2F%2Flive-resources-e2e-sales.ksb.com%2Fresource%2Fblob%2F1117884%2F8f2379ddf12503a548ff833dc472d8ef%2Fpic-power-measurement-1--data.jpg?ci_url_encoded=1&optipress=3.

You can always use this dual watt-meter method, even in cases when the power fluctuates or there is an inaccessible neutral conductor or phase imbalance. With a motor with a star configuration, if the neutral conductor is accessible and the phases are balanced, the total electrical power may be calculated using the formula $P_{eff} = 3 \cdot P_{eff1}$, which yields a single power value (P_{eff1}).

In order to determine the power output of a pump, one must have information on the efficiency of the motor (η_M) as well as the gearing.

To eliminate any uncertainties related to motor and gear efficiency, especially in frequency-controlled drives, a torque and speed-measuring instrument is typically installed in the drive train before the pump, particularly during testing in a test bay. Additionally, instruments for measuring the power input of a pump during testing in a test bay are also utilized.

Instruments for measuring the power input of a pump during testing in a test bay:

- Hub-type torque meter, *see* Figure 5.14 Power measurement.

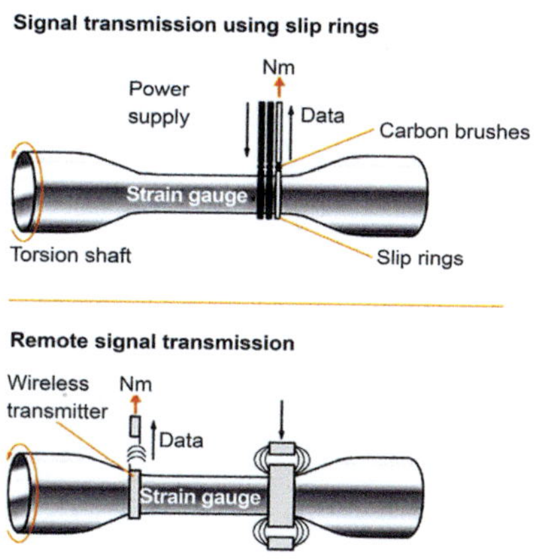

Figure 5.14. *Power measurement: hub-type torque meter.*

Source: https://acszigalen.cloudimg.io/v7/https%3A%2F%2Flive-resources-e2e-sales.ksb.com%2Fresource%2Fblob%2F1115860%2Ffcd562947bc304fffd6a91806b1fd1bd%2Fpic-power-measurement-2--data.jpg?ci_url_encoded=1&optipress=3.

- Disc-type torque meter, *see* Figure 5.15 Power measurement

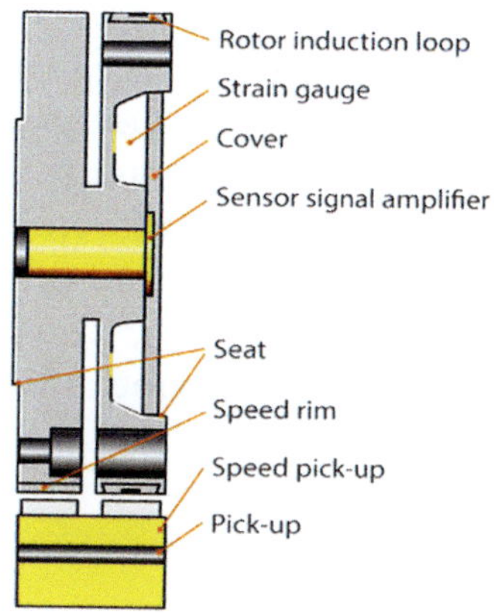

Figure 5.15. *Power measurement: disc-type torque meter.*

Source: https://acszigalen.cloudimg.io/v7/https%3A%2F%2Flive-resources-e2e-sales.ksb.com%2Fresource%2Fblob%2F1117176%2F7f61cb62a756c96e562a61c443a761f4%2Fpic-power-measurement-3--data.jpg?ci_url_encoded=1&optipress=3.

These instruments are installed between the motor and pump or between the gear and pump as part of the coupling or as a separate instrument. Rotational speed is generally also supported and measured.

Different systems, the majority of which are electronic, are available for transferring the signals. The main challenge with this technology is to reliably transfer the measuring signal from the rotating to the stationary system and to supply power to the measuring elements. Initially, slip rings were used for this purpose. However, they are very susceptible to malfunctions, which is why they have been replaced with contact-free transmission devices in precision systems. The latter transmit and display the measuring signals in the form of amplitude or frequency modulations, depending on the system.

These instruments can be mounted as a standalone instrument or as part of the coupling between the gear and the pump or between the motor and the pump. In most cases, rotational speed is also supported and quantified.

To transport the signals, there are various systems available, most of them are electrical. The primary problem with this technology is to give electricity to the measuring parts and transfer the measurement signal from the rotating to the stationary system with reliability. At first, this was done via slip rings. In precision systems, contact-free gearbox devices have taken their place because they are less prone to malfunctions. Depending on the system, the latter transmit and display the measuring signals as amplitude or frequency modulations.

ROLE MODEL

HANS CHRISTIAN ØRSTED:

A Danish physicist and chemist who discovered that electric currents create magnetic fields, which was the first connection found between electricity and magnetism.

Biography

A discovery by Hans Christian Ørsted forever changed the way scientists think about electricity and magnetism.

While preparing to perform an experiment during a lecture at the University of Copenhagen, he found that the magnetized needle of a compass was deflected whenever the electric current through a voltaic pile (an early form of the battery) was started or stopped. This surprising occurrence was solid evidence that electricity and magnetism are related phenomena. The announcement of Ørsted's discovery incited a tremendous outbreak of research in the nascent field of electromagnetics in the early 1820s. Though it is a matter of some debate, another scientist, Italian Gian Domenico Romagnosi, reportedly made the same discovery as Ørsted, but more than a decade and a half earlier. Romagnosi's finding was described in an Italian newspaper in 1802 but was never recognized by most of his contemporaries.

Ørsted was a native of Denmark, where he was born on August 14, 1777, in Rudkørsted bing, a city located on the island of Langeland. Ørsted's father worked as a pharmacist, and after helping out in the family business in his youth, the younger Ørsted would eventually decide to follow in his father's footsteps. Both he and his brother, Anders Sandørsted Ørsted, were directed in their early studies at home by their father and private tutors. That education enabled the brothers to gain entry into the University of Copenhagen in 1793. While Anders studied law, Hans concentrated his efforts in pharmacy. Hans Ørsted proved an exceptional student, writing prize-winning papers and passing the pharmacy examination with distinction. In 1799 he received a doctorate degree. His outlook on science and life at this time was greatly influenced by Immanuel Kant, and Ørsted's dissertation was an examination of his philosophies.

Following graduation, Ørsted served as a lecturer and pharmacist for a short time before grant and scholarship money enabled him to embark on the first of several foreign journeys that he would take over the course of his life. While visiting Germany, Ørsted became acquainted with Johann Wilhelm Ritter, whose views on electricity and magnetism reportedly aroused the possibility in Ørsted's mind that there was some sort of connection between the two (the notion of a unifying force underlying all of nature was widely held among adherents to the German Romantic movement). At that time, without any experimental evidence, such a connection was pure speculation. Ørsted was notably less influenced by the French scientists he met, whose approach to science was more mathematical.

In 1806 Ørsted gained a professorship at the University of Copenhagen. For a few years prior, he had been lecturing without an academic post to audiences paying a fee to hear him speak. The increasing popularity of his lectures did not go unnoticed by the university, resulting is the official offer of a faculty position. Ørsted became the first faculty member at the university to teach physical science. In addition to his teaching duties, Ørsted carried out research, chiefly in acoustics and electricity.

Ørsted left Copenhagen for about two years in 1812 to travel through Europe again. As did the earlier trip, this journey allowed him to exchange ideas with other notable scientists, helping broaden his views of the world. When Ørsted returned to Denmark he married Inger Birgitte Ballum. The couple remained together for the rest of his life and reared eight children.

1820 was a particularly important year for Ørsted's career. In April he made the discovery of the connection between electrical and magnetic phenomena for which he is chiefly known. He wrote a short treatise on the discovery, but much of the important subsequent work relating to his finding was carried out by others, such as François Arago and André-Marie Ampère, the latter of whom made his greatest contribution to science by rigorously applying mathematics to the study of electromagnetism. 1820 was also the year that Ørsted became the first person to isolate piperine (a component of pepper), thus making his mark in chemistry as well as physics. A few years later, in 1825, Ørsted again experienced a notable success in the field of chemistry when he produced an impure form of metallic aluminum.

As a man of broad interests, Ørsted enjoyed philosophizing and writing, as well as teaching and experimenting. He composed both poetry and prose. Ørsted was compiling his philosophical musings into a book, The Soul in Nature, when he passed away on

March 9, 1851. During his lifetime, he also encouraged the literary efforts of others, most significantly the Danish author Hans Christian Andersen, and was a strong proponent for the popularization of scientific knowledge. He was the founder of an organization expressly dedicated to the latter purpose. This Society for the Dissemination of Natural Science has awarded a prize named in his honor (the Ørsted Medal) since the early 1900s for remarkable Danish contributions to physics, chemistry or the popularization of science in general.

The unit of magnetic field strength in the CGS (centimeter-gram-second) system of physical units was also named in honor of Ørsted in the 1930s. One oersted is equivalent to the field strength one centimeter from a unit magnetic pole under vacuum conditions.

CLASS ACTIVITY

UNDERSTANDING ELECTRICAL MEASUREMENT

Objective: To help students grasp the fundamentals of electrical measurements, including voltage, current, and resistance, using a multimeter and other basic tools.

Materials Needed: Digital multimeters, Batteries (AA or 9 V), Resistors, LED bulbs, Breadboards, Connecting wires, Small DC motors.

Assessment: Evaluate students based on their participation, accuracy of measurements, and understanding of the concepts. Provide a short quiz at the end of the session to reinforce the concepts learned.

SUMMARY

- Measurement means to compare to a fixed/standard value. Measurement can include measuring some of the physical quantities like length, mass, time, etc., which we use in our daily lives. Different measuring devices and scales are used to measure these quantities to make various instruments or for various scientific applications and research work.
- Measurement begins with a definition of the quantity that is to be measured, and it always involves a comparison with some known quantity of the same kind. If the object or quantity to be measured is not accessible for direct comparison, it is converted or "transduced" into an analogous measurement signal.
- Measurement theory is the study of how numbers are assigned to objects and phenomena, and its concerns include the kinds of things that can be measured, how different measures relate to each other, and the problem of error in the measurement process.
- The errors, which occur during measurement are known as measurement errors. Measurement errors refer to the discrepancies between the actual values of the quantities being measured and the values obtained from the measurement process.
- Random errors, on the other hand, are unpredictable and arise from various unpredictable factors, including environmental conditions, observer variability, and inherent fluctuations in the measurement process.
- Electrical measuring instruments are all the devices used to measure the magnitude of an electric current with different objectives. The values that are usually measured with this equipment are current, voltage, resistance, and power. Each is expressed in different units: amps, volts, ohms, and watts, respectively.
- The ability of a measuring system to maintain standard of performance over prolonged periods of time. Zero stability defines the ability of an instrument restore to zero reading after the input quantity has been brought to zero, while other conditions remain the same.

REVIEW QUESTIONS

1. What are the primary instruments used to measure electricity in circuits?
2. How do electrical measuring instruments behave under stable conditions?
3. How do electrical measuring instruments respond to changing conditions in circuits?
4. What are the standard units used to measure electricity in circuits?
5. What are the fundamental types of connections in electrical circuits, and how do they differ?
6. How is voltage measured in an electrical circuit, and what instrument is commonly used for this purpose?

7. How can resistance be measured using a digital multimeter, and what role does it play in circuits?

REFERENCES

1. Bell, D. A. (2006). *Electronic Instrumentation and Measurements*. Oxford University Press.
2. Bosworth, W. C. (2003). *Electrical Measurement and Instrumentation*. Prentice Hall.
3. Carr, J. J. (1987). *Elements of Electronic Instrumentation and Measurement*. Prentice Hall.
4. Cooper, W. D., & Helfrick, A. D. (1992). *Modern Electronic Instrumentation and Measurement Techniques*. Prentice Hall.
5. Deobeling, E. O. (2004). *Measurement Systems, Applications and Design*. Tata McGraw Hill Publishing Company Limited.
6. Doebelin, E. O. (2003). *Measurement Systems: Application and Design*. McGraw-Hill Education.
7. Gooday, G. (2004). *The Morals of Measurement: Accuracy, Irony, and Trust in Late Victorian Electrical Practice*. Cambridge University Press. ISBN 0-521-43098-4.
8. Kalsi, H. S. (2011). *Electronic Instrumentation*. Tata McGraw-Hill Education.
9. Oliver, B. M. (2016). *Electrical Measurements and Measuring Instruments*. Cambridge University Press.
10. Sawhney, A. K. (2006). *A Course in Electrical and Electronic Measurements and Instrumentation*. Dhanpat Rai Publications.
11. Schuler, C. A. (2005). *Instrumentation and Measurement in Electrical Engineering*. Prentice Hall.
12. Webster, J. G. (Ed.). (2014). *The Measurement, Instrumentation, and Sensors Handbook*. CRC Press.

CHAPTER

6 Magnetism, Magnetic Circuits, and Electromechanical Energy Conversion

LEARNING OBJECTIVES

After studying this chapter, you will be able to:

- Discuss the basic concept of magnetism
- Focus on magnetic materials
- Explain the magnetic circuits and electromechanical energy conversion

KEY TERMS FROM THIS CHAPTER

- Capacitors
- Electric current
- Electromagnetic
- Energy conversion
- Antiferromagnetic materials
- Electric dipoles
- Electromotive force (EMF)
- Magnetic flux
- Current flow
- Electrical energy
- Electrons
- Magnetic flux
- Curie-Weiss law
- Electrostatics
- Ferromagnetism
- Magnetomotive Force (MMF)

INTRODUCTION

Magnetism represents a fundamental natural force that emerges from the interplay of magnetic fields. These fields originate from the motion of electric charges, predominantly electrons, within atoms. One of magnetism's most distinctive characteristics is its capacity to draw specific materials such as iron, nickel, and cobalt towards it. Comprehending magnetism is of paramount importance in several areas, including electromechanical energy conversion and magnetic circuits. In the realm of electromechanical energy conversion, the synergy between electricity and magnetism allows for the conversion of electrical energy into mechanical energy, and subsequently, the reversal of this process. This fundamental process is integral to devices such as electric motors, generators, transformers, and inductors, which are indispensable components in numerous contemporary technologies.

An essential component of electromechanical energy conversion systems is magnetic circuitry. Although they deal with the flow of magnetic flux rather than electric current, these circuits are comparable to electrical circuits. They are made up of air gaps that allow magnetic flux to pass through them and magnetic materials like iron cores. Magnetic circuits govern the distribution of magnetic flux in electromechanical systems, much as resistors, capacitors, and inductors control the flow of electric current in electrical circuits. Magnetic reluctance, which is comparable to electrical resistance in electrical circuits, is one of the fundamental ideas in magnetic circuitry. Reluctance, which measures a material's resistance to the flow of magnetic flux, is dependent on a number of variables, including the material's permeability and geometric structure. Similar to how greater resistance in electrical circuits reduces current flow, higher reluctance results in lower flux.

Magnetic circuits are vital to the effective transfer of energy in electromechanical devices such as electric motors and transformers. Transformers, for example, use electromagnetic induction and magnetic circuits to move electrical energy from one circuit to another. A transformer's primary and secondary windings are coiled around a single magnetic core, which permits energy to move from the primary winding to the secondary winding via variations in magnetic flux.

In contrast, electric motors use the interaction of magnetic fields to transform electrical energy into mechanical energy. A magnetic field is created inside the motor by an electric current flowing through one or more coils. This magnetic field interacts with other coils or permanent magnets. The motor rotates and produces useful work as a result of this interaction's mechanical force production. Many technological applications rely on the interconnected fields of physics and engineering, including magnetism, magnetic circuits, and electromechanical energy conversion. Gaining an understanding of these ideas is crucial for developing effective electromechanical systems and streamlining energy transfer procedures across a range of industries.

6.1. BASIC CONCEPT OF MAGNETISM

As we all know, the electromagnetic industry uses a huge number of devices that require magnets, thus they are essential to modern life. When people used natural iron minerals, particularly magnetite, in ancient times, they were able to experience magnetic phenomena. Magnetic phenomena were not understood from the perspective of electromagnetics until recently, a field to which many physicists, including Oersted and Faraday, made significant contributions. Ampère's circuit law, in particular, introduced the idea of a magnetic moment or magnetic dipoles, which are comparable to electric dipoles. The physical equivalent of a circuit current in a wire and a bar magnet are shown in macroscopic electromagnetic phenomena. wherein the rotational motion of a microscopic electron and a magnetic moment or dipole are equivalent but not distinguished at all. But quantum mechanics, which emerged in the 20th century, has provided the real understanding of the origin of magnetism.

Large amounts of information about the magnetic characteristics of materials were gathered prior to the development of quantum mechanics, and by observing how each material responded to a magnetic field, a perfectly logical classification was established. These investigations were conducted with Gouy and Faraday's magnetic balances. A balance is used to measure the force applied to materials in a magnetic field, illustrating the principle of magnetic measurement. Generally speaking, materials can be divided into two groups based on the directions of the force: diamagnetic and paramagnetic substances.

6.1.1. Origins of Magnetism

Though few people are aware of how a magnet functions, almost everyone is aware of the capabilities of magnetic materials. In order to comprehend this phenomenon, one must first realize how closely magnetism and electricity are related. It is possible to create a basic electromagnet by coiling copper wire and connecting it to a battery. The coil generates a magnetic field, but it only lasts as long as current passes through it. The workings of an ordinary bar magnet are attributed to the motions and interactions of its electrons, which are tiny charged particles that orbit the nucleus of each atom. Since an ordinary bar magnet does not have a visible connection to electricity, how does it function? Since electricity is the flow of electrons, whether they are in an atom or a wire, every atom can be thought of as a tiny permanent magnet on its own. The orbital magnetic moment of the circulating electron is measured in Bohr magnetons (μB), and it is accompanied by a spin magnetic moment because the electron spins on its own axis, just like the earth does (Figure 6.1). Because electrons are grouped in pairs and the magnetic moment is canceled by its neighbor, most materials have resultant magnetic moments.

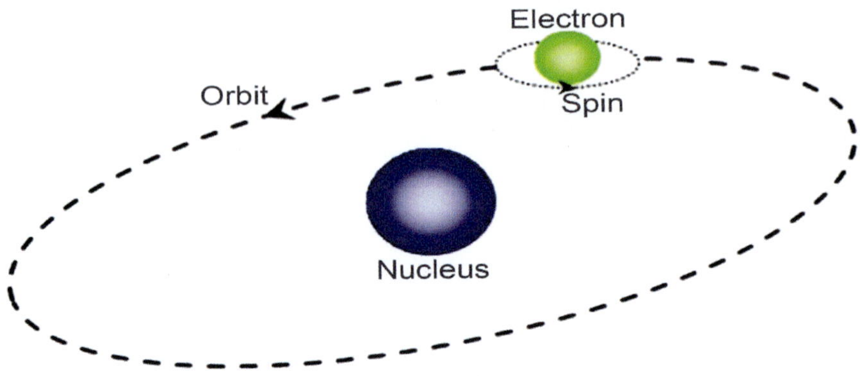

Figure 6.1. *The orbit of a spinning electron about the nucleus of an atom.*

Source: https://www.researchgate.net/profile/Salahudeen-Gene/publication/263785799/figure/fig9/AS:296084262932491@1447603397959/Orbit-of-a-spinning-electron-about-the-nucleus-of-an-atom.png.

A unified magnetic field is produced in some magnetic materials when a significant fraction of the electrons' magnetic moments line up. Much like a compass needle, any magnet will encounter resistance when attempting to align itself with an externally applied field because the field generated in the material (or by an electromagnet) has a direction of flow. These forces power electric motors, generate audio in speaker systems, regulate the voice coil in CD players, and other devices.

6.1.2. Magnetic Domains

Tiny magnets are what make up electrons. They revolve around an axis and also have north and south poles. There is a very small but very noticeable magnetic field created as a result of this spinning. Each electron's axis can be oriented in one of two ways.

Atoms are arranged in most materials so that the orientation of one electron's magnetic field cancels out the orientation of another. However, iron is not like other ferromagnetic materials (ferrum is Latin for iron). Because of the way their atoms are arranged, smaller groups of atoms band together to form regions known as domains, where every electron has the same magnetic orientation. The way these domains react to an external magnetic field is demonstrated in the interactive animation Figure 6.2.

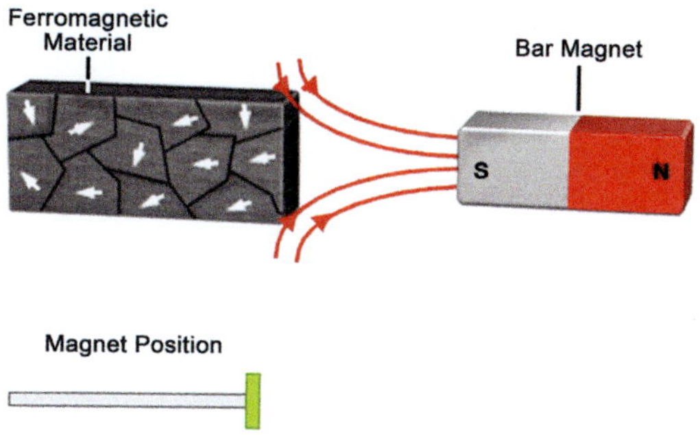

Figure 6.2. *Magnetic domains.*

Source: https://nationalmaglab.org/magnet-academy/watch-play/interactive-tutorials/magnetic-domains/.

The domains in the above image of Ferromagnetic Material are randomly aligned; the image illustrates the process rather than the actual domain size or shape. Red Magnetic Field Lines, which are normally invisible, can be seen extending from the bar magnet's poles. To get the magnet to interact with the field lines, move it closer to the ferromagnetic material using the Magnet Position slider. You'll notice that the domains gradually align with each other and with the bar magnet's field as you repeat the process.

When you're finished, the ferromagnetic material has transformed into a dipole with opposing north and south poles, a permanent magnet in and of itself. All ferromagnetic objects with all of their domains aligned in the same direction are equivalent to permanent magnets. Only four elements—iron, nickel, cobalt, and gadolinium—are ferromagnetic at room temperature and have the ability to become permanently magnetized.

Dysprosium, a fifth element, turns ferromagnetic at low temperatures. When exposed to an external magnetic field, ferromagnets can retain their magnetism for a very long time—sometimes for millions of years. Hysteresis is the term for this inclination to hold onto magnetism.

6.1.3. Domain Structure

6.1.3.1. Why Domains Form

Instead of existing in a state where magnetization is uniformly distributed throughout the material, a piece of magnetic material, like iron, naturally splits into distinct domains in order to reduce its internal energy. A substantial magnetic field that extends into the surrounding space is produced by a large area of ferromagnetic material that is continuously magnetized (diagram a, right). Much of the magnetostatic energy held in the field must be used for this. The sample can split into two domains, with the magnetization in each domain pointing in opposite directions, to lower this energy (diagram b right). The magnetic field outside the material is reduced as the magnetic field lines pass through each domain in loops that face opposite directions. Each of these domains has the ability to split further to further reduce the field energy. This will produce smaller parallel domains with alternating directions of magnetization and less field outside the material (Figure 6.3).

Large domains splitting into smaller ones, as described here, is not typically how the domain structure of actual magnetic materials forms. For instance, when a sample is cooled below its Curie temperature, the equilibrium domain configuration just manifests itself. However, domains can split, and the energy tradeoffs involved in domain formation are frequently revealed by describing how domains split.

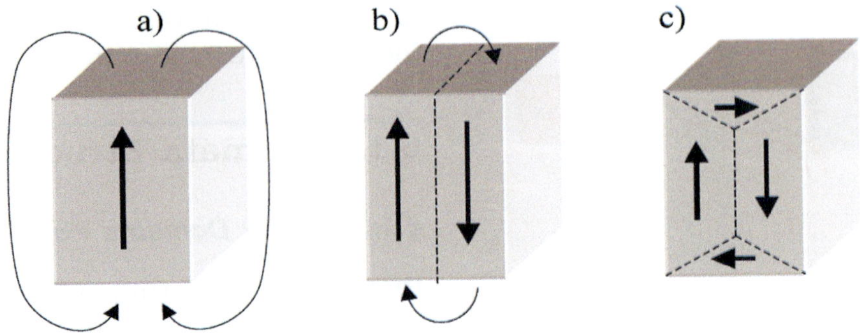

Figure 6.3. *Domain structure.*

Source: https://upload.wikimedia.org/wikipedia/commons/7/76/Powstawanie_domen_by_Zureks.png.

6.1.3.2. Size of Domains

As previously mentioned, an excessively large domain is unstable and will split up into smaller domains. The size of the domains formed in a material, however, is determined by how small a domain must be to remain stable and not split. The balance of various energies within the material determines this size. A domain wall is formed between two domains whenever a region of magnetization splits into two, allowing magnetic dipoles (molecules) with magnetization pointing in different directions to be next to each other. Neighboring dipoles tend to align so they point in the same direction due to the exchange interaction that generates the magnetization. It takes energy to move neighboring dipoles so they point in different directions. As a result, a domain wall needs additional energy, which is correlated with the wall's area and is known as the domain wall energy.

As a result, the net energy reduction that occurs when a domain splits is equal to the difference between the energy saved by the magnetic field and the energy needed to build the domain wall. There exists a relationship between the field energy and the cube of the domain size, and a relationship between the square of the domain size and the domain wall energy. Therefore, the net energy saved by splitting reduces as the domains get smaller. Until the energy required to build an additional domain wall is exactly equal to the energy saved in the field, the domains continue to split into smaller domains. Then the domains of this size are stable. In most materials the domains are microscopic in size, around 10^{-4}–10^{-6} m.

6.1.4. Magnetic Field and Its Properties

A field created by moving electric charges or a magnet is known as the magnetic field. The area in which the force of magnetism acts around a magnetic material or moving electric charge is known as the magnetic field. Magnetic field lines are used to depict magnetic fields. It is a visual aid for determining the magnetic field's strength and direction.

A compass needle can be used to draw lines through magnetic fields. Put the compass needle on a piece of paper and place it close to the magnet. Verify the direction indicated by the compass needle and record it. Mark the directions as you move the compass needle in various directions. The magnetic field lines can be seen by joining the points.

Some important properties of magnetic field lines are listed below:

- The tangent drawn to the magnetic field lines gives the direction of the magnetic field.
- The closeness or density of the field lines is directly proportional to the strength of the field.
- Magnetic field lines appear to emerge or start from the north pole and merge or terminate at the south pole.
- Inside the magnet, the direction of the magnetic field lines is from the south pole to the north pole.
- Magnetic field lines never intersect with each other.
- Magnetic field lines form a closed-loop.
- Field lines have both direction and magnitude at any point on the field. Therefore, magnetic field lines are represented by a vector.
- They denote the direction of the magnetic field.
- The magnetic field is stronger at the poles because the field lines are denser near the poles.

The imaginary lines that separate the two poles are known as magnetic field lines. There are various techniques for creating magnetic lines. One better way to test the existence of the magnetic field is to use a compass. One of the fundamental forces of nature, electromagnetic energy is characterized by magnetism and magnetic areas. The magnetic field line is impacted by the application of electromagnetic current flow. Since iron nails contain electromagnetic energy, they can be turned into magnets to measure magnetic fields.

The properties of magnetic field lines are:

- The lines of magnetic fields are curved lines and it has originated from the North pole and ended on the South pole.
- Stronger field lines need the lines to be closed and vice-versa.
- The magnetic field lines never intersect among each other.
- Near the South and North poles, the lines of magnetic fields are crowded and apart widely.

6.1.5. Description of Magnetic Field Lines

The imaginary lines surrounding a magnet are called magnetic field lines. The density of a field's lines indicates the field's magnitude. A magnet's magnetic field is strongest close to its North and South poles and becomes weaker as it travels away from them. You can better understand this idea by carrying out a quick experiment. Place a bar magnet in the middle of a white paper sheet that has been fixed to a table. Iron filings should be scattered all around the magnet. Tap the table gently. It is evident that the iron filings arrange themselves in a particular pattern, which symbolizes the magnet's field. Clear observation of these patterns reveals that iron filings are concentrated close to the poles, while their concentration is lower in the area farther from the poles.

A bar magnet, a compass, and chart paper can be used to draw magnetic field lines. Place the paper on a drawing board first. Mark the location with a pencil after positioning the bar magnet in the center. Maintain the compass close to any one of the magnet's poles. Verify that nothing else magnetic is in the area. The compass arrow is visible to be pointing in certain directions. Place a in that path. Reposition the compass so that the base of the arrow is at the by moving it from that location. Place a fresh where the compass's current arrow is pointing.

Repeat this process until the compass reaches the magnet's opposite pole. Connect the dots. Once more, return to the starting point and carry out the same procedures from there. Upon drawing several lines, it becomes evident that the lines are forming a closed loop, which appears to originate from one magnet pole and terminate at a different pole. This is how lines in a magnetic field are drawn. It is possible to see how similar the patterns are if these lines and the alignment of iron filings are compared. Magnetic field lines will also differ depending on the type of magnets.

Analyzing the magnetic field lines and their energy in the various poles indicates that space volume is essential. Magnetic fields between the earth's two poles are useful for exploring new areas. In the magnetic field line, various magnet shapes are essential for the approval statements. The earth's magnetic properties are related to the magnetic field that surrounds magnetic materials. A magnetic area is depicted by a magnetic field line. To find the lines, use a visual aid that provides information about the strength and direction of the magnetic field.

Ways to Measure Magnetic Lines: Various ways are applied in general to measure the magnetic lines among the two poles on the globe.

Some useful applied ways are illustrated below:

- A straight copper wire that carries the electric flow may be attached with the iron nail and can work as the instrument to measure the magnetic flow.
- Magnetic compasses are available to measure the magnetic field lines. In transportation systems such as the navy, aviation uses the magnetic compass to measure the magnetic lines and that helps get the right direction.
- With the flow of current, there are efficient ways to develop. There are different available ways for the development of the magnetic field's measurement.
- **Origin of the Magnetic Field:** Magnetic fields exist everywhere on Earth; their importance can be determined by observing the approaches of their strengths. According to various earlier definitions, the Sun is the source of the magnetic field. Since the Earth was assessed from a portion of the Sun, the magnetic energy contained within it was also assessed. Two different types of electron motion in atoms—or electron motion in an orbit within the solar system—are the source of magnetism.
- **Ways to Draw Magnetic Field Lines:** The magnetic field lines apply the force of magnetism through the

application of electrons within the atoms. Therefore, the number of lines of energy per unit dimension of space determines the strength of the magnetic field. Magnetic bars can be used to draw lines in a variety of situations and justify areas. They can also be useful for transportation purposes because they can incorporate a magnetic compass that is drawn to magnet poles placed in the North Pole. On a piece of paper that has been stabilized beforehand, a straight line can be drawn using the compass arrow and no magnetic material in the vicinity.

6.1.6. How Is the Magnetic Field Produced?

In addition to the magnet, other sources of magnetic field generation include electric currents and moving charges. It is common knowledge that matter is composed of minuscule units known as atoms. Protons and neutrons make up an atom's nucleus, and electrons orbit around it. The magnetic field is created by protons and neutrons spinning and orbiting around each other or the atom's nucleus. The orbital and spinal directions determine the magnetic field's direction. The mathematical symbol for the magnetic field is 'B.' Tesla is its unit (T).

6.1.6.1. Earth's Magnetic Field

Sir William Gilbert provided the first evidence of the earth's magnetic field in 1600. He discovered that the earth has a magnetic field and exhibits some magnetic properties based on some experiments. A magnet will automatically align itself in a north-south direction and come to rest if it is free to rotate in a horizontal plane while hanging from a thread. The magnet will be oriented so that its south pole is drawn toward the geographical north and its north pole is drawn toward the geographical south.

The existence of a few neutral points in the magnetic field lines serves as the second piece of evidence. The earth's magnetic field cancels out the magnetic field created by the magnet, which is used to draw field lines. These neutral points would not be visible without the earth's magnetic field. The third piece of evidence is that if a soft iron is buried in a north-south direction, it turns into a magnet.

6.1.6.2. The Hypothesis for the source of Earth's Magnetic Field

- The Earth's core is in the form of hot molten liquid and it contains ions. These ions are circulating in the form of current loops inside the liquid and as a result, the magnetic field is produced.
- The Earth is rotating about its axis and matter on the earth is made of charged particles. These charged particles also rotate about earth's axis in the form of current loops and are responsible for the production of the magnetic field.
- The Earth's outer layer consists of ionized gasses. When the earth is rotating, movement of ions produces electric current and a magnetic field is produced due to this.

6.1.6.3. Characteristics of the Earth's Magnetic Field

- Earth's magnetic field is uniform.
- The magnetic field strength at the surface of the earth is approximately 10–4 Tesla

- The magnetic field of the earth is extended up to a height of 5 times the radius of the earth.

6.1.6.4. Applications of Magnets in Real Life

- Magnets are used in electric bells.
- They are used in the construction of generators and electrical motors.
- Magnets are used to find the geographical directions.
- Magnets play an important role in the separation of magnetic and non-magnetic materials from the scrap.
- In the medical field also, magnets are widely used in treating pain of different body parts.

6.1.7. Operating the Magnetic Field

All magnets, no matter how big or small, have a field around them. To learn more about the magnetic field, place an iron nail at a specific distance from the magnet. The nail then starts to attract the magnet at that point. If the iron nail is far away from the magnet, it will not be able to affect it in any way. The B field is the name given to the magnetic field that surrounds magnets.

Electric charges or moving magnets produce the magnetic field. Within the magnetic field that envelops magnetic materials or moving charges, magnetism is at work. A magnetic field is shown by a magnetic field line. an indicator that shows the magnetic field's strength and direction visually.

With a compass needle, the lines of the magnetic field can be drawn. It is advised to position the compass needle on a piece of paper close to the magnet. Verify and mark the compass needle's direction. Mark the directions as you move the compass needle in different directions. The magnetic field lines are visible when the points are joined.

6.2. MAGNETIC MATERIALS

There are two types of magnetic materials: magnetically soft and magnetically hard. Although easily magnetized, magnetically soft materials typically only experience momentary induced magnetism. For instance, rubbing a permanent magnet against a nail or screwdriver will cause the object to momentarily become magnetized and produce a weak magnetic field of its own. This is as a result of the external magnetic field momentarily aligning a large number of their iron atoms in the same direction.

Strong external magnetic fields, like those produced by an electromagnet, have the ability to magnetize magnetically hard materials just like they can magnetically soft materials. The distinction is that, unless they are subjected to an opposing magnetic field, elevated above their curie temperature, or allowed to corrode, magnetically hard materials will always remain magnetized.

Permanent magnets are made from alloys that are magnetically hard and typically contain different amounts of iron, aluminum, nickel, cobalt, and rare earth elements like neodymium, dysprosium, and samarium. Neodymium magnets, the strongest permanent magnets, are composed of an iron, boron, and neodymium alloy. Unlike magnetically soft materials, permanent magnets have atomic magnetic domains that are difficult to align but once they are, they stay that way forever. This makes them challenging to magnetize.

6.2.1. Types of Magnetic Materials

The types of magnetism have been observed and classified on the basis of the magnetic behavior of materials in response to magnetic fields at different temperatures.

From this point of view, these types of materials are:

- Paramagnetic materials.
- Diamagnetic materials.
- Ferromagnetic materials.
- Ferrimagnetic materials or ferrites.
- Antiferromagnetic materials.

6.2.2. Paramagnetic Materials

Materials classified as paramagnetic have atomic dipoles that are aligned in the direction of an external field and are subject to individual action. Because their magnetizations are comparatively small and only last as long as an applied field is present, diamagnetic and paramagnetic materials are classified as nonmagnetic. Paramagnetic materials are those whose magnetic susceptibility, or ⊠, is positive. The induced magnetization reinforces the material's magnetic field. A wide range of chemical elements and compounds are considered paramagnetic materials. Their relative magnetic permeability is marginally higher than one (i.e., a negligible positive magnetic susceptibility) and are drawn to magnetic fields as a result. Since nonmagnetic materials lack permanent magnetization in the absence of an external magnetic field, they are typically referred to as para- or diamagnetic.

Envision a hydrogen atom where a negatively charged electron orbits a nucleus containing a single, stationary, positively charged proton. If we consider that orbiting electron to be a kind of current loop, which it is, one might assume that hydrogen would have a strong magnetic moment. Actually, the magnetic properties of ordinary hydrogen gas are very weak. Because the formula H_2 (which has a lower chemical energy (for H by an astounding 218 kJ mol^{-1}) than two isolated atoms, it should be remembered that each hydrogen atom is not isolated but rather is bonded to one another to form a molecule. It is not by accident that each electron in these molecules has an angular momentum that is opposite to that of its neighbor, giving the molecule little magnetic moment overall. Many substances exhibit this behavior, which is why it's said that they don't have a permanent magnetic moment.

The electrons in orbit planes at a right angle to the magnetic field will cause a very slight change in momentum when the molecule is exposed to a magnetic field. Faraday's Law, which states that an induced E-field will occur as the field is increased and that electrons, being charged particles, will experience this as a force, predicts this. As a result, the molecule gains an induced magnetic moment and the individual magnetic moments no longer entirely cancel. This phenomenon, known as diamagnetism, is the property shared by all materials in which the induced moment is opposite to the applied field. A nearby magnet repels hydrogen, ammonia, bismuth, copper, graphite, and other diamagnetic materials (though the effect is very weak). Consider it an example of Lenz's law in action. Materials classified as diamagnetic have only paired electrons in their atoms.

On the other hand, the electrons' magnetic moments are partially canceled in other molecules, like oxygen, where there are unpaired electrons. Even when no external field is applied, an O_2 molecule has a net or permanent magnetic moment. The electron orbits are still changed by an external magnetic field in the same way that diamagnets are, but the permanent moment typically has a stronger effect. The molecule's "poles" have a tendency to reinforce the field by aligning themselves parallel to it. These molecules are known as paramagnetic because they have a permanent magnetic moment. While paramagnetic materials such as copper sulfate, oxygen, tin, and aluminum are drawn to a magnet, the attraction is not nearly as strong as that of diamagnetism. The reason is that, at room temperature (liquid oxygen at −183 °C can be pulled around by a strong magnet), thermal vibration continuously knocks the permanent moments out of alignment with the field.

6.2.3. Diamagnetic Materials

Materials classified as diamagnetic are those that most people consider to be non-magnetic. A magnetic field applied to diamagnetic materials produces an induced magnetic field in them that is directed in the opposite direction, producing a repulsive force. Changes in electron orbital motion brought about by an external field give rise to diamagnetism. Water, wood, most organic substances like petroleum and certain polymers, and various metals like copper—especially the heavy ones with lots of core electrons like bismuth, gold, and mercury—are examples of diamagnetic materials. With susceptibilities on the order of -10^{-5}, the effect is minuscule and opposes the applied field. Diamagnetic materials, like water, or water-based materials, have a relative magnetic permeability that is less than or equal to 1, and therefore a magnetic susceptibility less than or equal to 0, since susceptibility is defined as ⊠$_v$ = $\mu_v - 1$. Because their magnetizations are comparatively small and only last as long as an applied field is present, diamagnetic and paramagnetic materials are classified as nonmagnetic. A material is diamagnetic if its magnetic susceptibility, or ⊠, is negative. The induced magnetization weakens the material's magnetic field. Magnetic fields repel diamagnetic materials. For instance, diamagnets like water have a magnetic susceptibility of ⊠v = -9.05×10^{-6}. Bismuth has the strongest diamagnetic property, with a ⊠$_v$ of -1.66×10^{-4}. Since nonmagnetic materials lack permanent magnetization in the absence of an external magnetic field, they are typically referred to as para- or diamagnetic.

6.2.4. Ferromagnetic Materials

Ferromagnetism is the fundamental process by which a substance transforms into a permanent magnet (i.e., e. materials that, when exposed to an external magnetic field, become magnetized and stay that way even after the field is removed. Ferromagnetism is the strongest type and is responsible for this common phenomenon. Materials classified as ferromagnetic, ferrimagnetic, or antiferromagnetic have a well-defined zero-field susceptibility and are permanently magnetized even in the absence of an external magnetic field. In ferromagnetic materials, atomic magnetic moments resulting from unpaired electron spins as a function of electron structure give rise to permanent magnetic moments. Additionally, there is a tiny orbital magnetic moment contribution compared to the spin moment. As a result, the magnetic moments of the electrons in the material align themselves parallel to one another even in the absence of an applied field. The Curie temperature, also known as the Curie point, is the specific temperature at which each ferromagnetic material loses its ferromagnetic characteristics. This is because the energy-lowering caused by ferromagnetic order is overpowered by the thermal tendency toward disorder. Few materials possess ferromagnetic properties. The most common ones are nickel, cobalt, iron, and most of their alloys; some rare earth metal compounds are also included. Ferromagnetism plays a major role in contemporary industry and technology. Magnetically "soft" materials, such as annealed iron, can become magnetized but do not typically retain it, and magnetically "hard" materials, on the other hand, do. In order to align their internal microcrystalline structure and make them extremely difficult to demagnetize, "hard" ferromagnetic materials like ferrite and alnico are processed differently in a strong magnetic field during the manufacturing process to create permanent magnets. The ferromagnets, which include compounds of iron, nickel, cobalt, and manganese as well as a few more unusual ones, are the most significant class of magnetic materials. When compared to a diamagnetic or paramagnetic material, the magnetization curve appears substantially different. It is worth mentioning that even though manganese is not ferromagnetic, the element's name, mágnes lithos, which means "stone from Magnesia" in Greek, has a common ancestor with magnetism.

It's crucial to understand that purity, heat treatment, and other variables have a significant influence on the magnetization curves for ferromagnetic materials. Two things about this curve, though, stand out right away: first, it is actually curved, not straight, as with non-ferromagnets, and second, the vertical scale is now expressed in Teslas.

Figure MPB is a normal magnetization curve because it starts from an unmagnetized sample and shows how the flux density increases as the field strength is increased. You can identify four distinct regions in most such curves. These can be explained in terms of changes to the material's magnetic 'domains':

- Close to the origin a slow rise due to 'reversible growth.'
- A longer, fairly straight, stretch representing 'irreversible growth.'
- A slower rise representing 'rotation.'
- An almost flat region corresponding to paramagnetic behavior and then μ_0 – the core can't handle any more flux growth and has *saturated*.

6.2.5. Ferrimagnetic Materials

While ferromagnets and ferrimagnets share similar macroscopic magnetic properties, they differ in the source of their net magnetic moments. Populations of atoms with opposing magnetic moments, as in antiferromagnetism, are what make a material ferrimagnetic; however, in ferrimagnetic materials, the opposing moments are unequal and spontaneous magnetization persists. Ferrites are typically ferrimagnetic ceramic compounds made from iron oxides, and they are widely used in household items like refrigerator magnets. Fe_3O_4 (magnetite) is a well-known example. The right material selection is essential in the main generator of nuclear power plants as well as power plants in general when there are strong magnetic fields present.

In general, a main generator consists of a rotating part and a stationary part:

- **Stator**: Stator is the stationary part of an electric generator, which surrounds the rotor. The stator has a wire winding in which the changing field induces an electric current
- **Rotor**: Rotor is the rotating part of an electric generator and generates a magnetic field.

Ferrimagnetic materials are used in almost all modern electronic devices, including microphones, motors, deflection yokes, interference suppressors, antenna rods, proximity sensors, recording heads, transformers, and inductors. The market is very large. Ferrimagnets have permeability to rival most ferromagnets, but because of the material's higher electrical resistivity, their eddy current losses are far lower. These are the properties that make ferrimagnets so ubiquitous. Also, using low-cost methods like extrusion or pressing to create various shapes is feasible, if not simple.

What are ferrimagnetic materials made of? Generally speaking, they are iron oxides mixed with one or more transition metals, like manganese, nickel, or zinc e.g., $MnFe_2O_4$. Frequently, barium is found in permanent ferrimagnets. The raw ingredient is ground into a powder, which is then sintered or fired in a kiln to create a hard, brittle, dark gray ceramic material with a cubic crystalline structure.

The interaction between the electrons attached to the metal ions determines the magnetic properties at the atomic level. Neighboring atomic magnetic moments become locked in anti-parallel with their neighbors (which contrasts with the ferromagnets). There is an overall magnetic moment, though, because the magnetic moments in one direction are weaker than the moments in the other.

6.2.5.1. Saturation

A limit that arises in inductors with a ferrimagnetic or ferromagnetic core is called saturation. At first, the flux rises proportionately to the increase in current. But eventually, additional current increases result in progressively smaller increases in flux. When the core can no longer contribute to flux growth, the only increase available is that given by μ_0, which is possibly three orders of magnitude smaller. While ferrites typically saturate between 200 and 500 mT, iron saturates at about 1.6 mT.

Since saturation is followed by a drop in inductance, it is typically imperative to avoid reaching saturation. In many circuits, inductance (I = V * T / L) and coil current growth rate are inversely correlated. As a result, any decrease in inductance accelerates the rise of the current, strengthening the field and pushing the core even closer to saturation.

Core manufacturers normally specify the saturation flux density for the particular material used. You can also measure saturation using a simple circuit. There are two methods by which you can calculate flux if you know the number of turns and either -

- The current, the length of the magnetic path and the B-H characteristics of the material.
- The voltage waveform on a winding and the cross sectional area of the core – see Faraday's Law.

Although saturation is mostly a risk in high power circuits it is still a possibility in 'small signal' applications having many turns on an ungapped core and a DC bias (such as the collector current of a transistor).

If you find that saturation is likely then you might:

- Run the inductor at a lower current
- Use a larger core
- Alter the number of turns
- Use a core with a lower permeability
- Use a core with an air gap

Some combination thereof – but you will need to re-calculate the design in any case.

6.2.6. Antiferromagnetic Materials

Among the transition metal compounds are frequently found the antiferromagnetic materials. Examples of antiferromagnetic materials are oxides of nickel, chromium, hematite, and iron-manganese alloys. The magnetic moments of atoms or molecules, which are typically connected to the spins of electrons, align in a predictable pattern with neighboring spins (on different sublattices) pointing in opposite directions in materials that display antiferromagnetism. This is an example of ordered magnetism, just like ferromagnetism and ferrimagnetism (Figure 6.4).

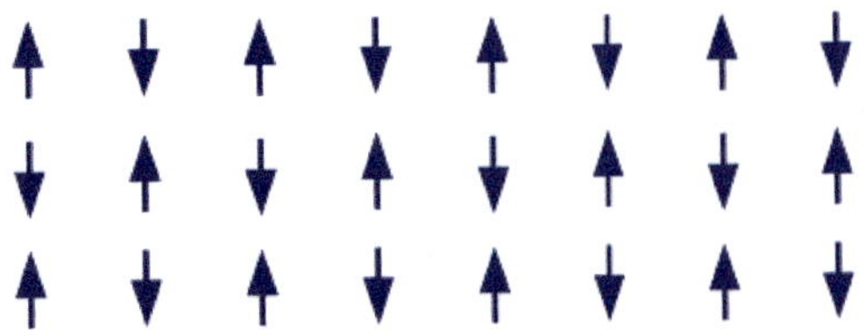

Figure 6.4. *Antiferromagnetic ordering.*

Source: https://upload.wikimedia.org/wikipedia/commons/thumb/5/5d/Antiferromagnetic_ordering.svg/665px-Antiferromagnetic_ordering.svg.png.

Generally speaking, antiferromagnetic order disappears at and above the Néel temperature, which is named for Louis Néel, the person who initially discovered this kind of magnetic ordering. It may exist at low enough temperatures. The material is usually paramagnetic above the Néel temperature. The first demonstration of antiferromagnetic structures was achieved by neutron diffraction of transition metal oxides, including oxides of nickel, iron, and manganese. Clifford Shull's experiments produced the initial evidence that magnetic dipoles could be arranged in an antiferromagnetic structure.

Transition metal compounds, particularly oxides, are frequently found to contain antiferromagnetic materials. Hematite, metals like chromium, alloys like iron manganese (FeMn), and oxides like nickel oxide (NiO) are a few examples. Additionally, there are a lot of examples in metal clusters with high nuclearity. Rarely, organic molecules can also display antiferromagnetic coupling; this is demonstrated by radicals like 5-dehydro-m-xylylene.

By growing a ferromagnetic film on top of the antiferromagnet or annealing it in an aligning magnetic field, the antiferromagnet and ferromagnet's surface atoms align, allowing antiferromagnets to couple to ferromagnets. This process is known as exchange bias. This makes it possible to pin a ferromagnetic film's orientation, which is one of the primary applications in "spin valves," the foundation of magnetic sensors that include the read heads of contemporary hard disk drives. The blocking temperature of a layer is the temperature, usually lower than the Néel temperature, at or above which the antiferromagnetic layer can no longer pin the magnetization direction of a neighboring ferromagnetic layer. The net moment is zero if the sublattice moments of A and B are exactly equal but opposite. Antiferromagnetism is the name given to this kind of magnetic ordering (Figure 6.5).

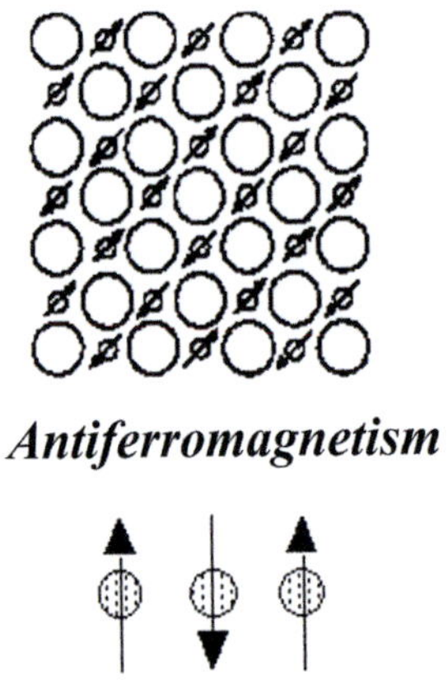

Figure 6.5. *Antiferromagnetic order may exist at sufficiently low temperatures.*

Source: https://files.askiitians.com/cdn1/images/201754-17551316-8857-8-ferri-magnetic-substance.gif.

The clue to antiferromagnetism is the behavior of susceptibility above a critical temperature, called the Néel temperature (T_N). Above T_N, the susceptibility obeys the Curie-Weiss law for paramagnets but with a negative intercept indicating negative exchange interactions (Figure 6.6; Table 6.1).

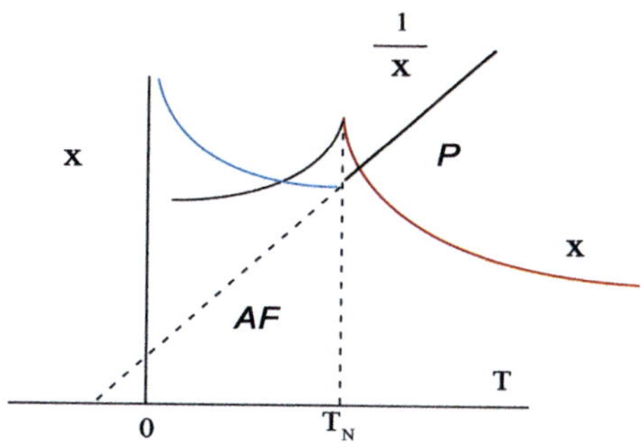

Figure 6.6. *Variation of susceptibility and inverse susceptibility with temperature in an AFM material.*

Source: https://archive.nptel.ac.in/content/storage2/courses/115103039/module1/lec1/images/7.png.

Table 6.1. Magnetic Properties of Minerals

Mineral	Composition	Magnetic Order	T_c(°C)	σ_s (Am^2/kg)
Oxides	–	–	–	–
Magnetite	Fe_3O_4	ferrimagnetic	575–585	90–92
Ulvospinel	Fe_2TiO_2	AFM	–153	
Hematite	αFe_2O_3	canted AFM	675	0.4
Ilmenite	$FeTiO_2$	AFM	–233	–
Maghemite	γFe_2O_3	ferrimagnetic	~600	~80
Jacobsite	$MnFe_2O_4$	ferrimagnetic	300	77
Trevorite	$NiFe_2O_4$	ferrimagnetic	585	51
Magnesioferrite	$MgFe_2O_4$	ferrimagnetic	440	21
Sulfides	–	–	–	–
Pyrrhotite	Fe_7S_8	ferrimagnetic	320	~20
Greigite	Fe_3S_4	ferrimagnetic	~333	~25
Troilite	FeS	AFM	305	
Oxyhydroxides	–	–	–	–
Goethite	αFeOOH	AFM, weak FM	~120	<1
Lepidocrocite	γFeOOH	AFM(?)	–196	–
Feroxyhyte	δFeOOH	ferrimagnetic	~180	<10
Metals & Alloys	–	–	–	–
Iron	Fe	FM	770	
Nickel	Ni	FM	358	55
Cobalt	Co	FM	1131	161
Awaruite	Ni_3Fe	FM	620	120
Wairauite	CoFe	FM	986	235

FM = ferromagnetic order

AFM = antiferromagnetic order

T_c = Curie or Néel Temperature

σ_s = saturation magnetization at room-temperature.

6.2.6.1. Measurement

The antiferromagnetic structure corresponds to a vanishing total magnetization in the absence of an external field. The antiferromagnetic phase may exhibit ferrimagnetic behavior in the presence of an external magnetic field if one of the sublattice magnetizations' absolute values differs from the other sublattice's, leading to a nonzero net magnetization. At absolute zero, the net magnetization should be zero, but as hematite demonstrates, spin canting frequently results in the development of a tiny net magnetization.

An antiferromagnetic material's magnetic susceptibility usually reaches its maximum at the Néel temperature. Conversely, the susceptibility will diverge as the ferromagnetic and paramagnetic phases merge. A divergence in the staggered susceptibility is noted in the antiferromagnetic case.

Antiferromagnetic structures could result from a variety of microscopic (exchange) interactions between the magnetic moments or spins. An Ising model on a bipartite lattice, e, can be considered in the most basic scenario. g. the basic cubic lattice, where spins are coupled at their closest neighboring sites. There will either be ferromagnetic or antiferromagnetic order depending on the sign of that interaction. Different and possibly more complex magnetic structures can result from geometrical frustration or competing ferro- and antiferromagnetic interactions.

Similar to ferromagnetic materials, the relationship between magnetization and the magnetizing field is non-linear. This is because ferromagnetic materials have a residual magnetization as a result of the hysteresis loop.

6.2.6.2. Geometric Frustration

Anti-ferromagnetic interactions, in contrast to ferromagnetism, can result in several optimal states (ground states, or states of minimal energy). The anti-ferromagnetic ground state in one dimension is an alternating sequence of spins, such as up, down, up, down, etc. However, there can be more than one ground state in two dimensions. Think about a triangle that is equilateral and has three spins—one on each vertex. As there are only two possible values for each spin (up or down), the system can exist in $2^3 = 8$ states, six of which are ground states.

When all three spins are up or all down, those are the two states that are not ground states. There will be two positive interactions and one negative interaction in each of the remaining six states. The system's inability to identify a single ground state serves as an example of frustration. It has been observed that minerals with crystal stacking structures like hexagonal or Kagome lattices exhibit this kind of magnetic behavior.

6.3. MAGNETIC CIRCUITS AND ELECTRO-MECHANICAL ENERGY CONVERSION

Modern technology relies heavily on magnetic circuits and electromechanical energy conversion, which are essential parts of many systems and devices. Like electrical circuits for the flow of electric current, magnetic circuits are the channels via which magnetic flux flows. They are made up of magnetic components that direct and regulate the distribution of magnetic flux, such as air gaps and iron cores.

Electrical energy can be converted into mechanical energy and vice versa in electromechanical energy conversion, which is made possible by the interaction of magnetic fields and electrical currents. This procedure is necessary for many commonplace devices found in daily life, such as electric motors, generators, transformers, and inductors.

One example of how magnetic circuits are used in electromechanical energy conversion is found in transformers. By using electromagnetic induction, these devices effectively move electrical energy from one circuit to another. Two or more coils twisted around a single magnetic core make up a transformer. A fluctuating magnetic flux is produced in the core when an alternating current passes through the primary coil. Electrical energy is transferred from the primary to the secondary circuit as a result of this fluctuating flux creating an electromotive force (EMF) in the secondary coil.

Another essential use for electromechanical energy conversion is in electric motors, which transform electrical energy into mechanical motion by means of magnetic circuits. Permanent magnets or electromagnets are the two main components that generate a magnetic field in an electric motor. The Lorentz force law states that this magnetic field interacts with the motor's current-carrying conductors, which are usually coils, to produce mechanical force. The motor shaft turns as a result, producing mechanical work from electrical energy.

In electromechanical devices, magnetic circuit design and optimization play a major role in device performance and efficiency. The behavior of magnetic circuits and, in turn, the effectiveness of energy conversion processes are influenced by various factors, including geometric configurations, magnetic permeability, and the characteristics of the core material.

Magnetic circuits are also necessary for managing magnetic fields in a variety of applications other than energy conversion. For example, magnetic circuits are used by inductors to store and release energy in electrical systems, and magnetic sensors use magnetic circuits to detect and measure magnetic fields for a variety of applications, such as proximity and position sensing.

Many technological advances are based on the related concepts of electromechanical energy conversion and magnetic circuits. Comprehending the fundamentals of magnetic circuits and their function in energy conversion procedures is essential for developing dependable and effective electromechanical systems in a variety of sectors.

6.3.1. Electromechanical Energy Conversion Principles

The apparatus that transforms electrical energy into mechanical energy or mechanical energy into electrical energy is known as an electromechanical energy conversion device. A magnetic field or an electric field can be used as the medium for electromechanical energy conversion; however, the majority of functional converters use a magnetic field as the coupling medium between electrical and mechanical systems because it has a much higher electric storing capacity than an electric field. Electromechanical energy converters can be classified as either electrical measuring instruments or loudspeakers, microphones, electromagnetic relays, and other gross motion devices.

The electromechanical energy conversion process makes extensive use of DC, induction, and synchronous machines. The device is referred to as a motor when it converts electrical energy into mechanical form, and as a generator when it converts mechanical energy into electrical form.

In these machines, conversion of energy from electrical to mechanical form or from mechanical to electrical from results from the following two electromagnetic phenomena:

- When a conductor is allowed to move in a magnetic field, a voltage is induced in the conductor.
- When a current-carrying conductor is placed in a magnetic field, then a mechanical force is experienced by the conductor.

Current passes through the conductors in a magnetic field during the monitoring action. There is a force applied to every conductor. The conductors are mounted on a rotor that has unrestricted motion. A mechanical load can be driven by the electromagnetic torque generated on the rotor, which is transmitted to the rotor shaft. Every conductor experiences an induced voltage as a result of the conductors rotating in the magnetic field. A prime mover powers the rotor during the generating action. Induced voltage occurs in the conductors of the rotor. A current will flow through the winding created by these conductors when an electrical load is connected to it, providing the load with power. Furthermore, the reaction torque that is created by the interaction between the conductors' current and the magnetic field tends to oppose the torque that the prime mover develops.

6.3.2. Magnetic Leakage and Fringing

The magnetic flux in a magnetic circuit that deviates from its intended path is called leakage flux. You can explain both the leakage flux and the fringing by using a solenoid as an example. A solenoid generates magnetic flux when current flows through it (Figure 6.7).

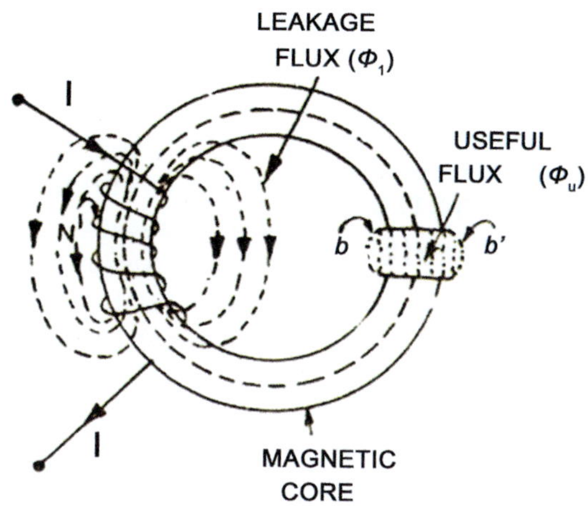

Figure 6.7. *Magnetic leakage and fringing.*

Source: https://circuitglobe.com/wp-content/uploads/2015/09/Leakage-flux-and-fringing-fig-compressor.jpg.

The majority of the flux is concentrated in the solenoid's core and travels along a specific path that goes through the air gap and is used in the magnetic circuit. We refer to this flux as useful flux φ_u. Since it is practically impossible for all of the flux in the circuit to take a specific path and build up in the magnetic core, some of the flux will instead build up around the coil or surround the coil's core without doing any work in the magnetic circuit. Leakage Flux is the type of flux that is not utilized for any work and is represented by φ_l.

Therefore, the total flux Φ produced by the solenoid in the magnetic circuit is the sum of the leakage flux and the useful flux and is given by the equation shown below:

$$\varphi = \varphi_u + \varphi_l \quad (1)$$

- **Leakage Coefficient:** The ratio of the total flux produced to the useful flux set up in the air gap of the magnetic circuit is called a leakage coefficient or leakage factor. It is denoted by (λ).

$$\lambda = \frac{\varphi}{\varphi_u} \quad (2)$$

- **Fringing:** When the useful flux builds up in the air gap, it tends to bulge outward at (b and b'), as seen in the above figure. As a result of this bulging, the air gap's effective area grows and its flux density falls. We call this effect Fringing. Fringing and air gap length are directly correlated, meaning that as the length grows, so will the fringing effect, and vice versa.

6.3.3. Magnetic Circuit – Series and Parallel Magnetic Circuit

6.3.3.1. Magnetic Circuit

A closed path that the magnetic flux follows is referred to as a magnetic circuit. A magnetic circuit is made up of a core made of highly conductive materials, such as soft steel, iron, etc. The reason for this is that these materials provide very little resistance to the magnetic flux flow.

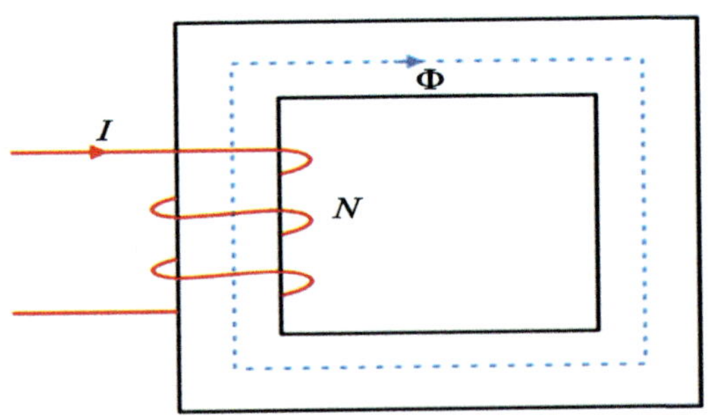

Figure 6.8. *Magnetic circuit.*

Source: https://www.tutorialspoint.com/assets/questions/media/54621/magnetic_circuit.png.

Examine an N-turn coil wound around an iron core (refer to the figure). Magnetic flux (ψ) is created in the core of the coil when an electric current I flows through it. Since the magnetic flux travels along the closed path ABCDA, the magnetic circuit is ABCDA (Figure 6.8).

In a magnetic circuit, the amount of magnetic flux in the core depends upon the current (I) and the number of turns (N). The product NI is known as the Magnetomotive Force (MMF).

$$MMF = NI = Ampere-Turns$$

The opposition offered by the magnetic circuit to the flow of magnetic flux is known as reluctance (S) of the magnetic circuit. The *reluctance* of the magnetic circuit depends upon the length of magnetic circuit, cross-sectional area of the circuit and nature of the material that makes up the magnetic circuit.

6.3.3.2. Types of Magnetic Circuit

There are two types of magnetic circuits:

- **Series Magnetic Circuit:** A magnetic circuit is referred to as series when the same magnetic flux, represented by ψ, flows through every component of the circuit. Examine a composite series magnetic circuit, which consists of two distinct magnetic materials with varying relative permeabilities. A composite series magnetic circuit is a series magnetic circuit with components of varying sizes and materials. Every component in this magnetic circuit in series will provide resistance to the magnetic flux ψ. The total reluctance of the magnetic circuit is equal to the sum of the reluctances of its individual components because the components are connected in series.

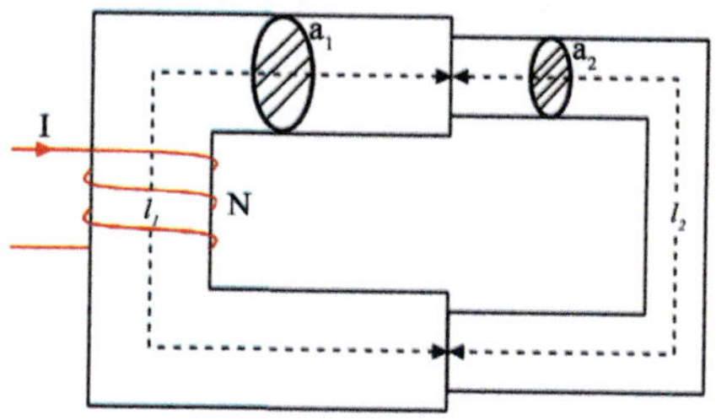

Figure 6.9. *Series magnetic circuit.*

Source: https://www.tutorialspoint.com/assets/questions/media/54621/series_magnetic_circuit.png.

Referring the Figure 6.9 of series magnetic circuit, we have;

$$\text{Total Reluctance}, S_T = \frac{l_1}{\mu_0\mu_{r1}a_1} + \frac{l_2}{\mu_0\mu_{r2}a_2}$$

$$\text{Total MMF} = \text{magnetic flux} \times \text{Total Reluctance}$$

$$\Rightarrow \text{Total MMF} = \psi\left(\frac{l_1}{\mu_0\mu_{r1}a_1} + \frac{l_2}{\mu_0\mu_{r2}a_2}\right)$$

$$\Rightarrow \text{Total MMF} = \left(\frac{B_1}{\mu_0\mu_{r2}}\right)\times l_1 + \left(\frac{B_2}{\mu_0\mu_{r2}}\right)\times l_2$$

$$\Rightarrow \text{Total MMF} = H_1 \times l_1 + H_2 \times l_2 \qquad (3)$$

Therefore, total MMF required to set up the magnetic flux in a series magnetic circuit is the sum of MMF required by individual parts of the circuit.

- **Parallel Magnetic Circuit:** A magnetic circuit which has more one path for the magnetic flux is called as parallel magnetic circuit (Figure 6.10).

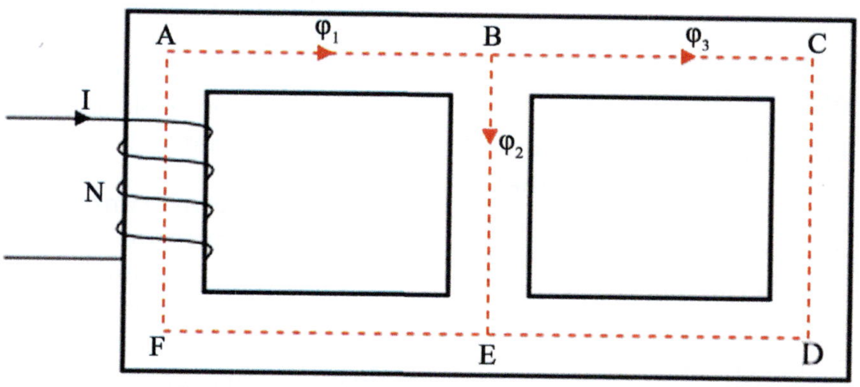

Figure 6.10. *Parallel magnetic circuit.*

Source: https://www.tutorialspoint.com/assets/questions/media/54621/parallel_magnetic_circuit.png.

Consider a coil of N turns wound on limb AF carries an electric current of I amperes. The magnetic flux φ_1 set up by the coil divides at B into two paths viz:

i. The magnetic φ_2 passes along the path BE.

ii. The magnetic flux φ_3 passes along the path BCDE.

Therefore, the total flux is;

$$\varphi_1 = \varphi_2 + \varphi_3 \qquad (4)$$

The path BE and BCDE are in parallel and hence form a parallel magnetic circuit. In a parallel magnetic circuit, the MMF required for the whole parallel magnetic circuit is equal to MMF required for any one of the parallel paths.

Let;

S_1=Reluctance of magnetic path ABEF

S_2=Reluctance of magnetic path BE

S_3=Reluctance of magnetic path BCDE

Total MMF=MMF for path ABEF+ MMF or path BE or BCDE

⊠Total MMF=$\varphi_1 S_1$+ ⊠$_2 S_2$+ $\varphi_3 S_3$

6.3.4. Attractive Force of Electromagnets

Particles with electric charge can interact with one another through electromagnetic fields, a phenomenon known as electromagnetic phenomena. Among the four basic forces of nature is the electromagnetic force. In the interactions between atoms and molecules, it is the dominant force. Electrostatics and magnetism are two separate but related phenomena that can be combined to form electromagnetism. Any two charged particles can experience electromagnetic forces. Magnetism is the interaction that happens between charged particles in relative motion, whereas electric forces cause an attraction between particles with opposite charges and a repulsion between particles with the same charge. Electromagnetic fields are used to explain these two forces. Microscopic charged particles are described by the Lorentz force, while macroscopic charged objects are described in terms of Coulomb's law for electricity and Ampère's force law for magnetism.

The electromagnetic force is the cause of numerous chemical and physical phenomena that are encountered in everyday existence. Atoms are held together by the electrostatic attraction between their electrons and atomic nuclei. Electric forces facilitate the amalgamation of distinct atoms to form molecules, encompassing the macromolecules like proteins that serve as the foundation of life. Chemical reactivity is also influenced by magnetic interactions between the spin and angular momentum magnetic moments of electrons; spin chemistry studies these relationships. Additionally, electromagnetic is essential to many aspects of contemporary technology, including the generation, transformation, and distribution of electrical energy, the production and detection of light, heat, and sound, fiber optic and wireless communication, computation, sensors, electrolysis, electroplating, and mechanical motors and actuators.

Since ancient times, scientists have studied electromagnetic. Numerous historical societies, such as the Greeks and the Mayans, developed intricate theories to account for phenomena such as lightning, static electricity, and the attraction between iron ore fragments that are magnetized. But scientists didn't start creating a mathematical foundation for comprehending the nature of electromagnetic interactions until the late 1700s. Prominent mathematicians and scientists like Coulomb, Gauss, and Faraday created namesake laws in the 18th and 19th centuries that contributed to our understanding of how electromagnetic fields interact and form. The discovery of Maxwell's equations, a system of four partial differential equations that offer a thorough explanation of classical electromagnetic fields, marked the culmination of this process in the 1860s. The existence of self-sustaining electromagnetic waves was predicted by Maxwell's equations, which also gave a solid mathematical foundation for the relationships between electricity and magnetism that scientists had been examining for centuries. It was subsequently demonstrated that Maxwell's theory—that such waves comprise visible light—was accurate. It has been established that all forms of electromagnetic radiation, including microwaves, radio waves, x-rays, ultraviolet, visible, and infrared radiation, are the same and only vary in their frequency range.

The modern era has seen scientists continue to improve the electromagnetism theorem to account for the effects of contemporary physics, such as relativity and quantum mechanics. Einstein's theory of special relativity was influenced by the theoretical implications of electromagnetism, specifically the determination of the speed of light based on the permeability and permittivity of the medium of propagation in 1905. In the meantime, Maxwell's equations have been altered by the study of quantum electrodynamics (QED) to better reflect the quantized nature of matter. The electromagnetic field's variations are described in terms of discrete excitations in QED, which are particles called photons, or light quanta.

6.3.5. Principles of Electromechanical Energy Conversion

Electric motors, generators, transformers, and other devices that run on electricity are powered by the fundamentals of electromechanical energy conversion. Fundamentally, electromechanical energy conversion is the process of converting electrical energy into mechanical energy or vice versa, made possible by the interaction of electric currents and magnetic fields.

The electromagnetic induction phenomenon, which Michael Faraday discovered in the 19th century, is fundamental to these ideas. According to Faraday's law, an electric current is produced in a conductor when an electromotive force (EMF) is induced by a fluctuating magnetic field. This idea underlies the operation of transformers and generators, which produce voltage and current through the relative motion of conductors and magnetic fields.

The transformation of mechanical energy into electrical energy is best illustrated by generators. A generator rotates a coil of wire inside a magnetic field using mechanical energy, which is frequently obtained from sources like turbines or engines. According to Faraday's law, an electromagnetic field (EMF) is produced by the coil's changing magnetic flux as it rotates. Mechanical energy is transformed into electrical energy by this induced voltage, which flows through the circuit linked to the generator.

On the other hand, electric motors show how electrical energy can be transformed into mechanical energy. The Lorentz force law states that an electric current flowing through a coil (or coils) in an electric motor interacts with a magnetic field to produce a mechanical force. By rotating the motor's shaft, this force transforms electrical energy into mechanical work. Electric motors power conveyor belts, fans, pumps, and other devices in a wide range of applications, from home appliances to industrial machinery.

Another essential application of the principles of electromechanical energy conversion is found in transformers. Through mutual electromagnetic induction, these devices exchange electrical energy between circuits. A common magnetic core is encircled by two or more primary and secondary coils to form a transformer. A voltage is produced in the secondary coil when an alternating current passes through the primary coil, causing a change in the magnetic flux in the core. This enables effective impedance matching in electronic circuits, as well as power distribution and voltage transformation across electrical grids.

The design of magnetic circuits, the characteristics of core materials, and the optimization of electrical and mechanical components are some of the variables that affect the efficiency and performance of electromechanical energy conversion systems. The electromechanical energy conversion technologies that power our modern world are becoming more efficient, compact, and versatile due to advancements in materials science, electronics, and control systems. The fundamental technologies that power the industrial, commercial, and residential sectors operate on the principles of electromechanical energy conversion.

These concepts facilitate the effective transfer and use of mechanical and electrical energy in transformers, generators, and motors, among other devices. This has led to advancements in a variety of industries, including manufacturing, automation, renewable energy, and transportation.

CASE STUDY

ELECTROMECHANICAL ENERGY CONVERSION IN WIND TURBINES

In the realm of renewable energy, wind turbines stand as iconic examples of electromechanical energy conversion. These structures harness the kinetic energy of wind and convert it into electrical power through a sophisticated process involving electromagnetism.

Introduction

Wind turbines are pivotal in the global shift towards sustainable energy sources. They exemplify how electromechanical systems can efficiently transform natural resources into usable electricity, contributing to clean energy goals and reducing carbon footprints.

Technological Overview

1. **Rotor and Blades:** The rotor of a wind turbine consists of multiple blades designed to capture wind energy. As wind flows over the blades, it causes them to rotate.
2. **Generator and Conversion Process:** Connected to the rotor is an electrical generator, typically a three-phase synchronous generator. This generator converts the mechanical energy of the rotating blades into electrical energy through the principles of electromagnetic induction.
3. **Power Transmission:** The electricity generated is then transmitted through power lines to homes, businesses, or the electrical grid, where it can be used immediately or stored for future use.

Key Components and Functionality

- **Blade Rotation:** The kinetic energy from the wind causes the blades to rotate. This rotational motion is critical as it drives the generator.
- **Electromagnetic Generator:** Inside the generator, the rotating motion induces a magnetic field within the stator coils, creating an alternating current (AC) output.
- **Power Regulation:** Control systems ensure the turbine operates efficiently by adjusting blade pitch and rotor speed based on wind conditions, optimizing energy capture.

Environmental Impact and Benefits

- Wind turbines produce clean, renewable energy without emitting greenhouse gases or other pollutants associated with fossil fuels.
- They contribute to energy independence and security by diversifying energy sources and reducing reliance on finite resources.

Challenges and Future Directions

- Challenges include intermittency (variation in wind speed), site selection, and visual impact, which require careful planning and technological advancements in energy storage and grid integration.
- Future innovations focus on increasing turbine efficiency, reducing costs, and expanding offshore wind farms to capitalize on stronger and more consistent wind patterns.

Conclusion

Wind turbines exemplify the successful integration of electromechanical energy conversion technologies into sustainable energy solutions. As the world transitions towards cleaner energy sources, understanding and optimizing these systems will be crucial for meeting global energy demands while mitigating environmental impacts.

CLASS ACTIVITY

Today, we dive into the fascinating world of magnetism. Students will work in pairs to investigate magnetic fields using bar magnets, iron filings, and compass needles. Each pair will hypothesize and test how magnetic fields interact, exploring concepts like magnetic poles, field lines, and the influence of materials. Through hands-on experimentation and guided discussion, students will grasp the principles of attraction, repulsion, and magnetic induction. This activity not only enhances understanding of magnetic phenomena but also encourages critical thinking and collaborative problem-solving skills among peers.

SUMMARY

- Magnetism is a fundamental force of nature that manifests through the interaction of magnetic fields. These fields arise from the movement of electric charges, primarily electrons, within atoms.
- Almost everyone is familiar with what a magnetic material can do but very few know how a magnet works. To understand this phenomenon one must first grasp the inextricable connection that exists between magnetism and electricity.
- Electrons are teeny tiny magnets. They have a north and a south pole, too, and spin around an axis. This spinning results in a very tiny but extremely significant magnetic field.
- The magnetic field is a field produced by the magnet or electric charges in motion. The magnetic field is the region around a magnetic material or moving electric charge within which the force of magnetism acts.
- Magnetic materials are categorized as magnetically hard, or magnetically soft materials. Magnetically soft materials are easily magnetized but the induced magnetism is usually temporary.
- Paramagnetic materials are those having permanent atomic dipoles, which are acted on individually and aligned in the direction of an external field.
- Ferromagnetism is the basic mechanism by which a material form permanent magnet (i.e., materials that can be magnetized by an external magnetic field and remain magnetized after the external field is removed).
- The antiferromagnetic materials are commonly found among the transition metal compounds. Hematite, chromium, alloys of iron manganese and oxides of nickel are the examples of antiferromagnetic material.
- Magnetic circuits and electromechanical energy conversion are integral components in various devices and systems, playing a fundamental role in modern technology.
- An electromechanical energy conversion device is the device that converts electrical energy into mechanical energy or, mechanical energy into electrical energy.
- Electromagnetism is an interaction that occurs between particles with electric charge via electromagnetic fields. The electromagnetic force is one of the four fundamental forces of nature. It is the dominant force in the interactions of atoms and molecules.

REVIEW QUESTIONS

1. Define magnetic flux density and magnetic field strength. How are they related?
2. Explain the difference between diamagnetic, paramagnetic, and ferromagnetic materials. Provide examples of each.

3. What is reluctance in the context of magnetic circuits? How is it analogous to electrical resistance?
4. Derive the expression for the magnetic field BBB inside a long solenoid carrying current III with nnn turns per unit length.
5. State Faraday's Law of Electromagnetic Induction and explain its significance in generating electrical energy.
6. How does a transformer work? Describe the principles behind step-up and step-down transformers.
7. Compare and contrast the operation of a DC motor and a DC generator. What principles do they share?
8. Define hysteresis loss and eddy current loss in magnetic materials. How are these losses minimized in practical applications?
9. Explain the significance of the magnetic permeability of a material. How does it affect the performance of magnetic devices?
10. Discuss one practical application of magnetic circuits and one application of electromechanical energy conversion in everyday life.

REFERENCES

1. Du Trémolet de Lacheisserie, É., Gignoux, D., & Schlenker, M. (2005). *Magnetism: Fundamentals* (pp. 3–6). Springer.
2. Furlani, E. P. (2001). *Permanent Magnet and Electromechanical Devices: Materials, Analysis and Applications*. Academic Press.
3. Jiles, D. (2015). *Introduction to Magnetism and Magnetic Materials* (3rd ed.). Boca Raton, FL: CRC Press.
4. Kronmüller, H. (2007). *Handbook of Magnetism and Advanced Magnetic Materials* (5 Volume Set). John Wiley & Sons.
5. Merzouki, R., Samantaray, A. K., & Pathak, P. M. (2012). *Intelligent Mechatronic Systems: Modeling, Control and Diagnosis* (pp. 403–405). Springer Science & Business Media.
6. Purcell, E. M. (2012). *Electricity and Magnetism* (3rd ed.). Cambridge University Press.
7. Tipler, P. (2004). *Physics for Scientists and Engineers: Electricity, Magnetism, Light, and Elementary Modern Physics* (5th ed.). W.H. Freeman.

CHAPTER

7

Transformer

LEARNING OBJECTIVES

After studying this chapter, you will be able to:

- Understand the overview of transformer
- Discuss the types of transformers
- Explain the applications of transformers
- Define EMF equation of ideal transformer
- Explain the transformer testing

KEY TERMS FROM THIS CHAPTER

Autotransformer
Distribution Transformer
Impedance
Load
Power Loss
Secondary Coil
Step-up Transformer
Transformer Oil
Core
Electromagnetic Induction
Isolation Transformer
Magnetic Flux
Primary Coil
Step-down Transformer
Tap Changer
Turns Ratio

INTRODUCTION

A transformer is a frequently utilized device in electrical systems that connects circuits operating at varying voltages. They are commonly employed in situations where there is a requirement to convert AC voltage from one level to another. Transformers can be used to either lower or raise the voltage and current in AC circuits, depending on the needs of the electrical equipment, device, or load. Power, instrumentation, and pulse transformers are examples of the wide range of applications that use transformers.

Transformers can be divided into two main types: electronic transformers and power transformers. Electronic transformers have low operating voltages and are used in consumer electronics such as televisions, computers, CD/DVD players, and other devices. Power transformers, on the other hand, have high power and voltage ratings and are commonly used in power generation, transmission, distribution, and utility systems to adjust voltage levels.

7.1. OVERVIEW OF TRANSFORMER

A transformer is a basic device that is used to transfer electrical energy from one alternating-current circuit to another, or to multiple circuits, through the process of electromagnetic induction. It operates by increasing or decreasing voltage based on the principle of electromagnetic induction. A transformer can either step up (Step-up transformer) or step down (Step-down transformer) AC voltage. Typically used in the transmission and distribution of alternating current power, transformers act as voltage control devices. They serve a variety of purposes, such as increasing voltage from electric generators for long-distance power transmission and decreasing voltage for low-voltage devices like doorbells and toy electric trains. (Chow Tai, 2006).

7.1.1. Components of Electric Transformers

For an efficient process, the electric transformer consists of certain components, which include the ones named and described below.

- **Transformer Core:** This part provides support for the primary and secondary windings, as well as supports the electromagnetic flux by creating a low reluctance path. The core is constructed from thin stacked sheets made of high-grade oriented steel, with a thin insulating material separating them. To reduce hysteresis and eddy currents in the system, the carbon content of the core steel is kept below 0.1%. Alloying with silicon can help reduce eddy currents. In a three-phase transformer, each limb holds the primary and secondary windings for each phase. The yokes connect these limbs magnetically. There are two types of constructions for the core: core and shell. The windings are surrounded by the core in shell-type construction.
- **Transformer Winding:** Each transformer phase contains two sets of windings: the primary winding and the secondary winding. The windings consist of multiple turns of copper or aluminum conductors that are insulated from each other and the transformer core. The arrangement and type of winding used are determined by factors such as current rating, short circuit strength, temperature rise, impedance, and voltage surges.

The winding that is rated for a higher voltage is known as an HV winding, while the one rated for a lower voltage is called an LV winding (Figure 7.1).

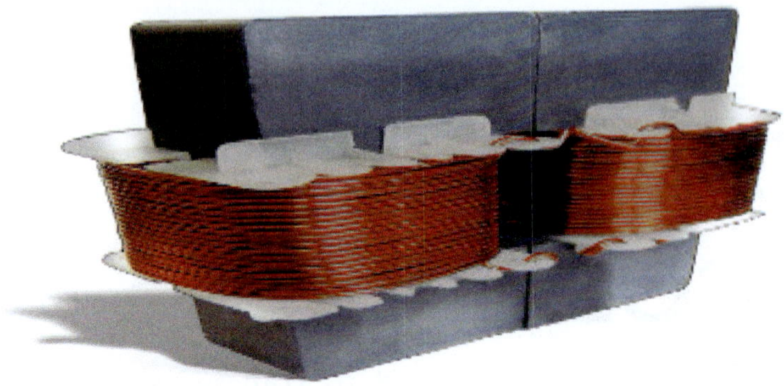

Figure 7.1. *Transformer winding.*

Source: https://www.iqsdirectory.com/articles/electric-transformer.html.

The high voltage winding conductors are typically thinner than the low voltage conductors in transformers, as they surround the low voltage winding from the outside. The low voltage windings are closer to the core, while the high voltage windings are sandwiched between the low voltage coils in shell type transformers. In core type transformers, the windings are classified into four types and the choice of winding type depends on the current carrying capacity and the number of turns needed (Figure 7.2).

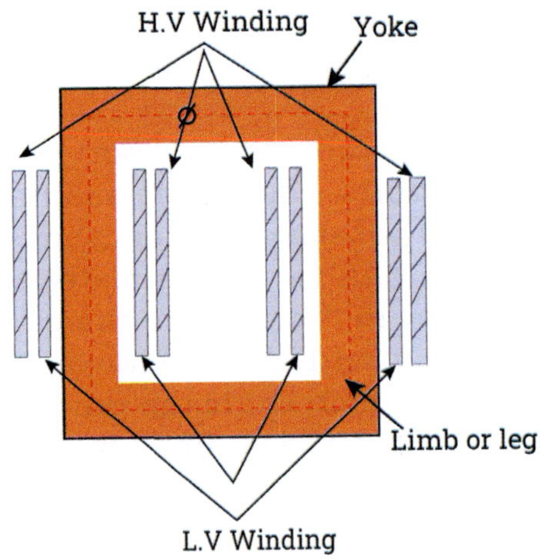

Figure 7.2. *Core and winding.*

Source: https://www.iqsdirectory.com/articles/electric-transformer.html.

- **Insulation of a Transformer:** Insulation is crucial for transformers to prevent severe damage in the event of a failure. It is essential for the insulation material to have high dielectric strength, thermal resistance, and mechanical properties. Common materials used for insulation in transformers include cotton, synthetic materials, and papers. Insulators must be strategically placed between windings, between the core and windings, and between parts carrying current (Figure 7.3).

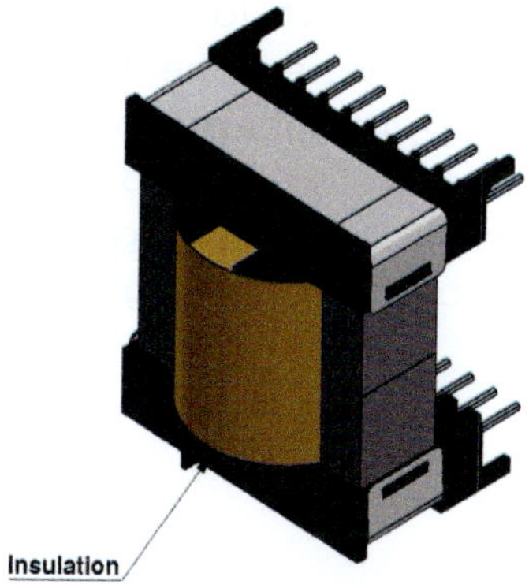

Figure 7.3. *Insulation of a transformer.*

Source: https://www.iqsdirectory.com/articles/electric-transformer.html.

- **Transformer Tank:** The tank has a dual function: shielding the core and windings from external conditions, and holding oil as well as supporting additional transformer accessories. To create these tanks, steel plates are rolled into containers. In circumstances where weight must be minimized, aluminum sheets may be substituted for steel sheets. Nonetheless, this can lead to higher costs, as aluminum sheets are pricier than steel sheets (Figure 7.4).

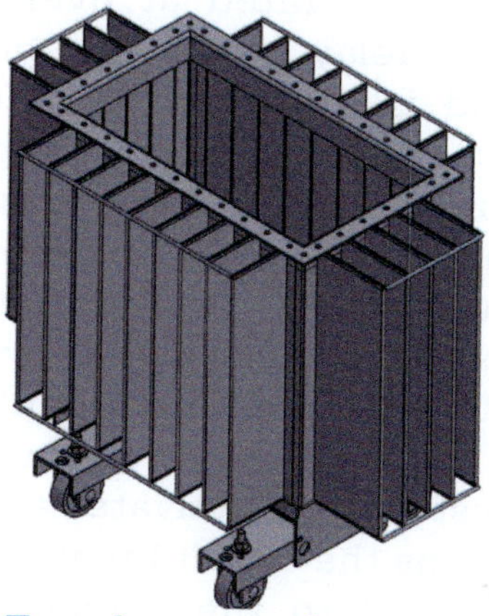

Figure 7.4. *Transformer tank.*

Source: https://www.iqsdirectory.com/articles/electric-transformer.html.

- **Transformer Oil:** Oil immersed transformers have the added benefit of using oil for insulation, which helps to cool the transformer and identify any potential issues with faulty components.
- **Terminal and Bushings:** Terminals found inside transformers serve to connect incoming and outgoing cables. These terminals are linked to bushings, which are in turn connected to the end of the windings. Bushings act as insulators, creating a barrier between the terminals and the tank. They also provide a safe passage for the conductor that links the terminals to the windings. Bushings are commonly made from materials such as epoxy resins and porcelain.
- **Oil Conservators:** The oil conservator sits positioned above the tank and bushings. Some oil conservators contain a rubber bladder to accommodate oil expansion from temperature changes. This allows for proper oil expansion and contraction within the transformer. A level indicator on the conservator shows the oil level inside. A pipe connects the oil conservator to the main tank.
- **Electric Transformer Breather:** Breathers are installed in oil-immersed transformers to keep the oil free from moisture. As the temperature changes, the oil expands and contracts, causing air to flow in and out of the conservator tank. The breather ensures that this air is free from moisture by removing it as it enters and exits the tank. Positioned at the end of the pipe carrying the air, the breather contains silica gel which

absorbs the moisture, resulting in moisture-free air for the conservator.

- **Radiators and Fans:** The heat generated in transformers is a byproduct of the energy conversion process. Dry transformers are typically cooled by natural air, while oil-immersed transformers utilize a variety of cooling methods. These can include cooling fans and radiators attached to the transformer tank, which are chosen based on factors such as power rating, power loss, and required cooling level for each specific transformer. The process begins with the heat generated in the core and windings being transferred to the transformer oil, which is then dissipated at the radiator. Larger transformers use forced cooling, where cooling radiators are installed in conjunction with the main radiators.
- **Explosion Vent:** The explosion vent is designed to act as an emergency release for oil and air gases trapped inside a transformer. Positioned above the conservator tank, it is a metal pipe with a diaphragm at one end. When faults in the oil lead to a rise in pressure inside the tank, the diaphragm ruptures at a low pressure threshold, allowing the built-up forces to escape into the surrounding atmosphere.
- **Tap Changers:** Tap changers are used to adjust the secondary voltage of electric transformers. They are designed to change the turns ratio of the transformer as needed. There are two types of tap changers: on load tap changers and off load tap changers. Off load tap changers are used when the transformer is not supplying any loads. On load tap changers allow for adjustments without interrupting the current flow to the load (Figure 7.5).

Figure 7.5. *Tap changer.*

Source: https://www.iqsdirectory.com/articles/electric-transformer.html.

- **Buchholz Relay:** The Buchholz relay is utilized in oil-filled transformers that are rated at over 500 kVA. This relay consists of oil and gas and is designed to detect faults occurring when the components of the transformer are submerged in oil. By detecting the gases released during short circuits that generate excessive heat and decompose the oil into gas, the Buchholz relay triggers an alarm system. This alarm system then activates a circuit that opens the circuit breaker supplying current to the primary winding, thus disrupting the flow of current. (Kothari & Nagrath, 2010).

7.1.2. Working Principle of Transformer

The working of the transformer is based on the principle of mutual inductance between two coils which are magnetic coupled (Figure 7.6).

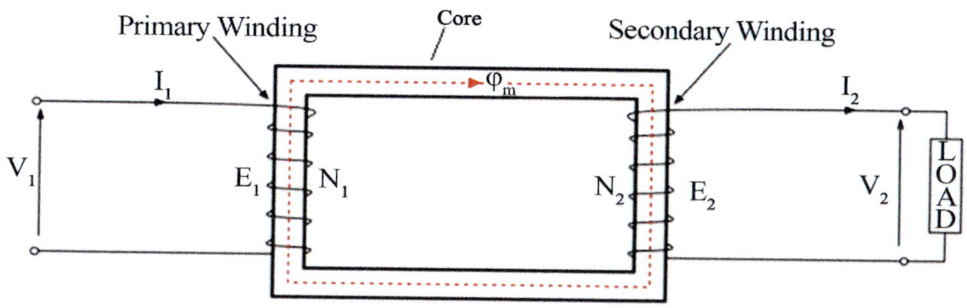

Figure 7.6. *Working principle of transformer.*

Mutual inductance states that when an alternating voltage is applied to the primary winding of a transformer, it produces an alternating flux ϕ_m, known as the mutual flux, in the core. This flux links both windings magnetically, inducing EMFs E_1 in the primary winding and E_2 in the secondary winding according to Faraday's law. E_1 is referred to as the primary EMF, while E_2 is the secondary EMF, given by:

$$E_1 = -N_1 \frac{d\varphi_m}{dt} \quad (1)$$

And

$$E_2 = -N_2 \frac{d\varphi_m}{dt} \quad (2)$$

Therefore;

$$\frac{E_2}{E_1} = \frac{N_2}{N_1} \quad (3)$$

From the above expression it can be seen that the magnitude of EMFs E_1 and E_2 depend upon the number of turns in the primary and secondary windings of the transformer, respectively, i.e., if $N_2 > N_1$, then $E_2 > E_1$, thus the transformer will be a step-up transformer and if $N_2 < N_1$, then $E_2 < E_1$, thus the transformer will be a step-down transformer.

When a load is connected to the secondary winding, the EMF E_2 will generate a load current I_2. In this way, a transformer facilitates the transmission of power from one electrical circuit to another while altering the voltage level.

7.1.3. Advantage and Disadvantage of Electric Transformer

Here are the advantages and disadvantages of electric transformers:

Advantages

1. Voltage Regulation:
   - Step-Up and Step-Down Voltage: Transformers can increase (step-up) or decrease (step-down) voltage levels, facilitating the efficient transmission of electricity over

long distances and reducing power loss.

- Stable Voltage Supply: They help maintain a stable voltage level, ensuring that electrical devices and appliances operate within their optimal voltage range.

2. Energy Efficiency:
   - High Efficiency: Modern transformers can have efficiencies exceeding 98%, meaning very little energy is lost in the form of heat during the transformation process.
   - Reduced Power Losses: By stepping up the voltage for transmission and stepping it down for distribution, transformers minimize power losses in the electrical grid.
3. Isolation and Safety:
   - Electrical Isolation: Transformers provide electrical isolation between circuits, enhancing safety by preventing direct contact with high voltage.
   - Protection: They can help protect electrical devices from voltage spikes and surges.
4. Cost-Effectiveness:
   - Long Lifespan: Transformers are durable and have a long operational life, often spanning several decades with proper maintenance.
   - Low Maintenance: Once installed, transformers require minimal maintenance compared to other electrical components.
5. Scalability:
   - Adaptable to Different Loads: Transformers can be designed to handle varying power loads, making them suitable for a wide range of applications from small electronics to large industrial equipment.

Disadvantages

1. Initial Cost:
   - High Capital Expenditure: The initial cost of purchasing and installing transformers can be significant, especially for large power systems.
   - Infrastructure Costs: Associated infrastructure, such as cooling systems and housing, can add to the overall cost.
2. Size and Weight:
   - Bulky and Heavy: Transformers, especially high-power ones, are large and heavy, requiring substantial space and support structures for installation.
   - Transportation Challenges: Moving and installing large transformers can be logistically challenging and expensive.
3. Energy Losses:
   - Iron and Copper Losses: Although efficient, transformers do experience energy losses due to hysteresis (iron losses) and resistance (copper losses) in the windings.
   - Heat Generation: Energy losses result in heat generation, which may require cooling systems to prevent overheating.

4. Environmental Concerns:
   - Oil Spills: Oil-filled transformers pose a risk of oil spills, which can be environmentally harmful and require specialized containment and clean-up measures.
   - Noise Pollution: Transformers can generate noise, particularly at high power levels, which may be a concern in residential areas.
5. Dependence on Grid:
   - Grid Reliability: Transformers are reliant on a stable grid infrastructure. Failures or disruptions in the grid can impact the performance and reliability of transformers.
   - Limited Flexibility: Once installed, transformers are not easily adaptable to significant changes in load demand or grid configuration without additional infrastructure changes.

7.2. TYPES OF TRANSFORMERS

There are several transformer types used in the electrical power system for different purposes, like in power generation, distribution and transmission, and utilization of electrical power. The transformers are classified based on voltage levels, Core medium used, winding arrangements, use and installation place, etc. Here we discuss different types of transformers are the step up and step down Transformer, Distribution Transformer, Potential Transformer, Power Transformer, 1-ϕ, and 3-ϕ transformer, Autotransformer, etc. (Horning, Kelly, et al., 2004).

7.2.1. Types of Transformers Based on Voltage Levels

The most commonly used transformer types are for all applications and are categorized as step-up or step-down transformers based on the voltage ratios between primary and secondary windings.

- **Step-Up Transformer:** As indicated by its name, the secondary voltage is increased in relation to the primary voltage by using more windings in the secondary than in the primary, as illustrated in the diagram. A step-up transformer is commonly used in power plants to connect the generator to the grid (Figure 7.7).

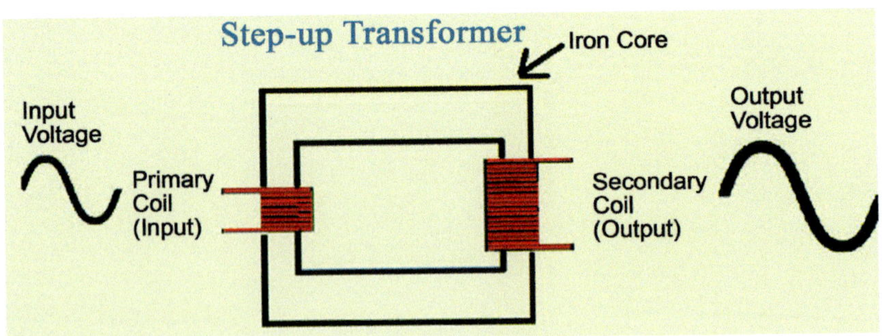

Figure 7.7. *Step-up transformer.*

Source: https://www.elprocus.com/various-types-of-transformers-applications/.

- **Step-Down Transformer:** The transformer reduces the voltage from a higher level at the primary side to a lower level at the secondary side, making it a step-down transformer. There are more winding turns on the primary side than on the secondary side (Figure 7.8).

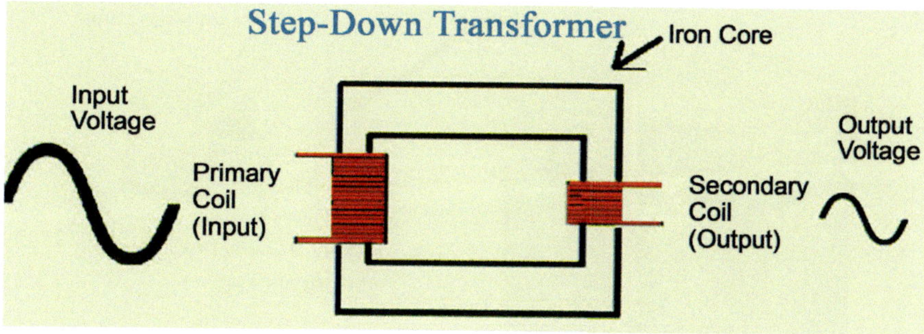

Figure 7.8. *Step-down transformer.*

Source: https://www.elprocus.com/various-types-of-transformers-applications/.

In distribution networks, the step-down transformer is commonly used to convert the high grid voltage to low voltage that can be used for home appliances.

7.2.2. Types of Transformers Based on the Core Medium Used

Transformers are classified as air core or iron core based on the type of medium placed between the primary and secondary winding.

- **Air-Core Transformer:** The primary and secondary windings are both wound on a non-magnetic strip in an air-core transformer, where the flux linkage between them occurs through the air. In comparison to an iron core transformer, the mutual inductance is lower in an air-core transformer due to the higher reluctance offered by the air medium. However, hysteresis and eddy current losses are eliminated in an air-core transformer (Figure 7.9).

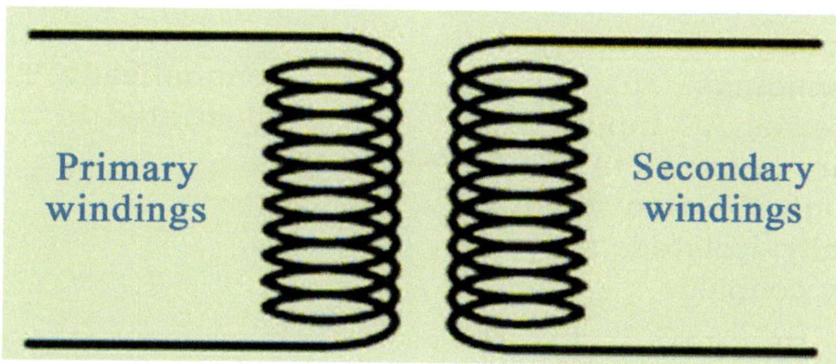

Figure 7.9. *Air-core transformer.*

Source: https://www.elprocus.com/various-types-of-transformers-applications/.

- **Iron Core Transformer:** The primary and secondary windings are wrapped around multiple iron plates, creating an ideal path for the generated flux. The iron's conductive and magnetic properties result in lower reluctance to the flux linkage. These transformers are commonly used for their high efficiency compared to air-core transformers (Figure 7.10).

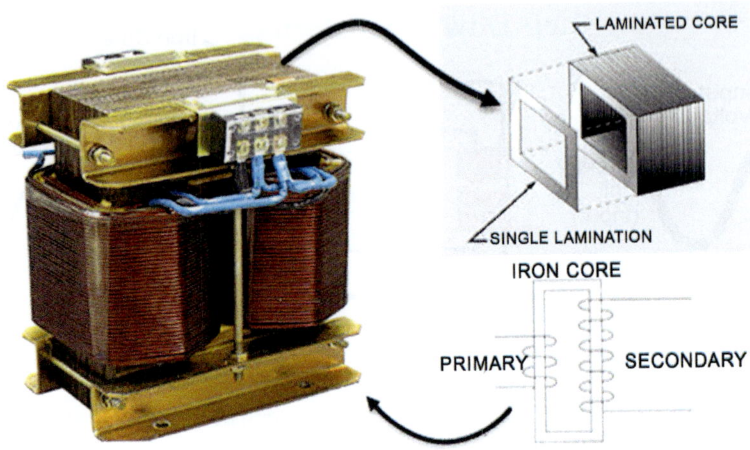

Figure 7.10. *Iron core transformer.*

Source: https://www.elprocus.com/various-types-of-transformers-applications/.

7.2.3. Types of Transformers Based on Winding Arrangement

Transformers based on winding arrangement are discussed below.

- **Two Winding Transformer:** A transformer, such as a two-winding transformer, is composed of separate windings for each phase, including the primary and secondary windings. The primary winding receives AC input, while the secondary is connected to the load. Although the two windings are electrically isolated, they are magnetically coupled.

  The EMF generated in the secondary winding is a result of the changing magnetic flux caused by the fluctuating current in the primary winding, known as mutual induction. Therefore, the output voltage is a direct result of this induction process. The magnitude of this voltage is primarily determined by the ratio of the number of windings in the primary and secondary coils and can either increase or decrease the input voltage.

- **Autotransformer:** In standard transformers, the primary and secondary windings are oriented in separate directions. However, in autotransformers, the primary and secondary windings are connected in series both physically and magnetically. This configuration is illustrated in the Figure 7.11.

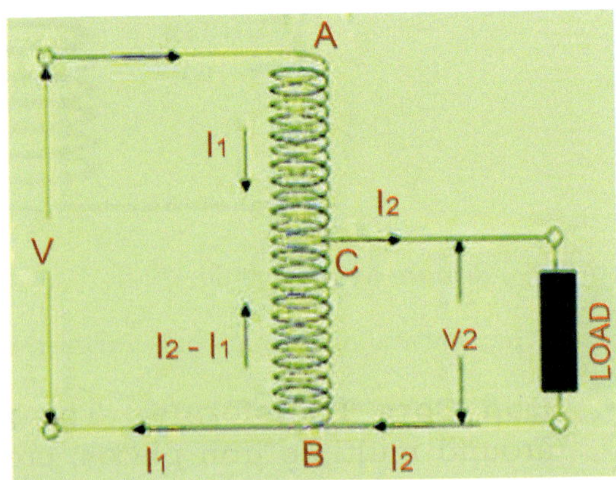

Figure 7.11. *Autotransformer.*

Source: https://www.elprocus.com/various-types-of-transformers-applications/.

A common coil with both primary and secondary windings is used, with voltage adjusting based on the location of the secondary tapping on the coil.

7.2.4. Transformers Based on Usage

These transformers are classified based on their necessity as power transformer, distribution transformer, measuring transformer, and protection transformer.

- **Power Transformer:** Power transformers are large in size and designed for high voltage applications (greater than 33 KV), commonly used in power generation stations and transmission substations. They have a high level of insulation (Figure 7.12).

Figure 7.12. *Power transformer.*

Source: https://www.elprocus.com/various-types-of-transformers-applications/.

- **Distribution Transformer:** Transformers are utilized to distribute power from power generation plants to faraway destinations. They are primarily used for distributing electrical energy at low voltages of less than 33 KV for industrial purposes and 440 V–220 V for domestic use (Figure 7.13).
  - It works at low efficiency at 50–70%
  - Small size
  - Easy installation
  - Low magnetic losses
  - It is not always fully loaded

Figure 7.13. *Distribution transformer.*

Source: https://www.elprocus.com/various-types-of-transformers-applications/.

Distribution transformers are categorized based on several factors, including their mounting location, insulation type, number of phases, voltage class, and Basic Impulse Insulation Level (BIL).

- **Measurement Transformer:** These devices are utilized for measuring electrical quantities such as voltage, current, power, and are categorized into potential transformers, current transformers, etc.

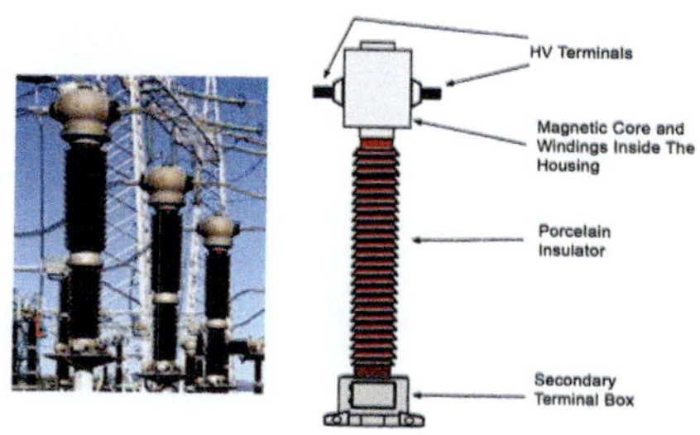

Figure 7.14. *Current transformer.*

Source: https://www.elprocus.com/various-types-of-transformers-applications/.

- **Potential Transformer:** The potential transformer, also referred to as a voltage transformer, allows for the measurement of high voltage lines by connecting the primary winding to the line. Instruments and meters are then connected to the secondary winding for measurement purposes. The primary function of this transformer is to reduce the high voltage to a safe level. The primary winding is typically grounded for safety precautions.

There are two main types of potential transformers: conventional wound and capacitor voltage. Conventional wound transformers are more expensive than capacitor voltage transformers because they require insulation.

- **Current Transformer:** The main purpose of the current transformer (CT) is to measure and also for safety purposes. When the current in the circuit is too high to be directly measured, the transformer is used to convert the high current into the desired value necessary for the circuit (Figure 7.14 and 7.15).

Figure 7.15. *Current transformer.*

Source: https://www.elprocus.com/various-types-of-transformers-applications/.

In this transformer, the primary winding is connected in series with the main supply and various measuring instruments like voltmeter, ammeter, protective relay coil, and wattmeter. These transformers are designed with specific current ratio, phase relation, and accuracy in order to precisely measure the load on the secondary side. The term ratio is of great importance in current transformers (CT).

Types of Current Transformers are three types like wound, toroidal, and bar type.

- **Wound Current Transformer:** The primary winding of the transformer can be physically connected in series using a conductor, which carries the current within the circuit. The magnitude of the secondary current is primarily determined by the transformer's turns ratio.
- **Toroidal Current Transformer:** This transformer does not have a primary winding. Instead, it uses a hole or window to thread the line that carries current in the circuit. Some current transformers have a split core that allows for easy opening, closing, installation, and connection without having to separate the network.
- **Bar-Type Current Transformer:** This transformer utilizes the main network's actual cable or bus-bar for the primary winding, which is a single turn. It is fully insulated from the system's high voltages and typically bolted to the current-carrying device.

- **Protection Transformers:** This type of transformer is utilized for safeguarding components. The key distinction between measuring and protection transformers lies in their accuracy, with protection transformers requiring greater precision compared to measuring transformers.
- **Instrument Transformer***:* An instrument transformer, also known as an isolation or measurement transformer, is an electrical device used to adjust voltage and current levels. Its primary function is to safely isolate the secondary winding when the primary winding receives a high voltage and current supply. This ensures that energy meters, relays, and measuring instruments connected to the secondary winding are protected from damage (Figure 7.16).

Figure 7.16. *Instrument transformers.*

Source: https://www.elprocus.com/various-types-of-transformers-applications/.

7.2.5. Transformers Based on Phase

Transformers based on phase are discussed below.

- **Single Phase Transformer:** The single phase transformer is a stationary device that operates based on Faraday's law of mutual induction. It functions by transmitting AC power from one circuit to another at a steady frequency and voltage level. This transformer consists of primary and secondary windings, with the primary winding receiving the AC supply and the load being connected to the secondary winding (Figure 7.17).

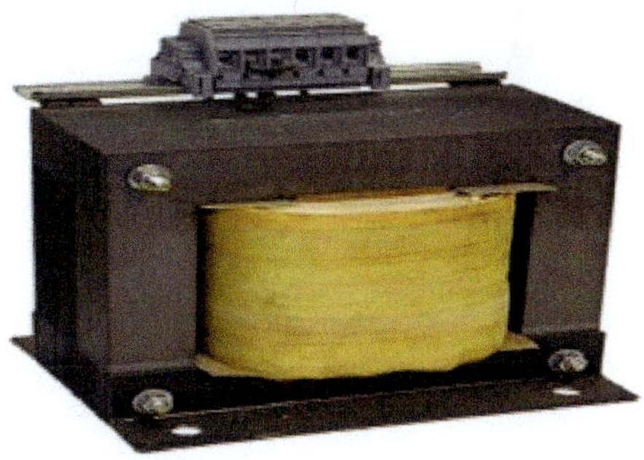

Figure 7.17. *Single phase transformer.*

Source: https://www.elprocus.com/various-types-of-transformers-applications/.

- **Three Phase Transformer:** When three 1-phase transformers are connected together with all 3 primary windings coupled as one and all 3 secondary windings connected as one, it creates a 3-phase transformer. This type of transformer is commonly used for generating, transmitting, and distributing electric power in industries. Assembling the transformer is cost-effective and it can be connected using Star or Delta connections. (Parker, Ula, &Webb, 2005) (Figure 7.18).

Figure 7.18. *Three phase transformer.*

Source: https://www.elprocus.com/various-types-of-transformers-applications/.

7.3. APPLICATIONS OF TRANSFORMERS

Here we have some applications of transformers given below:

- **Changing Electric Voltage:** By adjusting the coil's number of turns, the voltage can be altered. Transformers are primarily utilized to control electric voltage, either increasing it (step-up) or decreasing it (step-down). There are various types of transformers for each function, as previously mentioned.

  $$Vp/Vs = Np/Ns \quad (4)$$

  $$Vp = (Np/Ns)*Vs \quad (5)$$

  $$Vs = Vp(Ns/Np) \quad (6)$$

Even if the value of Vp remains constant as the input voltage, we can still adjust the output voltage (Vs) by altering the number of turns in both the primary and secondary coils.

- **Power Supply:** Electric transformers allow us to efficiently transmit electric power over long distances, reducing energy waste. Step up transformers help transmit power over long distances, while step down transformers reduce voltage for safe household use.
- **Current Conversion:** Transformers are utilized to convert Alternating Current to Direct Current, often in combination with rectifiers. These devices are commonly used in electronic power supplies.
- **Impedance Matching:** To achieve optimal power transfer between different circuits, impedance must be matched using transformers. This is commonly applied in audio equipment and telecommunication systems. Transformers facilitate not only power transfer, but also ensure maximum efficiency through impedance matching.
- **Isolation:** Transformers ensure that there is isolation between input and output devices, which makes them safe to use. They also protect sensitive equipment from potentially harmful voltage fluctuations.
- **Audio Transformers:** Audio transformers are utilized in audio equipment to combine and transfer audio signals between different stages, as well as to modify them, often between amplifiers and mixers. They play a role in maintaining signal fidelity and reducing external noise interference within the channel.
- **Tesla Coil:** Transformers are crucial components in Tesla coils, serving to generate high voltage and high frequency alternating electric current. They are primarily utilized for demonstrations and for generating powerful electric discharges (Figure 7.19).

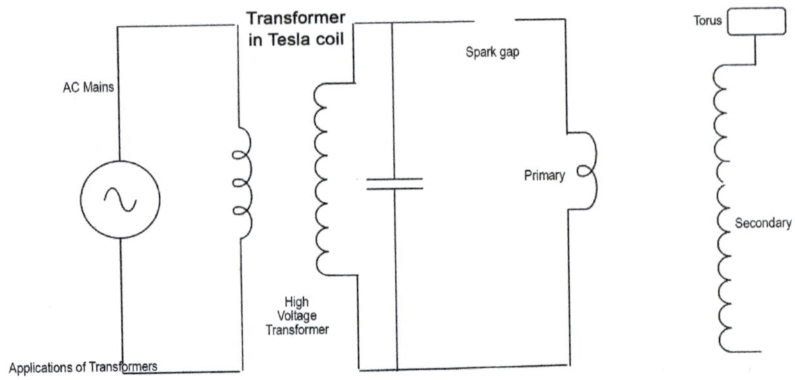

Figure 7.19. *Transformer in tesla coil.*

Source: https://www.geeksforgeeks.org/applications-of-transformers/.

- **Medical Usage:** Transformers are utilized in various applications beyond electric devices such as motors. They are also utilized in the medical field for machines like MRI and X-Ray, where they generate high voltage current to support imaging and diagnostics.
- **Railway Electrification:** Railway electrification uses transformers to reduce high voltage from overhead lines to a suitable level for trains. These transformers guarantee that trains are given the correct amount of power for optimal performance, enabling safe and dependable railway travel (Figure 7.20).

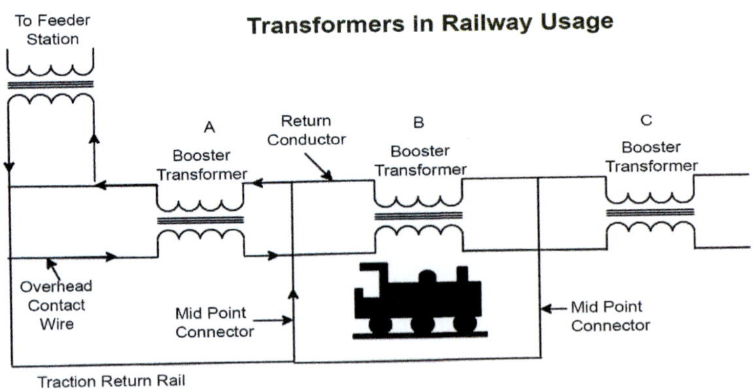

Figure 7.20. *Transformers in railway usage.*

Source: https://media.geeksforgeeks.org/wp-content/uploads/20240503154956/Untitled-Diagram---2024–05–03T154944850.webp.

- **Backup Power Systems:** Transformers are essential components in UPS systems, ensuring a constant power supply during outages. They adjust voltage levels to support key equipment such as computers, servers, and medical devices, preventing disruptions and protecting against data loss or damage. Their role is crucial in maintaining operational continuity and ensuring uninterrupted power flow.

- **Voltage Stabilization:** Tap changers are used in transformers to maintain stable voltage in circuits by automatically adjusting their turn ratio. This process allows for consistent voltage levels, even in the face of fluctuating loads. By constantly monitoring and making adjustments, transformers help ensure a steady voltage, improving the reliability and efficiency of electrical systems, no matter the changing demands.
- **Arc Furnaces:** Electric arc furnaces require transformers to provide the necessary high currents for melting metals. These transformers also supply power for a variety of industrial metal refining processes, enabling efficient operations in metalworking and manufacturing sectors (Figure 7.21).

Figure 7.21. *Transformers in arc furnaces.*

Source: https://internationalelectricalsuppliers.weebly.com/uploads/5/0/6/7/50679149/7541319_orig.jpg.

- **Application of Transformers in Power Generation:** Transformers play a crucial role in the electricity generation process by ensuring efficient delivery of electricity from power plants to homes and businesses. One of their primary functions is to increase the voltage of electricity for long-distance transmission and maintain a constant voltage for consumers. They also serve as protectors against electrical issues like short circuits and lightning. In cases where electricity needs to be transmitted over long distances, transformers switch to direct current for improved transmission. Upon reaching communities, transformers lower the voltage to safe levels for use in households and appliances.
- **Transmission and Distribution:** Transformers play a crucial role in the transmission and distribution of electricity, ensuring that power generated from different sources can reach homes and businesses safely and effectively. One key function of transformers in power transmission is to regulate voltage levels. By stepping up voltage levels at power plants, transformers aid in the efficient transmission of high voltage electricity over long distances, minimizing power loss. Additionally, transformers in substations along transmission lines help regulate voltage to maintain a steady flow of electricity. Moreover, transformers have a crucial role in providing electricity to households. Community distribution transformers lower the voltage to a safe level for domestic purposes and regulate electrical distribution, adjusting power supply to meet

demand and maintain stability in various locations including homes, schools, and hospitals. Overall, transformers are essential for preserving the efficiency of power transmission and distribution, guaranteeing the consistent and reliable supply of electricity to consumers.

- **Lighting:** Transformers are essential in lighting systems as they help to regulate voltage levels to achieve the desired output. They are particularly crucial in low-power lighting applications like landscape or street lighting, where they play a key role in reducing voltage to a stable level that meets the lighting requirements. Additionally, transformers are used for dimming control and for providing safety isolation.
- **Telecommunication:** Transformers are commonly utilized in telecommunication for receiving different types of signals. The primary role of transformers in this field is for installation and adjustments. They are also essential in other devices like modems and routers, where they convert digital signals to analog for transmission, as well as vice versa. Furthermore, transformers play a crucial role in power supply units and voltage regulators to guarantee a steady and dependable power supply to communication equipment, while also protecting against fluctuations and electrical interference.

7.4. EMF EQUATION OF IDEAL TRANSFORMER

Let N_p represent the number of turns in the main winding, and Ns represent the number of turns in the secondary winding of the transformer. When an AC voltage is applied to the main coil of the transformer, the resulting current produces a magnetic flux that links the secondary coil and induces an emf. The magnitude of this emf is determined by the number of turns in the secondary coil. In an ideal (lossless) transformer with no resistance in the primary coil and all flux contained within the core linking the primary and secondary windings, the voltage V_p applied to the primary coil results in a flux linkage in each turn of the core at time t due to the current flowing in the primary coil. (William R. Huber, 2022) (Figure 7.22).

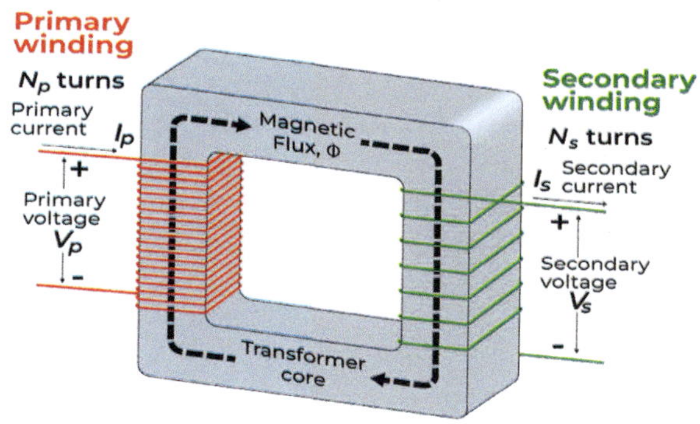

Figure 7.22. *EMF equation of ideal transformer.*

Source: https://www.geeksforgeeks.org/transformer/.

The induced emf or voltage (ε_s) in the secondary with N_s turns is then calculated.

$$\varepsilon_s = -N_s \times d\phi/dt \qquad (7)$$

In addition, the alternating flux generates a reverse emf in the main. This is it.

$$\varepsilon_p = -N_p \times d\phi/dt \qquad (8)$$

And for an ideal transformer, $\varepsilon_p = V_p$

By approximation, if the secondary is an open circuit or the current drawn from it is modest, $\varepsilon_s = V_s$.

The voltage across the secondary coil is V_s. As a result, Eqns (7) and (8) may be written as:

$$V_s = -N_s \times d\phi/dt \quad (9)$$

$$V_p = -N_p \times d\phi/dt \quad (10)$$

From Eqns (3) and (4), we have

$$V_s / V_p = N_s / N_p \quad (11)$$

The above equation is known as Transformer Equation or Transformer Formula.

7.5. TRANSFORMER TESTING

Testing of transformers includes several procedures to verify the specifications and functionality of transformers prior to and following their installation.

7.5.1. Type Test of Transformer

In order to demonstrate that the transformer meets the customer's specifications and design requirements, it must undergo various testing procedures at the manufacturer's facilities. Some of these tests are conducted to validate the fundamental design expectations of the transformer. These tests are typically performed on a prototype unit rather than on every unit in a production batch. The type test of the transformer verifies the main and essential design criteria for the entire production lot.

7.5.2. Routine Tests of Transformer

Routine tests on transformers are primarily conducted to verify the operational performance of each unit within a production batch. These tests are performed on every unit produced.

7.5.3. Special Tests of Transformer

Tests on transformers are tailored to meet the specific needs of customers, offering important data for the maintenance and operation of the equipment.

- **Pre-Commissioning Test of Transformer:** Besides these, the transformer undergoes additional tests before being officially commissioned at the site. These tests, known as pre-commissioning tests, are conducted to evaluate the transformer's condition post-installation and to compare the results of low voltage tests with the factory test reports.

Type tests of transformer include:

- Winding resistance test of transformer
- Transformer ratio test
- Transformer vector group test
- Measurement of impedance voltage/short circuit impedance (principal tap) and load loss (Short circuit test)
- Measurement of no-load loss and current (Open circuit test)
- Measurement of insulation resistance

- Dielectric tests of transformer
- Temperature rise test of transformer
- Tests on on-load tap-changer
- Vacuum tests on tank and radiators

Routine tests of transformer include:

- Winding resistance test of transformer
- Transformer ratio test
- Transformer vector group test
- Measurement of impedance voltage/ short circuit impedance (principal tap) and load loss (Short circuit test)
- Measurement of no load loss and current (Open circuit test)
- Measurement of insulation resistance
- Dielectric tests of transformer.
- Tests on on-load tap-changer.
- Oil pressure test on transformer to check against leakages past joints and gaskets

This means that routine tests include all type tests except for the temperature rise and vacuum tests. Additionally, an oil pressure test is performed to ensure there are no leakages at joints and gaskets.

Special Tests of transformer include:

- Dielectric tests.
- Measurement of zero-sequence impedance of three-phase transformers
- Short-circuit test
- Measurement of acoustic noise level
- Measurement of the harmonics of the no-load current.
- Measurement of the power taken by the fans and oil pumps.
- Tests on bought out components / accessories such as Buchhloz relay, temperature indicators, pressure relief devices, oil preservation system etc.

7.5.4. Transformer Winding Resistance Measurement

Transformer winding resistance measurements are conducted to determine the I^2R losses and calculate the winding temperature following a temperature rise test. This test is performed as both a type test and routine test. Additionally, it is conducted on-site to assess the condition of the transformer, including checking for loose connections, broken conductor strands, high contact resistance in tap changers, high voltage leads, and bushings.

There are different methods for measuring of the transformer winding, likewise:

- Current-voltage method of measurement of winding resistance.
- Bridge method of measurement of winding resistance.
- Kelvin bridge method of Measuring Winding Resistance.
- Measuring winding resistance by Automatic Winding Resistance Measurement Kit.

It is important to note that measurements of transformer winding resistance must be performed at every tap.

7.5.5. Transformer Ratio Test

The efficiency of a transformer is greatly influenced by the accuracy of its specific turns or voltage ratio. Therefore, conducting a transformer ratio test is a crucial step in testing a transformer. This test is not only essential, but also a routine procedure in the testing of transformers. To ensure that an electrical power transformer operates effectively, testing the voltage and turn ratio is essential.

Performing the transformer ratio test is straightforward. By connecting a three-phase 415 V supply to the HV winding while keeping the LV winding open, we can measure the induced voltages at both the HV and LV terminals to determine the actual voltage ratio of the transformer. This test is repeated for each tap position individually.

7.5.6. Magnetic Balance Test of Transformer

The magnetic balance test for transformers is specifically carried out on three-phase transformers to detect any discrepancies in the magnetic circuit.

7.5.6.1. Procedure of Magnetic Balance Test of Transformer

- Keep the tap changer of transformer in normal position.
- Now disconnect the transformer neutral from ground.
- Then apply single phase 230 V AC supply across one of the HV winding terminals and neutral terminal.
- Measure the voltage in two other HV terminals in respect of neutral terminal.
- Repeat the test for each of the three phases.

In case of an autotransformer, a magnetic balance test of transformer should be repeated for LV winding also.

In the core of the transformer, three limbs are arranged side by side with one phase winding wound in each limb. The voltage induced in different phases is determined by the position of the limb in the core. The voltage induced in each phase in relation to the neutral terminals is provided in the table below.

	Left Side Phase	**Central Phase**	**Right Side Phase**
	AN	BN	CN
Voltage applied at left phase	230 V	180 V	50 V
Voltage applied at central phase	115 V	230 V	115 V
Voltage applied at right phase	50 V	180 V	230 V

7.5.7. Magnetizing Current Test of Transformer

The magnetizing current test of a transformer is conducted to identify issues such as faults in the magnetic core structure, movement of windings, breakdown of insulation between turns, or malfunctions in tap changers. These problems can alter the magnetic circuit's effective reluctance, which in turn impacts the amount of current needed to create flux in the core.

- Keep the tap changer in the lowest position and open all IV and LV terminals
- Then apply three phase 415 V supply on the line terminals for three-phase transformers and single phase 230 V supply on single phase transformers
- Measure the supply voltage and current in each phase
- Now repeat the magnetizing current test of transformer test with keeping tap changer in normal position

- Repeat the test while keeping the tap at highest position

In most cases, two outer limb phases on a transformer core will typically read higher compared to the center limb phase in three phase transformers. Generally, an agreement within 30% of the exciting current measured in previous tests is deemed acceptable. However, if the measured exciting current value is 50 times greater than the factory test value, it may indicate a fault in the winding that requires additional analysis.

7.5.8. Vector Group Test of Transformer

It is crucial to conduct a vector group test on a 3-phase transformer in order to ensure proper vector grouping, which is necessary for the parallel operation of transformers. Various internal connections are available for three-phase transformers in the market, which offer different magnitudes and phases of the secondary voltage. While the magnitude can be adjusted for parallel operation by choosing the appropriate turn ratio, the phase difference cannot be compensated for.

In order to operate transformers in parallel, we must select one with matching phase sequence and divergence. Transformers with identical vector groups will have the same phase sequence and divergence between primary and secondary.

Before procuring an electrical power transformer, you should ensure the vector group of the transformer, whether it will be matched with his or her existing system or not. The vector group test of transformer confirms his or her requirements.

7.5.9. Insulation Resistance Test or Megger Test of Transformer

The insulation resistance test is a crucial type of test for transformers. It is conducted to verify the effectiveness of the insulation system in an electrical power transformer and ensure its proper functioning.

7.5.9.1. Procedure of Insulation Resistance Test of Transformer

- Disconnect all the line and neutral terminals of the transformer
- Megger leads to be connected to LV and HV bushing studs to measure insulation resistance IR value in between the LV and HV windings
- Megger leads to be connected to HV bushing studs and transformer tank earth point to measure insulation resistance IR value in between the HV windings and earth
- Megger leads to be connected to LV bushing studs and transformer tank earth point to measure insulation resistance IR value in between the LV windings and earth

NB: Testing the insulation resistance of each phase in a three-phase transformer is not required. This is because the windings on the high voltage side are connected together to form a star or delta configuration, as are the windings on the low voltage side. Therefore, insulation resistance values are taken collectively between all the windings.

Measurements are to be taken as follows:

- For autotransformer: HV-IV to LV, HV-IV to E, LV to E.
- For two winding transformer: HV to LV, HV to E, LV to E.
- Three winding transformers: HV to

IV, HV to LV, IV to LV, HV to E, IV to E, LV to E.

- Oil temperature should be noted at the time of insulation resistance test of the transformer, since the IR value of transformer insulating oil may vary with temperature.
- IR values to be recorded at intervals of 15 seconds, 1 minute and 10 minutes.
- With the duration of application of voltage, IR value increases. The increase in IR is an indication of dryness of insulation.
- Absorption coefficient = 1 minute value/15 secs. value.
- Polarization index = 10 minutes value/1 minute value.

7.5.10. Dielectric Tests of Transformer

The dielectric test of a transformer is a type of insulation test that is conducted to verify the anticipated insulation strength of the transformer. It is one of multiple tests carried out to ensure the appropriate quality of transformer insulation. The dielectric test is conducted in two separate stages.

The first test is known as the Separate Source Voltage Withstand Test for transformers. During this test, a single-phase power frequency voltage of a specified level is applied to the winding under examination for a period of 60 seconds, while the other windings and tank are grounded. The purpose of this test is to determine if any insulation failures occur. The second test is the induced voltage test on the Transformer. In this test, a three-phase voltage that is twice the rated secondary voltage is applied to the secondary winding for 60 seconds, while keeping the primary winding of the transformer open circuited. The frequency of the applied voltage should be twice that of the power frequency. If there is no failure of insulation, the test is considered successful. Apart from dielectric tests on transformers, there are other types of tests to check the insulation of the transformer, including lightning impulse tests, switching impulse tests, and partial discharge tests.

7.5.10.1. Induced Voltage Test of Transformer

The induced voltage test of the transformer is intended to check the inter-turn and line end insulation as well as main insulation to earth and between windings-

- Keep the primary winding of transformer open circuited.
- Apply three-phase voltage to the secondary winding. The applied voltage should be twice of the rated voltage of secondary winding in magnitude and frequency.
- The duration of the test shall be 60 seconds.
- The test shall start with a voltage lower than 1/3 the full test voltage, and it shall be quickly increased up to the desired value.

The test is successful if no breakdown occurs at full test voltage during the test.

7.5.11. Temperature Rise Test of Transformer

The temperature rise test is a crucial component of the transformer's type testing. This assessment involves verifying that the transformer's winding and oil temperatures do not exceed specified limits. During this test, both the oil and winding temperatures must be closely monitored to ensure they are within the allowable limits for the electrical transformer.

CASE STUDY

MODERNIZING POWER DISTRIBUTION WITH ELECTRIC TRANSFORMERS

Background

ABC Power Company, a leading utility provider, is faced with the challenge of modernizing its power distribution infrastructure to meet growing energy demands, improve efficiency, and integrate renewable energy sources into the grid. This case study explores how the company leverages electric transformers to upgrade its power distribution network.

Challenges

1. **Aging Infrastructure:** ABC Power Company's existing power distribution infrastructure consists of aging transformers and substations, which are prone to failures, inefficiencies, and voltage fluctuations.
2. **Capacity Constraints:** The current infrastructure is reaching its maximum capacity, limiting the company's ability to meet increasing electricity demands from residential, commercial, and industrial customers.
3. **Integration of Renewable Energy:** With a growing emphasis on renewable energy sources such as solar and wind, ABC Power Company needs to integrate these intermittent energy sources into its grid while maintaining stability and reliability.
4. **Efficiency and Reliability:** Improving the efficiency and reliability of the power distribution network is essential to minimize energy losses, reduce downtime, and ensure consistent electricity supply to customers.

Solution

ABC Power Company implements a comprehensive plan to modernize its power distribution infrastructure, with a focus on upgrading electric transformers to enhance efficiency, capacity, and flexibility.

Implementation Steps:

1. **Transformer Upgrades:**
   i. **Replacement of Aging Transformers:** The company replaces outdated

transformers with modern, high-efficiency units equipped with advanced monitoring and control capabilities.

ii. **Installation of Smart Transformers:** Smart transformers with built-in sensors and communication capabilities are deployed to enable real-time monitoring of voltage levels, load conditions, and performance metrics.

Grid Modernization:

i. **Distribution Automation:** Automation technologies are implemented to improve the responsiveness and flexibility of the distribution network, enabling automatic reconfiguration and optimization of power flow.

ii. **Remote Monitoring and Control:** Remote monitoring and control systems are deployed to enable centralized monitoring of transformer operations, fault detection, and predictive maintenance.

Integration of Renewable Energy:

i. **Grid-Connected Inverters:** Grid-connected inverters are installed to facilitate the integration of renewable energy sources such as solar and wind into the grid. These inverters help manage voltage fluctuations and maintain grid stability.

ii. **Advanced Grid Management Systems:** Advanced grid management systems are implemented to optimize the utilization of renewable energy resources, balance supply and demand, and ensure grid reliability.

Energy Efficiency Measures:

i. **Loss Reduction Techniques:** Loss reduction techniques, such as voltage optimization and load management, are employed to minimize energy losses and improve overall system efficiency.

ii. **Demand-Side Management Programs:** Demand-side management programs are introduced to encourage customers to reduce peak electricity demand through incentives and energy efficiency measures.

Results

1. **Enhanced Efficiency:** The deployment of modern electric transformers and grid automation technologies improves the efficiency of the power distribution network, reducing energy losses and optimizing power flow.
2. **Increased Capacity:** Upgrading transformers and integrating renewable energy sources increases the capacity and flexibility of the grid, allowing ABC Power Company to meet growing energy demands and support future growth.
3. **Improved Reliability:** Advanced monitoring and control systems enable proactive

maintenance, early fault detection, and rapid response to disruptions, enhancing the reliability and resilience of the power distribution network.

4. **Integration of Renewable Energy:** The integration of renewable energy sources into the grid reduces dependency on fossil fuels, lowers greenhouse gas emissions, and contributes to a more sustainable energy future.

CLASS ACTIVITY

Objective: To familiarize students with the architecture and functionality of Transformers in NLP tasks.

Materials Needed:

- Whiteboard/flip chart
- Markers
- Printed handouts or digital slides explaining Transformer architecture
- Computers or tablets with internet access (optional, for demonstrations)

Assessment:

Explain key components such as self-attention mechanisms, encoder and decoder layers, and positional encoding.

Discuss how the self-attention mechanism allows the model to weigh the importance of different words in a sentence when processing information.

Provide each group with a set of discussion questions related to Transformer architecture. Questions can include:

- What are the advantages of using self-attention mechanisms in NLP tasks?
- How does the Transformer model handle sequential data without using recurrent neural networks (RNNs)?
- What are the limitations of the Transformer architecture?

SUMMARY

- Transformer is the simplest device that is used to transfer electrical energy from one alternating-current circuit to another circuit or multiple circuits, through the process of electromagnetic induction. A transformer works on the principle of electromagnetic induction to step up or step down voltage.
- The core is made up of stacked thin sheets, which are assembled from high grade oriented steel and a thin insulating material separates them. The carbon content of the core steel is kept below a degree of 0.1% in order to minimize the hysteresis and the eddy currents in the system.
- Insulation is the most important requirement for transformers and in case they experience a failure, severe damages to the transformers can occur. These should have high dielectric strength and the ability to withstand high temperatures together with having high mechanical properties.
- The tank serves two purposes: to protect the core and the windings from the outside environment. The other purpose is to contain oil and later also provide support for other transformer accessories. These tanks are made by rolling steel plates to containers. In cases where weight has to be reduced aluminum sheets are used in place of the steel sheets.
- The oil conservator is located above the tank and the bushings. In some oil conservators, a rubber bladder is usually inserted. Its use is to provide enough space for oil expansion after the oil in the transformer had expanded and contracted due to an increase and decrease in temperature.
- The explosion vent serves the purpose of being an emergency exit for oil and air gasses that are held inside a transformer. It is a pipe made up of metal, contained above the conservator tank, and has a diaphragm at one of its ends.
- The power transformers are big in size. They are suitable for high voltage (greater than 33 KV) power transfer applications. It used in power generation stations and Transmission substations. It has a high insulation level.
- The potential transformer is also known as a voltage transformer. In this transformer, the primary winding can be connected across the HV (High voltage) line whose voltage is to be calculated, & all the instruments used for measuring & meters are allied to the secondary winding of the transformer.
- The current transformer (CT) is mainly used to gauge as well as also for safety. Once the current within the circuit is high to directly apply to the measuring instrument, this transformer is mainly used to change the high current into the preferred value of the current necessary within the circuit.

REVIEW QUESTIONS

1. What is the basic function of a transformer, and how does it operate?
2. Explain the difference between step-up and step-down transformers. Provide examples of where each type might be used.

3. Discuss at least three common applications of transformers in everyday life and industry.
4. Derive the EMF equation of an ideal transformer and explain the significance of each term.
5. What are the main types of transformer testing? Describe each type and its purpose.
6. Explain the importance of insulation testing in transformer maintenance and safety.

REFERENCES

1. Bedell, F. (1942). History of A-C wave form, its determination and standardization. *Transactions of the American Institute of Electrical Engineers, 61*(12), 864. https://doi.org/10.1109/T-AIEE.1942.5058456.
2. Chow, T. L. (2006). *Introduction to Electromagnetic Theory: A Modern Perspective* (p. 171). Jones and Bartlett Publishers. ISBN 978-0-7637-3827-3.
3. Hartley, W. H. (2003). Analysis of transformer failures. *36th Annual Conference of the International Association of Engineering Insurers* (p. 7). Archived from the original on October 20, 2013.
4. Horning, M., Kelly, J., Myers, S., & Stebbins, R. (2004). *Transformer Maintenance Guide* (3rd ed.). Transformer Maintenance Institute, S. D. Myers Inc.
5. Huber, W. R. (2022). *George Westinghouse: Powering the World* (p. 84). McFarland & Company. ISBN 9781476686929.
6. Kalita, P. K., Muduli, S. K., D'Antoni, L., Reps, T., & Roy, S. (2022). Synthesizing abstract transformers. *Proceedings of the ACM on Programming Languages, 6*(OOPSLA2), 1291–1319. https://doi.org/10.1145/3563334.
7. Knowlton, A. E. (Ed.). (1949). *Standard Handbook for Electrical Engineers* (8th ed., pp. 547–644). McGraw-Hill.
8. Kothari, D. P., & Nagrath, I. J. (2010). *Electric Machines* (4th ed.). Tata McGraw-Hill. ISBN 978-0-07-069967-0.
9. Kubo, T., Sachs, H., & Nadel, S. (2001). *Opportunities for New Appliance and Equipment Efficiency Standards* (p. 39, fig. 1). American Council for an Energy-Efficient Economy.
10. Landee, R. W., Davis, D. C., & Albrecht, A. P. (1957). *Electronic Designers' Handbook*. McGraw-Hill Book Company.
11. Nagy, Á. Z. (1996). Lecture to mark the 100th anniversary of the discovery of the electron in 1897 (preliminary text). Budapest. Archived from the original on November 25, 2012.
12. Parker, M. R., Ula, S., & Webb, W. E. (2005). Transformers & the ideal transformer. In J. C. Whitaker (Ed.), *The Electronics Handbook* (2nd ed., pp. 172, 1017). Taylor & Francis. ISBN 0-8493-1889-0.

13. Skilling, H. H. (1962). *Electromechanics* (p. 39). John Wiley & Sons, Inc.
14. Survilo, J., & Kutjuns, A. (2009). Operation modes of HV/MV substations. *Scientific Journal of Riga Technical University. Power and Electrical Engineering, 25*(25), 81–86.

CHAPTER

8 DC Machines

LEARNING OBJECTIVES

After studying this chapter, you will be able to:

- Understand the electromechanical energy conversion
- Describe the constructional features and classification of DC machines
- Explain the types of DC machines
- Understand about losses in DC machines
- Learn about advantages and applications of DC machines

KEY TERMS FROM THIS CHAPTER

Ac machines	AC motor	Armature
Armature winding	Brushes	Commutator
DC generator	DC motor	DC power
Electric machine	Electrical energy	Electrical motor
Electromagnetic force	Field System	Magnetic field
Mechanical energy	Pressure	Torque
Electromechanical energy conversion		

INTRODUCTION

Energy converters, which connect the electrical and mechanical energy systems, are revolving electrical machines. Electromechanical energy conversion is the continuous conversion of electrical energy to mechanical energy or vice versa. Thus, as Figure 8.1 illustrates, an electric machine is a connection between an electrical and mechanical system. The conversion process in these machines is reversible. The device is referred to as a generator if the conversion is from mechanical to electrical. When a machine converts from electrical to mechanical, it is referred to as a motor (Figure 8.1).

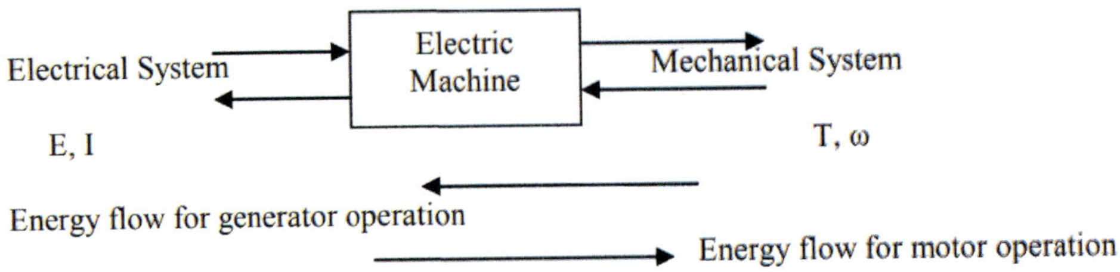

Figure 8.1. *Electromechanical energy conversion.*

Source: https://egyankosh.ac.in/bitstream/123456789/103885/1/unit%205.pdf.

The electrical machines can be classified as either dc or ac machines based on the kind of supply systems they are connected to. Single-phase, three-phase, or synchronous induction machines are the types of AC machines. Electrical devices that run on a direct current supply are categorized as dc devices. This chapter covers the topic of DC machines.

8.1. ELECTROMECHANICAL ENERGY CONVERSION

In any electromagnetic energy conversion process, energy is transferred between electrical and mechanical systems through a coupling medium, which in rotating electric machines is typically the magnetic field.

The principal electromagnetic phenomenon in dc machines are:

- When a conductor moves in a magnetic field, voltage is induced in the conductor, and
- Production of mechanical forces on current carrying conductors placed in a magnetic field.

8.1.1. Induced EMF

Any coil with dynamic magnetic flux linkages will eventually induce an electromagnetic field, according to Faraday's law of induction. The equation governing this law is;

$$e = \frac{Nd\varphi}{dt} = \frac{-d\lambda}{dt} \qquad (1)$$

Where; N is the number of turns of the coil; φ is the flux linking the coil in coil; λ is the flux linkage of the coil; and $\frac{d\varphi}{dt} =$ rate of change of flux

The negative sign indicates that the direction of the induced emf is to lessen the cause of the occurrence.

If a conductor of length l meters moves at the linear speed of V m/sec in a magnetic field of intensity B wb/m^2, than the magnitude of emf induced in the conductor is given by;

$$e = B\,l\,v \text{ volt} \qquad (2)$$

where vector B, l and v are mutually perpendicular. Fleming's right hand rule for emf and the right hand screw law both specify the polarity of induced emf. According to this rule, the middle finger of the right hand points in the direction of the induced electromagnetic field if the thumb, forefinger, and middle finger are held mutually at right angles with the forefinger pointing in the direction of magnetic field B and the thumb in the direction of motion (velocity).

8.1.2. Electromagnetic Force

The electromagnetic force F experienced by a conductor of length l meter carrying current I amp, placed in a magnetic field intensity of B wb/m^2 is given by;

$$F = B\,I\,l \text{ Newtons} \quad (3)$$

Force F, magnetic field B, and current I are all represented by vectors that are perpendicular to one another. The left hand rule indicates the direction of force F that the conductor is experiencing. According to this rule, the thumb indicates the direction of the force or motion if the thumb, forefinger, and middle finger are held perpendicular to each other, the forefinger pointing in the direction of the magnetic field, and the middle finger pointing in the direction of current flow.

8.1.3. Back EMF and Counter Torque

The armature windings of a generator induce electromagnetic fields. Eqn. (1) provides the value of the induced emf in the DC generator, which is determined by Faraday's Laws. In a similar manner, a DC motor draws current from its terminal after receiving a DC supply. Therefore, a DC motor's magnetic field also contains a current-carrying conductor that is rotating. Consequently, a dc motor's armature winding also experiences emf induction. This emf is roughly the same as the applied motor terminal voltage in terms of magnitude and direction. Therefore, this induced emf in a DC motor's armature winding is referred to as the back or counter emf. The armature resistance drop, brush voltage drop, and losses are the causes of the variation between the applied terminal voltage and back emf.

Motoring torque, which drives the mechanical load on the machine's shaft, is produced in an electrical motor by forces acting on current-carrying conductors in a magnetic field (Hughes Austin & Drury Bill, 2019). A mechanical torque will be produced even in a generator because the armature conductors are located in a magnetic field and carry current. But this torque will act in the opposite direction of the generator's prime mover torque. Consequently, this type of torque is known as a counter torque (William H. Yeadon & Alan W. Yeadon, 2001).

8.2. DC MACHINES

An electromechanical energy alteration device is called a DC machine. A DC machine rotates a DC motor by means of torque produced by the magnetic force acting on an electric current flowing through a coil in a magnetic field. The DC machines are divided into two categories: DC motors and DC generators (Wilson & Tricky, 1962).

A DC motor converts DC power into mechanical power, while the primary purpose of a DC generator is to convert mechanical power into DC electrical power. Industrial applications usually use AC motors to convert electrical energy to mechanical energy. On the other hand, DC motors are useful in situations requiring a wide range of speeds and effective speed regulation, such as electric-transaction systems.

DC machines can function as DC motors or generators. DC generators are no longer necessary because rectifiers are now used to produce DC. However, these are still in use in a few unique circumstances and applications. However, DC motors are still useful and the preferred machine for many tasks. Sometimes knowledge of dc motors requires knowledge of dc generator principles. As a result, the constructional features of a DC machine that can be utilized as a DC motor or generator will be briefly discussed in this section.

8.2.1. Construction of a DC Machine

The main parts of a dc machine are

- Armature – Rotating part
- Field – Stationary part
- Commutator – Rotating part
- Brush Assembly – Stationary part

A DC machine's electrical circuit consists of brush, armature and field winding, commutator, and terminals for the armature and field. The armature, field poles, yoke, and core are all connected by the magnetic circuit. Figures 8.2 and 8.3 show a cross section view of a typical DC machine showing its various parts.

(a) Stator

(b) Rotor

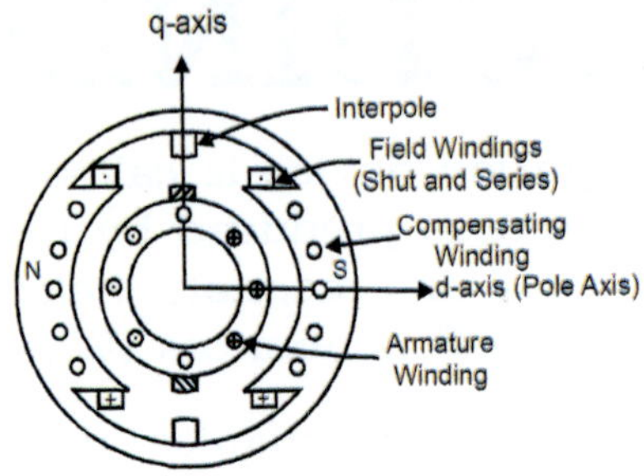

(c) Schematic Cross-sectional View for a 2-pole Machine

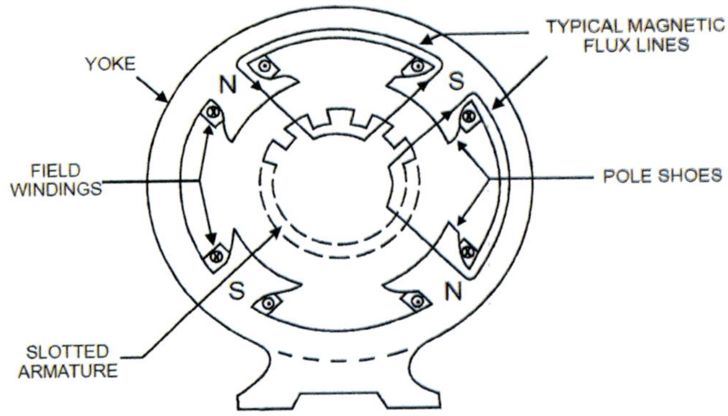

(d) Field System

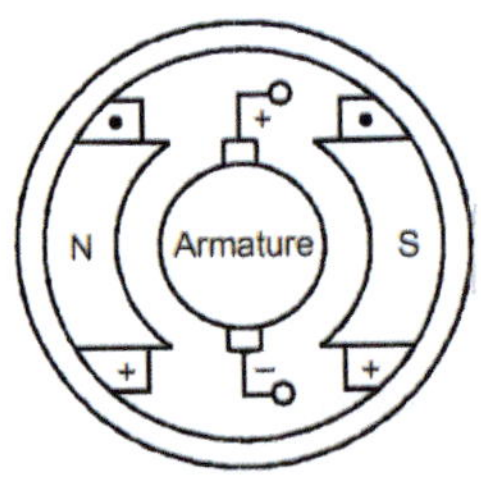

(e) Interpole, Armature etc.

Figure 8.2. *Different views of DC machine.*

Source: https://egyankosh.ac.in/bitstream/123456789/103885/1/unit%205.pdf.

i. **Armature:** The term "armature core," or simply "armature," refers to the primary rotating component of a DC machine. An armored core consists of a stack of laminations with slots around the outside. There is armature winding in these slots. There are venting ducts available to cool the armature. A cylinder is created by stacking the steel armature core laminations, as seen in Figure 8.3. Conductors of the armature winding are carried by axial slots on the armature's perimeter. Laminations are sealed off from the rotor iron and from one another. A wood or fibrous material wedge holds the conductors in place in the slot.

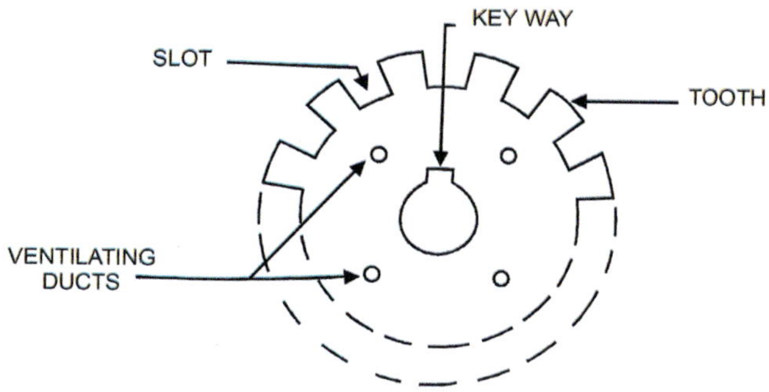

Figure 8.3. *Armature lamination of DC machine.*

Source: https://egyankosh.ac.in/bitstream/123456789/103885/1/unit%205.pdf.

When using a dc generator, the current is sent from the generator to the external electrical load circuit through the conductors, where the emf is induced in the armature winding.

When a dc motor is operating, the armature winding's current-carrying conductors exert force, which causes electromagnetic torque to develop and the rotor (armature) to rotate.

ii. **Field System:** Field system produces magnetic field. This is also known as the starter and is typically a stationary component of the machine. A DC machine's field winding is mounted on field poles. These field poles, also known as yoke, are projecting poles on the inner rim of the stator frame. The poles are composed of steel laminations, and the yoke is composed of solid cast steel. A magnetic field is created by dc field current flowing through field coils that are fixed on field poles. Between the rotor's cylindrical iron surface and external pole shoe, there is an air gap. Flux is established by the field current and flows through the armature, yoke, pole shoe, and air gap. Figure 8.2 displays the flux path and the various components of a DC machine.

iii. **Commutator and Brush:** A DC machine's armature winding experiences alternating induced emf. This alternating voltage is changed to direct current, or dc, by the commutator. The commutator is composed of several tapered copper segments that are mounted on the shaft and are insulated from one another by mica sheets. The commutator segments mounted on the shaft are seen in Figure 8.4. The two ends of the coils are connected to two commutator segments, and the number of coils on the armature equals the number of commutator segments.

Carbon brushes, which ride on commutator segments, lead current into and out of the armature conductors. Two fixed brush sets are placed diametrically opposed to one another on two pole machines. The number of brush sets in machines with four or more poles is equal to the number of poles. These brushes surround the commutator at equal intervals. The polarities of adjacent brush sets alternate between being positive and negative. Positively polarized brushes are brought out to the positive terminal after being connected to one another. Similar to this, brushes with opposite polarity are linked to the negative terminal. We refer to these two terminals as armature terminals.

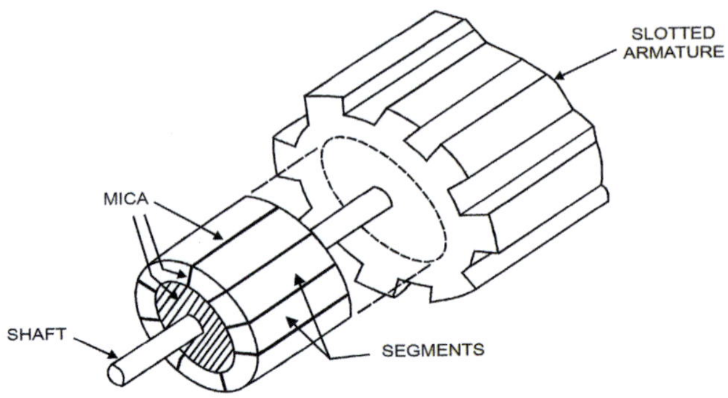

Figure 8.4. *Commutator segment mounted on a shaft.*

Source: https://egyankosh.ac.in/bitstream/123456789/103885/1/unit%205.pdf.

The commutator rotates in tandem with the armature, allowing the segments to brush past the stationary brushes. Carbon is used to make the brushes because it is soft enough and has good electrical conductivity, which prevents wear on the commutator. Brush holders hold the brushes in place and use adjustable springs to press them up against the commutator. One can manually change the pressure. Inadequate pressure causes sparking and imperfect contact. Over pressurized conditions lead to increased friction, heating, and premature brush wear.

iv. **Armature Winding:** On the rotor of a DC machine is the armature winding. The following is a definition of some common terms used in armature winding. A turn is made up of two conductors that are joined at one end by a connector. Forming a coil involves joining multiple turns in series. A winding is created by joining multiple coils in series. Figure 8.5 illustrates the turn, coil, and winding. S indicates the start of a turn, and F indicates its conclusion.

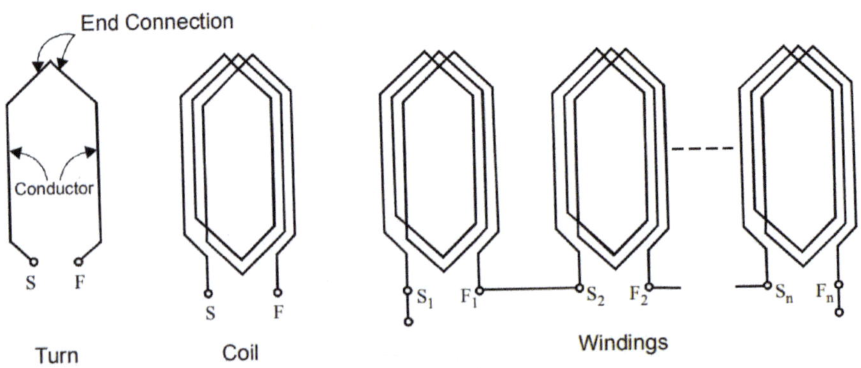

Figure 8.5. *Turn, coil and winding of DC machine.*

Source: https://egyankosh.ac.in/bitstream/123456789/103885/1/unit%205.pdf.

The armature winding's coils can be connected to one another in several different ways. Armature coils connected by lap winding and wave winding are the most popular types of interconnections.

When it comes to lap winding, the number of parallel paths (a) and the number of brushes (P) are always equal. Large current machines frequently use this kind of connection. Figure 8.6 depicts a typical lap winding schematic along with the corresponding coil representation.

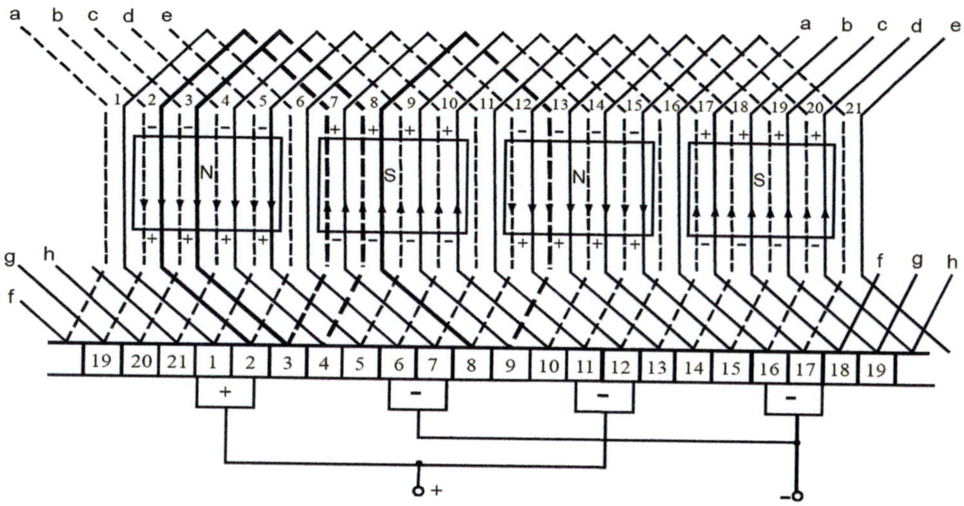

(a) Unrolled Winding

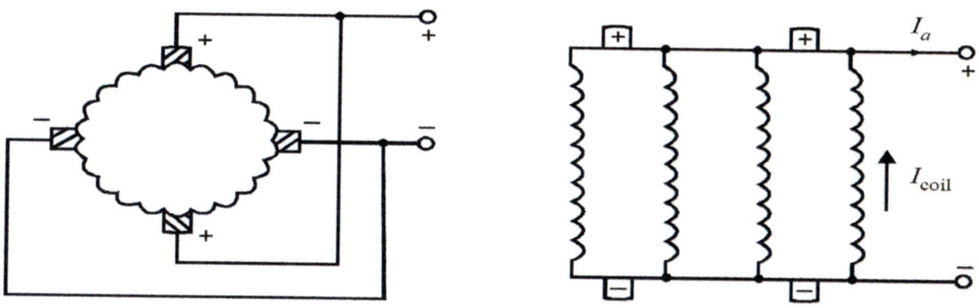

(b) Equivalent Coil Representation

Figure 8.6. *Lap winding.*

Source: https://egyankosh.ac.in/bitstream/123456789/103885/1/unit%205.pdf.

The number of parallel paths "a" in the case of wave windings is always two. Usually, high voltage, low current rated DC machines employ this scheme. Figure 8.7 displays a typical wave winding schematic along with the corresponding coil representation.

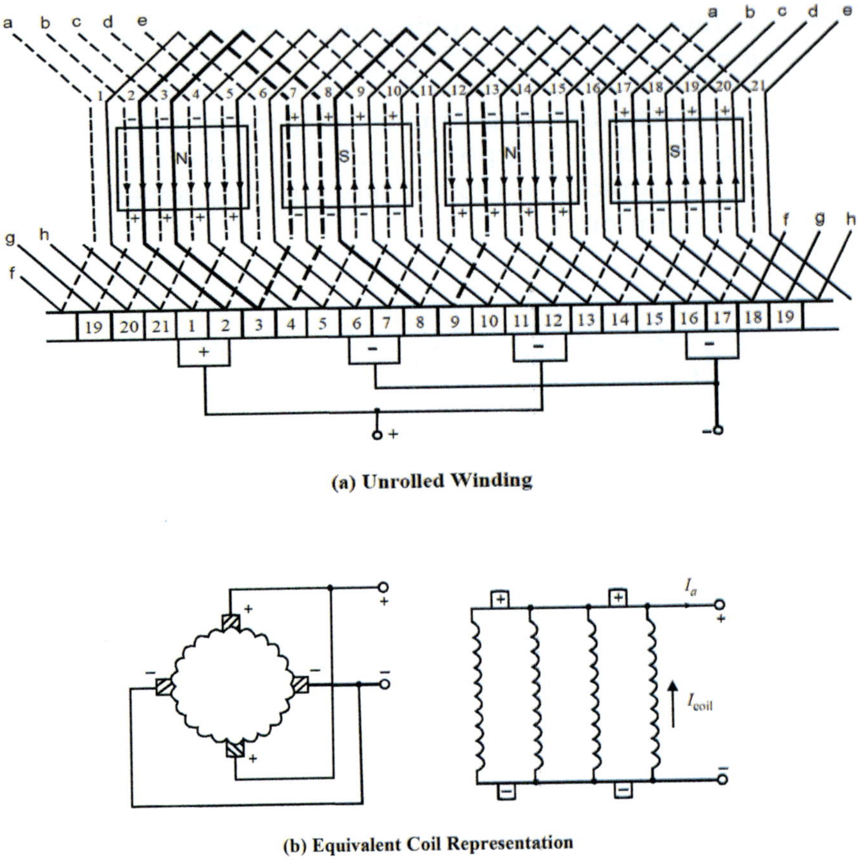

Figure 8.7. *Wave winding.*

Source: https://egyankosh.ac.in/bitstream/123456789/103885/1/unit%205.pdf.

8.2.2. Armature Voltage

As the armature rotates in the magnetic field produced by the starter poles, emf is induced in the armature winding.

Let P = number of poles,

Z = total number of armature conductors;

N = armature speed in rpm;

φ = flux per pole in wb; and

A = number of parallel paths through the armature winding.

The average emf induced in the conductor is given by;

$$E_{av} = \frac{\text{flux cut}}{\text{Time of cutting the flux}} = \frac{d\phi}{dt} \quad (4)$$

Since the armature is rotating at N rpm (= N/60 rps), time for one revolution is 60/N secs. Time taken for a conductor to move one pole pitch (i.e., 1/P of a circumference) or 1/P of a revolution will be $\Delta t = \left(\frac{1}{P}\right)\left(\frac{60}{N}\right)$ secs. (5)

Average emf induced in a conductor

$$E_{av} = \frac{\left(\frac{\phi}{P}\right)}{\left(\frac{60}{N}\right)} = \frac{NP\,\phi}{60} \text{ volts} \quad (6)$$

$$E_{av} = \frac{\phi\, NP}{60} \text{ volts} \quad (7)$$

There are a total of Z conductors arranged in A parallel paths on the armature. In other words, number of conductors in series per parallel path is Z/A. Thus, the actual emf induced in one parallel path (and, therefore, for the whole winding) is given by;

$$E = \left\{\frac{NP\,\phi}{60}\right\}\left\{\frac{Z}{A}\right\} \quad (8)$$

$$E = \frac{NPZ\,\phi}{60\,A} \text{ Volts} \quad (9)$$

A = 2 for wave winding, and

A = P for lap winding.

For a given dc machine, Z, P and A are constants, so that, emf induced

$$E \propto N\phi \quad (10)$$

or

$$E_{av} = K_a\, N\phi = K_a\, \omega\, \phi \quad (11)$$

Induced armature emf is dependent on rotational speed and flux (i.e., field current). Whether the machine runs as a generator or a motor has no bearing on this expression of induced emf. When referring to a generator, it is called generated emf (or voltage), and when it comes to motor operation, it is called back emf.

8.2.3. Electromagnetic Torque

There are various methods by which an expression for torque can be developed.

Let r = average radius of armature

I_a = armature current

L = effective length of each conductor

Current in each conductor $i = \frac{I_a}{A}$

Average flux density $B = \frac{\phi}{\text{area}}$

$\text{Area} = \frac{2\pi\, r\, l}{P} =$ Cross sectional area of flux path at radius r

$$\therefore \quad B = \frac{\phi\, P}{2\pi\, r\, l}$$

Force on each conductor F = Bil

Torque on one conductor T = F . r Nm

Total armature torque $T = Z.F.r$

$$= Z\,(\text{Bil})\,r$$

$$= Z\left\{\left(\frac{\phi P}{2\pi r l}\right)\left(\frac{I_a}{A}\right)1\right\}r$$

$$T = Z\,(\phi\, I_a)\left(\frac{P}{2\pi A}\right)$$

$$\left(\frac{1}{2\pi}\right)Z\,(\phi\, I_a)\left(\frac{P}{A}\right)$$

$$T = 0.159\, Z\,\phi\, I_a\left(\frac{P}{A}\right)\text{ Nm} \quad (12)$$

Hence, torque developed by the armature of a dc machine depends upon flux and armature current. Flux depends upon field current. For a dc machine Z, P and A are constants, therefore $T \propto \varphi I_a$.

Example 8.1

Calculate the emf generated by a 4-pole wave wound armature having 45 slots with 18 conductors per slot when driven at 1,200 rpm. The flux per pole is 0.016 wb.

Solution

Let number of armature conductors be Z. Then

Z = No. of slots × No. of conductors per slot

= 45 × 18

= 810

Given

P = 4

A = 2 (wave winding)

Speed N = 1200 rpm

Flux φ = 0.016 wb

Emf generated

$$E = \left(\frac{NPZ\,\phi}{60\,A}\right)\text{ volt} \quad (13)$$

$$E = \frac{(1200 \times 4 \times 810 \times 0.016)}{(60 \times 2)} \quad (14)$$

$$E = 518.4 \text{ volts} \quad (15)$$

Example 8.2

The armature of a 4-pole dc machine has a lap winding accommodated in 60 slots, each slot having 16 conductors. The flux per pole is 0.06 wb. What is the total torque developed by the motor when the armature current is 500 amps.

Solution

Given

φ = 0.06 wb

I_a = 500 A

P = 4

Z = 60 × 16

= 960 conductors

A = 4 (lap)

$$\text{Torque} = 0.159\,\phi\, Z\left(\frac{P}{I}\right)I_a\text{ Nm}$$

$$= 0.159 \times 0.06 \times 960 \times \frac{4}{4}\,500$$

$$= 458\text{ Nm} \quad (16)$$

8.2.4. Classification of DC Machines

One great benefit of dc machines is their ability to link their armature and field windings in different ways to offer a wide range of performance characteristics. Field windings provide energy to the field poles. These field windings can be connected in series with the armature winding, which is known as a series field winding, or they can be connected across the armature winding, which is known as a shunt field winding. A shunt field winding is one that is connected across the armature voltage, has a higher resistance, more turns, and draws a smaller current (less than 5% of the rated armature current). Because the series field winding must supply the full rated armature current, it has a low resistance and few turns. A DC machine that has both series and shunt field windings will have either a differentially or cumulatively compound connection between the armature and field winding.

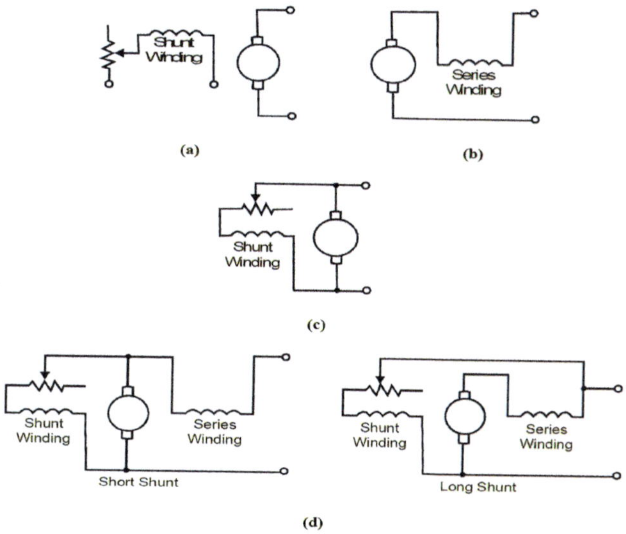

Figure 8.8. *Different connections of DC machines (a) separately excited Dc machine; (b) series DC machine; (c) Shunt DC machine; and (d) Compound DC machine.*

Source: https://egyankosh.ac.in/bitstream/123456789/103885/1/unit%205.pdf.

When a DC machine's field winding receives stimulation from an independent DC source, the device is referred to as a separately excited DC machine. Conversely, a self-excited DC machine is a DC machine that generates its own field current. It is a shunt machine if the field winding is positioned across the armature winding of the self-excited DC machine.

Similar to this, a series machine's armature and field windings are connected in series. The compound machine can be connected as a long or short shunt and has a series and shunt field. It is a cumulatively compounded machine if the shunt winding mmf is assisted by the series winding mmf. It is a differentially compounded DC machine if the series winding mmf is in opposition to the shunt winding mmf. Figure 8.8 depicts the various connections and classifications of DC machines that were previously discussed.

In order to control the filed current and alter the field mmf for voltage or speed control, a rheostat is typically included in the circuit of the shunt field.

Permanent magnets can also be used to provide field excitation. This could be a type of continuously mmf machine that is separately excited. The functioning and features of several DC machines—first as generators, then as motors—will be examined in the ensuing sections.

8.3. TYPES OF DC MACHINES

A DC machine is a device that transforms mechanical energy into electrical energy in the form of direct current or electrical energy into mechanical energy. Consequently, an electromechanical energy conversion device is essentially what a DC machine is.

Based on the energy conversion, DC machines can be classified into the following two types:

- DC Motor
- DC Generator

Both DC motors and DC generators share a nearly identical basic construction. However, the underlying ideas that drive the functioning of a DC generator and motor are distinct.

In essence, every DC machine is made up of a system of conductors and an electromagnet system. In DC machines, the conductor system is referred to as the armature and is fixed to a movable shaft. The field system, which consists of magnets or electromagnets, generates the necessary working magnetic flux.

8.3.1. DC Motor

The block diagram of a DC motor is shown in Figure 8.9.

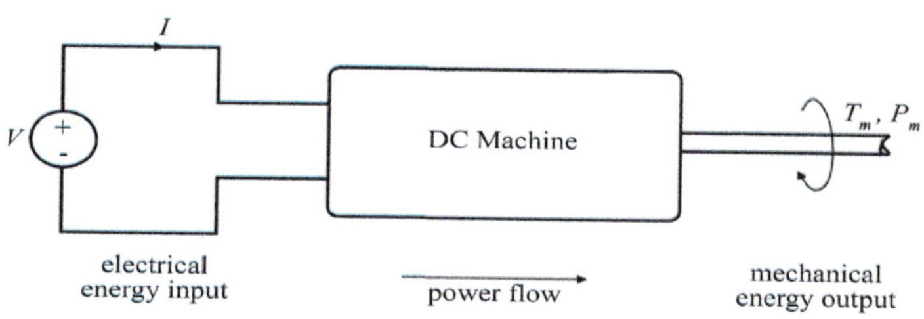

Figure 8.9. *DC machine as motor.*

Source: https://www.tutorialspoint.com/electrical_machines/electrical_machines_types_of_dc_machines.htm.

A DC motor is a DC machine that is intended to transform DC electrical energy into mechanical energy through rotation. Consequently, in the case of a DC motor, mechanical energy is extracted from the shaft in the form of shaft rotation, and electrical energy is supplied to the device through input terminals.

8.3.2. DC Generator

The block diagram of a DC generator is shown in Figure 8.10.

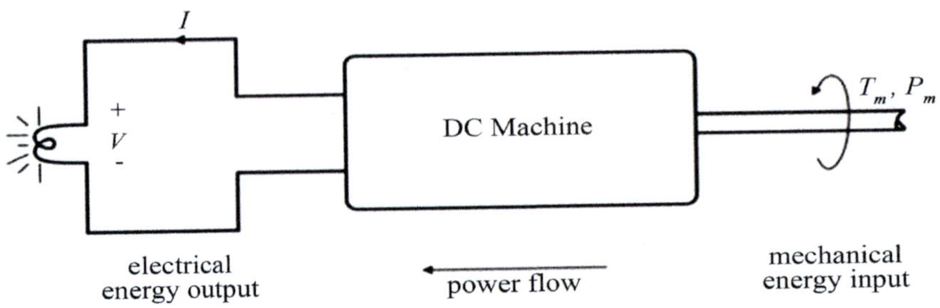

Figure 8.10. *DC machine as generator.*

Source: https://www.tutorialspoint.com/electrical_machines/electrical_machines_types_of_dc_machines.htm.

A DC generator is a device that can transform mechanical energy input into electrical energy output. As a result, mechanical energy from an engine, turbine, etc. can be used in a DC generator. is delivered to the DC machine as the shaft's rotational energy, and the armature terminals provide the DC electrical energy as an output.

8.4. LOSSES IN DC MACHINES

In DC machines (generator or motor), the losses may be classified into three categories namely,

- Copper losses
- Iron or core losses
- Mechanical losses

All these losses appear as heat and hence raise the temperature of the machine. They also reduce the efficiency of the machine.

8.4.1. Copper Losses

Copper losses are the losses that happen in DC machines as a result of the resistance of the machine's various windings. Since the copper losses are caused by current passing through the windings' resistance, they are also referred to as I^2R losses.

The major copper losses that occur in dc machines are as;

$$\text{Armature copper loss} = I_a^2 R_a \quad (17)$$

$$\text{Series field copper loss} = I_{se}^2 R_{se} \quad (18)$$

$$\text{Shunt field copper loss} = I_{sh}^2 R_{sh} \quad (19)$$

Brush contact resistance causes brush contact loss in DC machines as well. This loss is typically included in the armature copper loss in practical calculations.

8.4.2. Iron Losses

Because the armature rotates in the magnetic field, iron losses happen in the armature core of DC machines. We also refer to these losses as core losses because they take place in the armature's core. Iron or core losses come in two flavors: eddy current loss and hysteresis loss.

8.4.3. Hysteresis Loss

Hysteresis loss is the term used to describe the core loss that happens in the armature core of a DC machine as a result of magnetic field reversal in the armature core as it passes under successive magnetic poles with different polarities. The hysteresis loss is given by the following empirical formula,

$$\text{Hysteresis loss}, P_h = k_h B_{max}^{1.6} f V \quad (20)$$

Where, k_h is the Steinmetz's hysteresis coefficient, B_{max} the maximum flux density, f is the frequency of magnetic reversal, and V is the volume of armature core.

By using materials like silicon steel for the armature core, which have a low value of Steinmetz's hysteresis coefficient, the hysteresis loss in DC machines can be decreased.

8.4.4. Eddy Current Loss

An electromagnetic field (EMF) is created in the armature core of a DC machine when it rotates in the magnetic field of the poles, causing eddy currents to flow through it. Eddy current loss is the term for the power loss brought on by these eddy currents. The eddy current loss is given by,

$$\text{Eddy current loss}, P_e = k_e B_{max}^2 f^2 t^2 V \quad (21)$$

Where, K_e is a constant of proportionality, and t is the thickness of lamination.

It is evident from the eddy current loss expression that the eddy current loss is dependent on the square of the lamination thickness. Thus, the armature core is constructed from thin laminations separated from one another by a thin varnish layer in order to minimize this loss.

8.4.5. Mechanical Losses

Mechanical losses are the power losses in a DC machine caused by windage and friction. The friction loss in a DC machine takes the form of brush friction, bearing friction, etc. whereas the spinning armature's air friction causes windage loss. The machine's speed affects the mechanical losses. But for a certain speed, these losses are essentially constant.

Note: Iron or core losses and mechanical losses together are known as stray losses.

8.4.6. Constant and Variable Losses

In DC machines, we may group the above discussed losses in the following two categories:

- Constant Losses
- Variable Losses

Constant losses are those losses in a DC machine that don't change regardless of the load. – iron losses, shunt field copper loss, and mechanical losses are some of these losses.

Variable losses are those losses in a DC machine that change depending on the load. In a DC machine, the two variable losses are the series field copper loss and the armature copper loss.

Total losses in a DC machine = Constant losses + Variable losses

8.5. DC MACHINE ADVANTAGES

The advantages of this machine include the following.

- DC machines like dc motors have various advantages like starting torque is high, reversing, fast-starting & stopping, changeable speeds through voltage input
- These are very easily controlled as well as cheaper when compared with AC
- Speed control is good
- Torque is high
- Operation is Seamless
- Free from harmonics
- Installation and maintenance is easy

8.6. APPLICATIONS OF DC MACHINES

Currently, alternating current is the primary source of electrical energy generation. As a result, there is now less use of DC machines (motors or generators). Nevertheless, dc machines are still employed in a number of processes, such as welding, electrolysis, and providing excitation systems for alternators.

8.6.1. Applications of DC Generators

The applications of different kinds of DC generators are listed below:

- Separately-Excited DC Generator
  - Generators are mainly used in laboratories for experiments and testing purposes.
  - They are also used as a source of DC power for supplying DC motors.
- Series DC Generator
  - Series DC generators are used in DC locomotives for regenerative braking for supplying field excitation current.

- Series DC generators are also used as a booster in distribution systems.
- Differentially compounded series generators are used for arc welding.
- Over compounded cumulative series generators are used for lighting and heavy power supplies.
- Flat compounded series generators are used for supplying power to offices, houses, and the other commercial buildings.

- Shunt DC Generator
  - Shunt DC generators are mainly used for lighting purposes.
  - Shunt DC generators are used for charging batteries.
  - These generators are also used for supplying field excitation current to alternators.

8.6.2. Applications of DC Motors

The applications of different types of DC motors are given below:

- Series DC Motor
  - Series DC motors are used in applications where high starting torque is required.
  - Series DC motors are used in cranes and hoists.
  - Series DC motors are used in electric tractions.
  - They are used in air compressors.
  - They are also used in vacuum cleaners.
  - Series DC motors are also used in sewing machines, etc.
- Shunt DC Motor
  - Shunt DC motors are used in applications that require constant speed.
  - Shunt DC motors are used for driving lathe machines.
  - These are also used in centrifugal pumps and blowers.
  - These are used in fans, conveyors, and spinning machines.
  - These DC motors are also used in lifts.
- Compound DC Motor
  - Compound DC motors are used in those applications which require high starting torque and constant speed.
  - Used in printing presses.
  - They are also used in shears.
  - They are used in elevators and lifts.
  - Compound DC motors are also used in rolling mills and heavy planners, etc.

ROLE MODEL

WILLIAM STURGEON

William Sturgeon (22 May 1783 – 4 December 1850) was an English physicist and inventor who made the first electromagnets, and invented the first practical electric motor.

Source: https://encrypted-tbn0.gstatic.com/images?q=tbn:ANd9GcSmt3uPg6GKWZdcFWFp-WihfS3H9i2gQ6limQ&s.

Early Life:

Sturgeon was born on 22 May 1783 in Whittington, near Carnforth, Lancashire, and became apprenticed to a shoemaker.

Career:

Sturgeon joined the army in 1802 and taught himself mathematics and physics. In 1824 he became lecturer in Science and Philosophy at the East India Company's Military Seminary at Addiscombe, Surrey, and in the following year he exhibited his first electromagnet. He displayed its power by lifting nine pounds with a seven-ounce piece of iron wrapped with wire through which a current from a single battery was sent.

In 1832 he was appointed to the lecturing staff of the Adelaide Gallery of Practical Science in London, where he first demonstrated the DC electric motor incorporating a commutator.

In 1836 he established the journal *Annals of Electricity, Magnetism and Chemistry,* and in the same year he invented a galvanometer.

Sturgeon was a close associate of John Peter Gassiot and Charles Vincent Walker, and the three were instrumental in founding the London Electrical Society in 1837.

In 1840 he became superintendent of the Royal Victoria Gallery of Practical Science in Manchester. He formed a close social circle with John Davies, one of the Gallery's promoters, and Davies's student James Prescott Joule, a circle that eventually extended to include Edward William Binney and the surgeon John Leigh. The Gallery closed in 1842, and he earned a living by lecturing and demonstrating.

In 1843 he started the monthly journal, The Annals of Philosophical Discovery and Monthly Reporter of the Progress of Practical Science; issue 1 of volume 1 is dated July 1843. Each month's issue contains a mixture of original "long" papers (over 5 pages long), republished papers from foreign journals (translated where necessary) and shorter articles. However, the journal did not prove successful, and ceased publication at the end of volume 1, in December 1843. This single volume is archived at Internet.org.

Death and Burial:

Sturgeon died in Prestwich, Lancashire (now in Greater Manchester) on 4 December 1850. He is buried there, in the churchyard of the St Mary the Virgin: he is identified on his grave slab as "William Sturgeon – The Electrician."

CLASS ACTIVITY

The armature of a 6 pole dc machine 0.75 meter in diameter has 664 conductors each having an effective length of 0.30 meter and carrying a current of 100 A. If 70% of total conductors lie simultaneously in field of average flux density of 0.85 Wb/m2, student calculate the armature torque developed.

SUMMARY

- Rotating electrical machines are energy converters, which link electrical energy system to mechanical energy system. The process of continuous translation of electrical energy to mechanical energy or vice versa is called electromechanical energy conversion.
- A DC machine is an electromechanical energy alteration device. The working principle of a DC machine is when electric current flows through a coil within a magnetic field, and then the magnetic force generates a torque that rotates the dc motor. The DC machines are classified into two types such as DC generator as well as DC motor.
- The main function of the DC generator is to convert mechanical power to DC electrical power, whereas a DC motor converts DC power to mechanical power. The AC motor is frequently used in industrial applications for altering electrical energy to mechanical energy. However, a DC motor is applicable where good speed regulation & an ample range of speeds are necessary like in electric-transaction systems.
- The main rotating part of a dc machine is called as armature core or simply armature. Armature core is a stack of laminations with slots at its external periphery. These slots carry armature winding. Ventilating ducts are provided for cooling of armature.
- Magnetic field is produced by field system. This is normally a stationary part of the machine and is also referred to as starter. The field winding of a dc machine is placed on field poles. These field poles are projecting poles on the inner periphery of stator frame, also called as yoke. Yoke is made of solid cast steel and poles are made up of steel laminations.
- The induced emf in the armature winding of a dc machine is alternating in nature. The commutator converts this alternating voltage to unidirectional or direct current or dc. Commutator is made up of a number of tapered copper segments insulated from each other by mica sheets and mounted on the shaft and insulated from it.
- Armature winding of a dc machine are placed on rotor. Some basic terms related to armature winding are described below. A turn consists of two conductors connected to one end by the connector. Connecting several turns in series forms a coil. Several coils connected in series forms a winding.
- When the DC machine is designed to convert DC electrical energy into rotational mechanical energy, it is called a DC motor. Therefore, in case of a DC motor, the electrical energy is supplied to the machine through input terminals and mechanical energy output is taken from the shaft in the form of rotation of the shaft.
- A DC machine which can convert mechanical energy input into electrical energy output is known as a DC generator. Thus, in a DC generator, the mechanical energy from a source like engine, turbine, etc. is supplied to the DC machine in the form of rotational energy of the shaft and the DC electrical energy is received as output from the armature terminals.

REVIEW QUESTIONS

1. What is a DC Machine?
2. List the parts of a DC machine.
3. Name the two types of DC machines.
4. What are the classification of losses in DC machines?
5. What are the advantages of DC machine?

REFERENCES

1. Acarnley, P. P., & Al-Tayie, J. K. (1997). Estimation of speed and armature temperature in a brushed DC drive using the extended Kalman filter. *IEE Proceedings–Electric Power Applications, 144*(1), 13–20. https://doi.org/10.1049/ip-epa:19970927.
2. Herman, S. L. (2009). *Electric Motor Control* (9th ed.). Delmar, Cengage Learning.
3. Hughes, A., & Drury, B. (2019). *Electric Motors and Drives: Fundamentals, Types and Applications* (5th ed.). Newness. ISBN 978-0-08-102615-1.
4. Wilson, T. G., & Trickey, P. H. (1962). D.C. machine with solid state commutation. AIEE paper I. CP62-1372.
5. Yeadon, W. H., & Yeadon, A. W. (2001). *Handbook of Small Electric Motors*. McGraw-Hill Professional.

CHAPTER

9

Three Phase Induction Motors

LEARNING OBJECTIVES

After studying this chapter, you will be able to:

- Understand the overview of three phase induction motor
- Discuss on single phase induction motors
- Explain the special machines

KEY TERMS FROM THIS CHAPTER

AC electric motor
Auxiliary winding
Electromotive force (EMF)
Induction motor
Rotating Magnetic Field
Stator Casing
Stator Winding
Variable-frequency drives (VFD)
Asynchronous motor
Electromagnetic induction
Faraday's law
Magnetic field
Single-phase induction motor
Stator Core
Three-phase induction motor
Winding

INTRODUCTION

An induction motor, also known as an asynchronous motor, is an AC electric motor where the electric current in the rotor is created through electromagnetic induction from the stator winding's magnetic field. This eliminates the need for electrical connections to the rotor. The rotor of an induction motor can be either wound type or squirrel-cage type. Three-phase squirrel-cage induction motors are commonly used in industrial settings due to their self-starting capability, reliability, and cost-effectiveness. Single-phase induction motors are more commonly used for smaller loads like garbage disposals and stationary power tools. While traditionally used for constant-speed operations, both single- and three-phase induction motors are now being utilized in variable-speed applications through the use of variable-frequency drives (VFD). VFDs provide energy-saving opportunities for induction motors in applications with variable loads, such as fans, pumps, and compressors.

9.1. OVERVIEW OF THREE PHASE INDUCTION MOTOR

A three-phase induction motor is a specific type of AC induction motor that operates on a three-phase power supply, as opposed to a single-phase induction motor which operates on a single-phase power supply. The three-phase supply current creates an electromagnetic field in the stator winding, resulting in the generation of torque in the rotor of the motor due to the magnetic field.

9.1.1. History

French physicist François Arago first proposed the concept of rotating magnetic fields in 1824, which later became known as Arago's rotations. Walter Baily further demonstrated this phenomenon in 1879 by manually controlling switches, creating what is considered the first early form of an induction motor (Thompson Silvanus Phillips, 1895). Hungarian engineer Ottó Bláthy is credited with inventing the first single-phase AC induction motor without a commutator. Bláthy utilized this motor to power his creation, the electricity meter.

The first AC commutator-free polyphase induction motors were independently invented by Galileo Ferraris and Nikola Tesla, a working motor model having been demonstrated by the former in 1885 and by the latter in 1887. Tesla applied for US patents in October and November 1887 and was granted some of these patents in May 1888. In April 1888, the Royal Academy of Science of Turin published Ferraris's research on his AC polyphase motor detailing the foundations of motor operation. In May 1888 Tesla presented the technical paper A New System for Alternating Current Motors and Transformers to the American Institute of Electrical Engineers (AIEE) describing three four-stator-pole motor types: one having a four-pole rotor forming a non-self-starting reluctance motor, another with a wound rotor forming a self-starting induction motor, and the third a true synchronous motor with a separately excited DC supply to the rotor winding (Alger, & Arnold, 1976).

George Westinghouse, who was developing an alternating current power system at that time, licensed Tesla's patents in 1888 and purchased a US patent option on Ferraris' induction motor concept. Tesla was also employed for one year as a consultant. Westinghouse employee C. F. Scott was assigned to assist Tesla and later took over development of the induction motor at Westinghouse. Steadfast in his promotion of three-phase development, Mikhail Dolivo-Dobrovolsky invented the cage-rotor induction motor in 1889 and the three-limb transformer in 1890. Furthermore, he claimed that Tesla's motor was not practical because of two-phase pulsations, which prompted him to persist in his three-phase work. Although Westinghouse achieved its first practical induction motor in 1892 and developed a line of polyphase 60 hertz induction motors in 1893, these early Westinghouse motors were two-phase motors with wound rotors until B. G. Lamme developed a rotating bar winding rotor.

The development of three-phase induction motors by the General Electric Company (GE) began in 1891, leading to significant advancements in the industry. In 1896, General Electric and Westinghouse collaborated on a cross-licensing agreement for the bar-winding-rotor design, later known as the squirrel-cage rotor. Arthur E. Kennelly was the first to recognize the importance of using complex numbers (using j for the square root of minus one) to analyze AC problems, specifically the concept of a 90° rotation operator. Charles Proteus Steinmetz, an engineer at GE, further refined the application of AC complex quantities and introduced an analytical model called the induction motor Steinmetz equivalent circuit. Thanks to these innovations, a modern 100-horsepower induction motor today shares the same mounting dimensions as a 7.5-horsepower motor back in 1897.

9.1.2. Faraday's Law of Induction Motor

The theory of electromagnetic induction, developed by Michael Faraday, explains the key principles behind the operation of a three-phase induction motor. It shows how a changing magnetic field relates to the generation of an electromotive force (EMF).

The basic idea of Faraday's regulation is summarized as follows:

Whenever there is a change in the magnetic field surrounding a conductor, it generates an electromotive force (EMF) within the conductor. The magnitude of the induced EMF is directly proportional to the rate of change of magnetic flux passing through the conductor.

- **Rotating Magnetic Field:** In a 3-phase induction motor, three arrangements of windings are set 120° separated in space. At the point when a three-phase AC voltage is applied to these windings, it delivers a rotating magnetic field.
- **Interaction with Rotor Conductors:** The rotating magnetic field induces an EMF in the rotor conductors because of Faraday's law. As the magnetic field rotates, it cuts across the rotor conductors, causing an adjustment of attractive motion.
- **Generation of Current in Rotor:** The induced EMF in the rotor conductors causes the progression of current in the rotor. This current in the rotor conductors connects with the magnetic field, creating a torque that makes the rotor turn.
- **Rotor Movement and Synchronization:** The rotor begins to follow the rotating magnetic field, endeavoring to find its rotational speed. The asynchronous development (slip) between the rotor and the turning magnetic field permits the motor to ceaselessly create torque and rotate.

- **3 Phase Power:** Three-phase power is a form of electrical power distribution in which three sinusoidal voltages of the same frequency are generated or used. It is a common technique used in electrical and industrial systems, as well as in large electrical motors. The three-phase system is more efficient and provides a more stable power supply compared to a single-phase system (Figure 9.1).

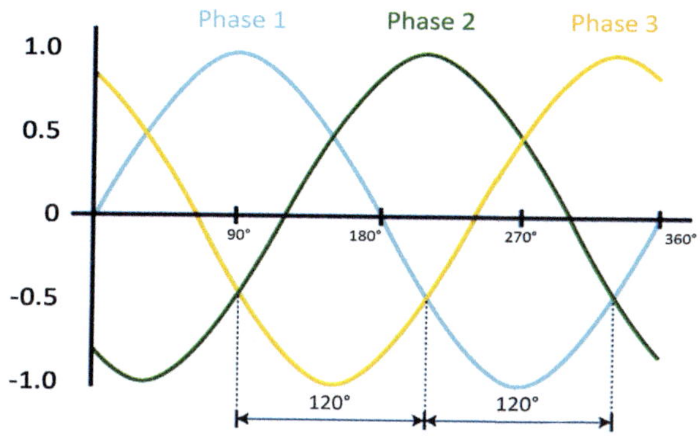

Figure 9.1. *Three phase wave-forms.*

Source: https://www.geeksforgeeks.org/three-phase-induction-motor/.

9.1.3. Construction of Three Phase Induction Motor

The construction of an induction motor is exceptionally basic and robust. It has predominantly two sections (Figure 9.2):

- Stator
- Rotor

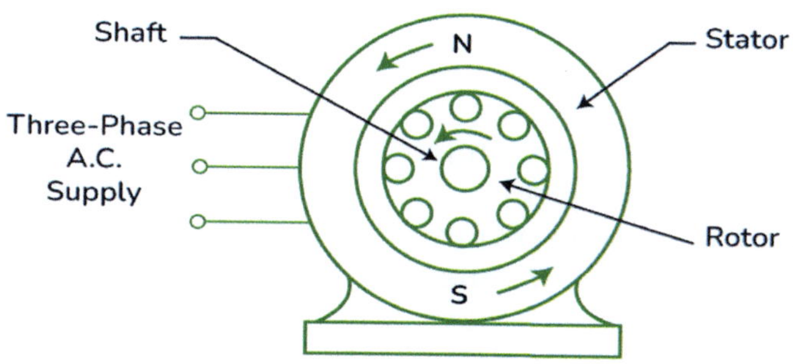

Figure 9.2. *Construction of three-phase induction motor.*

Source: https://www.geeksforgeeks.org/three-phase-induction-motor/.

1. **Stator of Three Phase Induction Motor:** As the name recommends, the stator is a fixed piece of the motor. The stator of the induction motor comprises of three primary parts;

- Stator Casing
- Stator Core
- Stator Winding
- **Stator Casing:** The stator housing is the external component of the motor, designed to provide support for the stator core and winding. Its main function is to provide mechanical reinforcement for the internal components of the motor. The housing is equipped with fins on its exterior for effective heat dissipation and motor cooling.

The enclosure is designed for small machines but can also be used for larger machines. It can be made from die-cast or fabricated steel, aluminum or aluminum alloys, or stainless steel, depending on the specific application.

- **Stator Core:** The main purpose of the stator core is to transport the alternating magnetic flux, which results in hysteresis and eddy current losses. To minimize these losses, the core is made of high-quality steel stampings with a thickness typically between 0.3 and 0.6 mm. These stampings are shielded from each other by a protective coating. The stampings are stacked together to form the stator core, which is then firmly attached to the stator frame. The inner layer of the stator core contains a varying number of slots.
- **Stator Winding:** The stator winding is located in the available gaps in the stator core and is made up of three phases, with a three-phase power supply being connected to it. The number of poles in the motor is determined by the arrangement of the stator winding, which in turn controls the motor's speed. A higher number of poles results in a lower speed, while a lower number of poles results in a higher speed. The poles are typically grouped in pairs, resulting in an even total number of poles. The connection between simultaneous speed and number poles is as displayed in the below equation;

$$N_s = 120f / P$$

Where;

f = Supply Frequency

P = Total Number of Poles

N_s = Synchronous Speed

As the end of winding connected with the terminal box. Consequently, there are six terminals (two of each phase) in the terminal box.

As per the application and sort of turning over techniques for motors, the stator winding is associated in star or delta and it is finished by the association of terminals in the terminal box.

2. **Rotor of Three Phase Induction Motor:** As the name proposes, the rotor is an rotating part of the motor. As indicated by the type of rotor, the induction motor is delegated;
   - Squirrel Cage Induction Motor
   - Phase Wound (Wound Rotor) induction motor / Slip-ring induction motor

The development of the stator is same in the two kinds of induction motor. We will examine the kinds of rotors utilized in 3-phase induction motor in the accompanying part of types of three phase induction motor.

9.1.4. Types of Three Phase Induction Motor

There are two main classes of three-phase motors based on the type of rotor winding: squirrel cage and slip ring (wound rotor motor).

- Squirrel Cage Induction Motor
- Slip-ring or Wound Rotor Induction Motor
- **Squirrel Cage Induction Motor:** The rotor of this motor is comparable to a squirrel's cage enclosure, earning it the nickname of an "induction squirrel cage motor." This type of rotor is both simple to manufacture and highly durable. As a result, approximately 80% of induction motors utilize a squirrel cage design for their rotors (Figure 9.3).

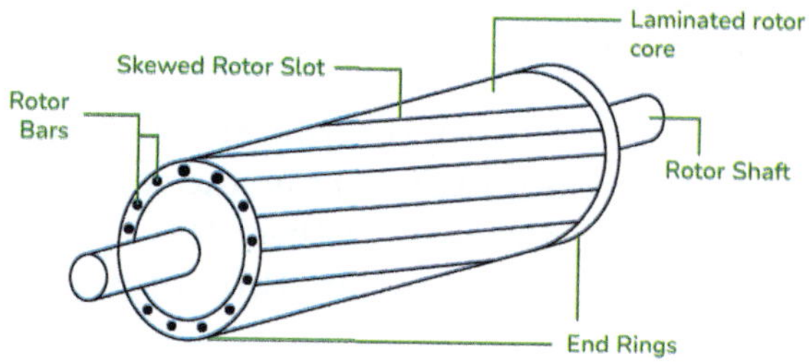

Figure 9.3. *Squirrel cage induction motor rotor.*

Source: https://www.geeksforgeeks.org/three-phase-induction-motor/.

The rotor of the motor has a tubular shape at its center, with spaces along its outer edges that are not evenly distributed. This design helps create a magnetic locking effect between the stator and rotor teeth, leading to smoother operation and reduced noise. However, this design also increases the resistance of the rotor due to the longer rotor guide. Unlike traditional rotor windings, the squirrel cage rotor is made up of rotor bars typically constructed from aluminum, steel, or copper. These bars are connected to end rings, effectively shortening the rotor circuit. The rotor bars are either welded or supported by the end rings for mechanical stability. The short-circuited rotor bars eliminate the need for slip rings and brushes, making the motor simpler and more durable in construction. This design also means there is no external protection for the rotor circuit.

9.1.5. Working Principle of a Three-Phase Induction Motor

The working principle of a 3-phase induction motor can be explained by considering a portion of it as follows (Figure 9.4):

- When the 3-phase stator winding is fed from a balanced 3-phase supply, a rotating magnetic field (RMF) is produced in the motor. This RMF rotates around the stator at *synchronous speed* which is given by;

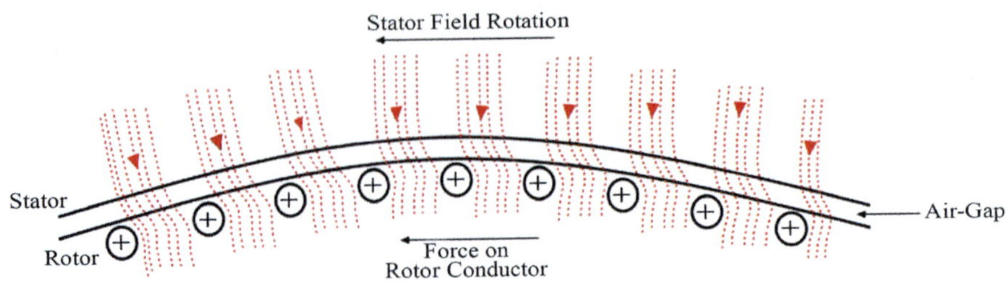

Figure 9.4. *Working principle of a 3-phase induction motor.*

Source: https://www.tutorialspoint.com/3-phase-induction-motor-definition-working-principle-advantages-and-disadvantages.

Synchronous Speed, N_S=120f / P

- The rotating magnetic field (RMF) travels through the air gap and intersects with the stationary rotor conductors. This movement creates electromagnetic forces (EMFs) in the rotor conductors. When the rotor circuit is completed with a short circuit, currents begin to flow in the rotor conductors.
- The rotor conductors, which carry current, are positioned in the magnetic field generated by the stator winding. This interaction causes the rotor conductors to undergo a mechanical force. The combined forces on all rotor conductors create a torque that propels the rotor in line with the rotating magnetic field. Therefore, the three-phase electric power input is converted into mechanical power output in a 3-phase induction motor.
- Additionally, as per Lenz's law, the rotor is expected to move in the direction of the stator field. This means that the rotor currents will flow in a way that opposes the action causing them. In this case, the relative speed between the rotating magnetic field (RMF) and the rotor conductors is what causes the rotor currents. Therefore, in order to decrease this relative speed, the rotor begins moving in the same direction as the RMF.

9.1.6. Advantages of Three Phase Induction Motor

Following are the chief advantages of a 3-phase induction motor:

- It has simple and rugged construction.
- It requires less maintenance.
- It has high efficiency and good power factor.
- It is less expensive.
- It has self-starting torque.

9.1.7. Disadvantages of Three Phase Induction Motor

The disadvantages of a 3-phase induction motor are given as follows:

- The 3-phase induction motors are constant speed motors; hence their speed control is very difficult.
- 3-phase induction motors have poor starting torque and high inrush currents (about 4 to 8 times of the rated current).
- They always operate under lagging power factor and during light loads, they operate at very worst power factor (about 0.3 to 0.5 lagging).

9.2. SINGLE PHASE INDUCTION MOTORS

A single-phase power source can be used to operate a three-phase motor, but it will not start on its own. It can be manually started in either direction and will reach full speed within a few seconds. However, it will only produce 2/3 of the 3-φ power rating because one winding is not used (Figure 9.5).

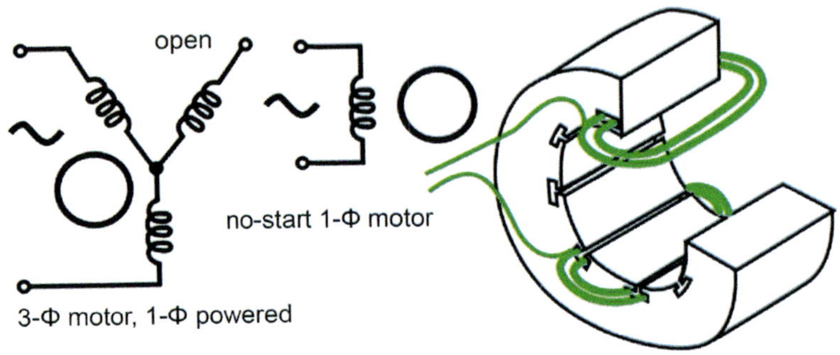

Figure 9.5. *3-φ motor runs from 1-φ power but does not start.*

Source: https://www.allaboutcircuits.com/textbook/alternating-current/chpt-13/single-phase-induction-motors/#:~:text=A%20three%2Dphase%20motor%20may,one%20winding%20is%20not%20used.

9.2.1. Single Coil of a Single Phase Motor

The single coil in a single-phase induction motor generates a pulsating magnetic field that peaks at 0° and 180° electrical, rather than a rotating magnetic field.

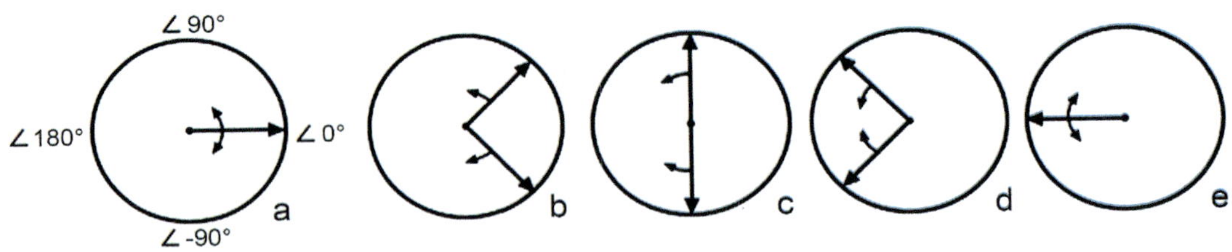

Figure 9.6. *Single-phase stator produces a nonrotating, pulsating magnetic field.*

Source: https://www.allaboutcircuits.com/textbook/alternating-current/chpt-13/single-phase-induction-motors/#:~:text=A%20three%2Dphase%20motor%20may,one%20winding%20is%20not%20used.

Another perspective suggests that a single-coil, excited by a single-phase current, generates two magnetic field phasors that rotate in opposite directions. These phasors align twice during each revolution at 0° (Figure 9.6a) and 180° (Figure 9.6e). At 90° and –90°, the phasors nullify each other (Figure 9.6c).

At angles of 45° and –45° (as shown in Figure 9.6b), the phasors are partially additive along the x-axis and cancel out along the y-axis. A similar scenario is observed in Figure 9.6d. The combined result of these two phasors is a stationary phasor in space, but with alternating polarity over time, preventing the development of any starting torque. However, if the rotor is rotated slightly below the synchronous speed, it will reach maximum torque at 10% slip compared to the forward rotating phasor. Torque generated will be less at slip values above or below 10%. The rotor will experience a slip of 200%–10% in relation to the counter-rotating magnetic field phasor. Minimal torque, with a double frequency ripple, is produced by the counter-rotating phasor. Consequently, the single-phase coil will generate torque once the rotor is in motion. Starting the rotor in the opposite direction will result in a similarly strong torque as it approaches the speed of the backward rotating phasor. Single-phase induction motors are equipped with a copper or aluminum squirrel cage placed within a core of steel laminations, a common feature in polyphase induction motors.

9.2.2. Permanent-Split Capacitor Motor

A solution to the single phase problem involves constructing a 2-phase motor that generates 2-phase power from a single phase source. This involves using a motor with two windings that are 90° apart electrically, powered by two phases of current that are 90° displaced in time. This type of motor is known as a permanent-split capacitor motor (Figure 9.7).

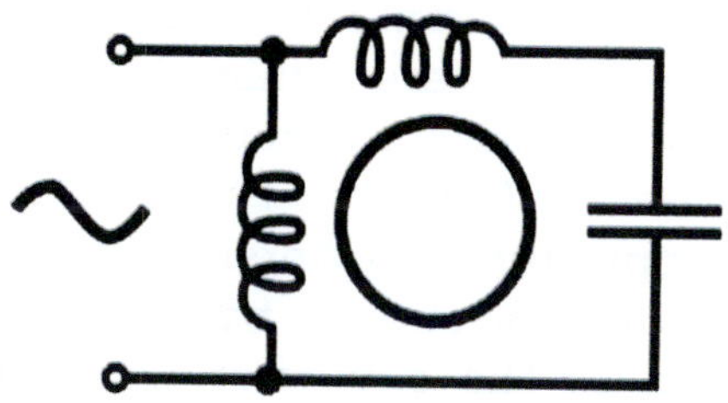

Figure 9.7. *Permanent-split capacitor induction motor.*

Source: https://www.allaboutcircuits.com/textbook/alternating-current/chpt-13/single-phase-induction-motors/#:~:text=A%20three%2Dphase%20motor%20may,one%20winding%20is%20not%20used.

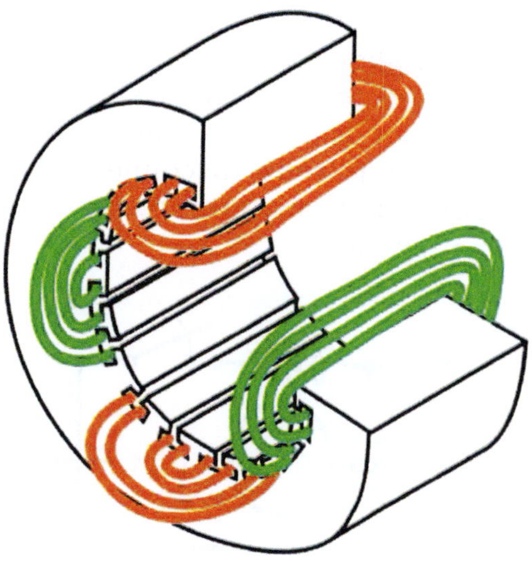

Figure 9.8. *Single-phase induction motor with embedded stator coils.*

Source: https://www.allaboutcircuits.com/textbook/alternating-current/chpt-13/single-phase-induction-motors/#:~:text=A%20three%2Dphase%20motor%20may,one%20winding%20is%20not%20used.

This particular motor experiences higher current levels and a delay in reaching full speed, resulting in torque fluctuations once at maximum speed. To address this issue, it is recommended to use a small capacitor to reduce losses. Compared to a shaded pole motor, the losses are minimal. This motor design is effective for motors up to 1/4 horsepower (200 watts), typically used for smaller applications. Reversing the direction of the motor is simple by changing the placement of the capacitor in relation to the other winding (Figure 9.8).

Single-phase induction motors may have coils embedded in the stator for larger models, while smaller sizes typically use simpler concentrated windings with salient poles.

9.2.3. Capacitor-Start Induction Motor

The diagram shows that a larger capacitor can be utilized to kickstart a single-phase induction motor using the auxiliary winding. This capacitor is then disconnected by a centrifugal switch once the motor reaches its optimal speed.

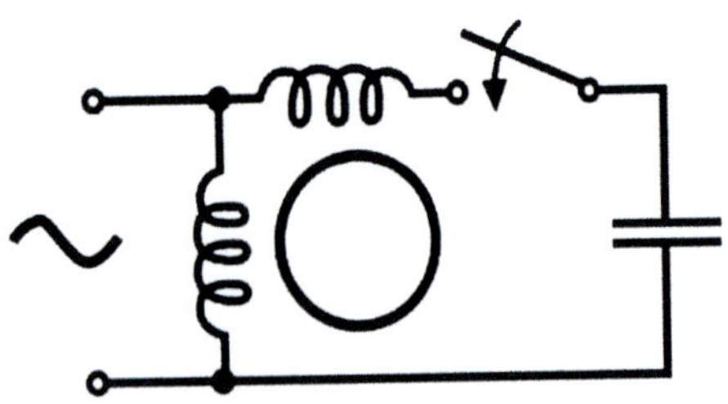

Figure 9.9. *Capacitor-start induction motor.*

Source: https://www.allaboutcircuits.com/textbook/alternating-current/chpt-13/single-phase-induction-motors/#:~:text=A%20three%2Dphase%20motor%20may,one%20winding%20is%20not%20used.

Additionally, the auxiliary winding may consist of many more turns of thicker wire compared to a resistance split-phase motor in order to prevent excessive temperature increase. As a result, greater starting torque is accessible for heavy loads such as air conditioning compressors. This motor setup is so effective that it is offered in a variety of horsepower (kilowatt) options (Figure 9.9).

9.2.4. Capacitor-Run Motor Induction Motor

Another option for starting a motor is using a capacitor-start motor with a large capacitor for high starting torque. After starting, a smaller capacitor can be put in place to improve running characteristics without drawing too much current. The capacitor-run motor is more complex but is beneficial for larger motors (Figure 9.10).

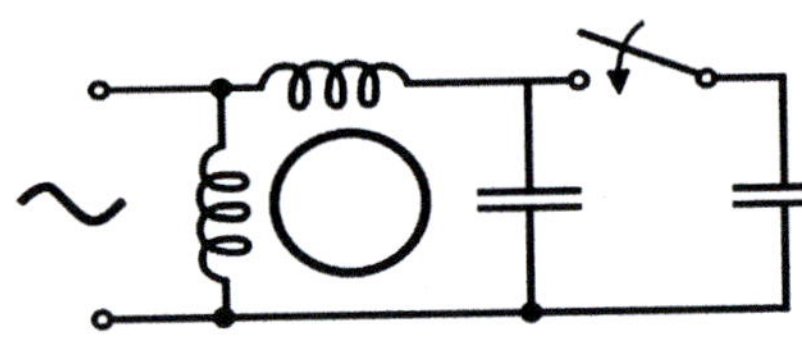

Figure 9.10. *Capacitor-run motor induction motor.*

A motor starting capacitor may be a double-anode non-polar electrolytic capacitor which could be two + to + (or – to -) series-connected polarized electrolytic capacitors. Such AC rated electrolytic capacitors have such high losses that they can only be used for intermittent duty (1 second on, 60 seconds off) like motor starting.

A capacitor for motor running must not be of electrolytic construction, but a lower loss polymer type.

9.2.5. Resistance Split-Phase Motor Induction Motor

A smaller wire auxiliary winding with much fewer turns placed at 90° to the main winding can help start a single-phase induction motor. This auxiliary winding has lower inductance and higher resistance, causing less phase shift compared to the main winding. This results in approximately 30° of phase difference. The coil provides moderate starting torque, which is disconnected by a centrifugal switch at 3/4 of synchronous speed. This uncomplicated setup, without a capacitor, works well for motors up to 1/3 horsepower (250 watts) that drive easily started loads (Figure 9.11).

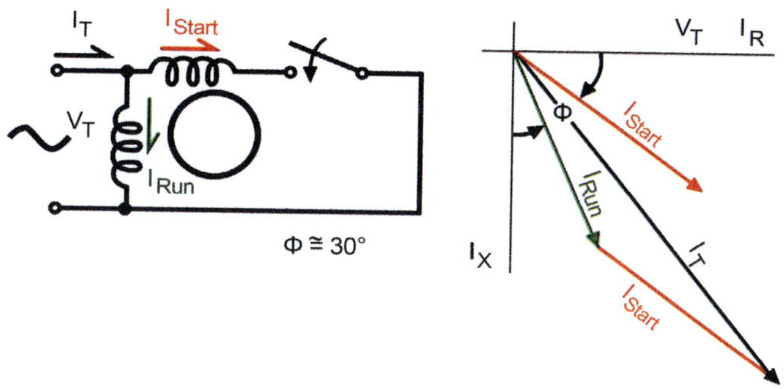

Figure 9.11. *Resistance split-phase motor induction motor.*

Source: https://www.allaboutcircuits.com/textbook/alternating-current/chpt-13/single-phase-induction-motors/#:~:text=A%20three%2Dphase%20motor%20may,one%20winding%20is%20not%20used.

This motor has a higher starting torque compared to a shaded pole motor, but lower than a two-phase motor with the same parts. The high current density in the auxiliary winding during startup causes a rapid temperature increase, making frequent or slow startups impractical.

9.2.6. Nola Power Factor Corrector

In the mid-1970s, Frank Nola from NASA introduced a power factor corrector designed to enhance the efficiency of AC induction motors. The concept behind the device stemmed from the understanding that induction motors operate inefficiently when not at full load, leading to a low power factor. This low power factor is primarily caused by the magnetizing current needed by the stator. As the motor load decreases, this fixed current makes up a greater proportion of the overall motor current. By decreasing the applied voltage at lighter loads, the power factor and efficiency could be improved by reducing the need for the full magnetizing current.

The power factor corrector senses power factor, and decreases motor voltage, thus restoring a higher power factor and decreasing losses. Since single-phase motors are about 2 to 4 times as inefficient as three-phase motors, there are potential energy savings for 1-φ motors. There are no savings for a fully-loaded motor since all the stator magnetizing current is required.

Reducing the voltage is not an option, but potential cost savings can be achieved by operating a motor below its full load capacity. A motor rated for 117 VAC can safely operate between 104–127 VAC, meaning it is not fully loaded when operating above 104 VAC – such as in the case of a 117 VAC refrigerator. By decreasing the line voltage to 104–110 VAC, the power factor controller can improve efficiency and save on energy costs. Motors operating at higher voltages have greater potential for savings. Additionally, single-phase induction motors that are idle or operate at 25% or less of their full load capacity are suitable candidates for a power factor controller. These motors should run for a significant number of hours per year, such as in lumber saws, punch presses, or conveyors, in order to justify the cost of the controller. It is recommended to invest in a power factor controller for single-phase induction motors as they are more cost-effective compared to three-phase motors. However, the cost of the controller may not be recovered if the motor operates only a few hours per day.

9.3. SPECIAL MACHINES

There a various types of special motors, which are as follows:

9.3.1. Permanent Magnet Synchronous Motors (PMSM)

The Permanent Magnet Synchronous Motor (PMSM) is an AC synchronous motor that utilizes permanent magnets for field excitation and produces a sinusoidal back EMF waveform. It combines features of both induction motors and brushless DC motors, with a permanent magnet rotor and stator windings. The stator windings are designed to generate a sinusoidal flux density in the air gap similar to an induction motor. The PMSM has higher power density compared to induction motors of the same rating due to not needing stator power for magnetic field production. Thanks to the permanent magnets, the PMSM can generate torque at zero speed, but requires a digitally controlled inverter for operation. PMSMs are commonly used in high-performance and high-efficiency motor applications, as they allow for smooth rotation across all speeds, full torque control at zero speed, and rapid acceleration and deceleration. In order to achieve this level of control, vector control techniques like field-oriented control (FOC) are employed for PMSMs. This algorithm breaks down the stator current into separate components for generating magnetic fields and torque, allowing for individual control of each component after decomposition.

- **Construction:** Permanent magnet synchronous motors, like other rotating electric motors, consist of a rotor and a stator. The construction of a permanent magnet synchronous motor is similar to a basic synchronous motor, with the main difference being the rotor. In this motor, permanent magnets are attached to the rotor instead of field windings. These magnets create field poles and are typically made of materials like samarium-cobalt, medium iron, and boron for their high permeability. Neodymium-boron-iron magnets are the most commonly used due to their cost-effectiveness and widespread availability.
- **Working Principle:** The permanent magnet synchronous motor operates similarly to a synchronous motor, relying on the interaction between the rotating magnetic field of the stator and the constant magnetic field of the rotor. The motor generates electromotive force at synchronous speed by utilizing the rotating magnetic field. When the stator winding is powered by a 3-phase supply, a rotating magnetic field is created in the air gaps, producing torque when the rotor field poles align with the rotating magnetic field at synchronous speed. As these motors require a variable frequency power supply to start, they are not self-starting.
- Characteristics
  - No sparks, safer in explosive environments
  - Clean, fast, and efficient
  - More compact, efficient, and lighter than an ACIM
  - Designed for high-performance servo applications

- Runs with/without position encoders
- Low audible noise and EMI
- Smooth low- and high-speed performance
- Produces optimal torque when coupled with Field-Oriented Control (FOC)

- Types

Depending on how magnets are attached to the rotor and the design of the rotor, permanent magnet synchronous motor can be classified into two types:

- Surface permanent magnet synchronous motor (SPMSM)
- Interior permanent magnet synchronous motor (IPMSM).

SPMSM mounts all magnet pieces on the surface, and IPMSM places magnets inside the rotor.

Also, depending on the stator design, a permanent magnet synchronous motor can be:

- With distributed winding
- With concentrated winding

- Permanent Magnet Synchronous Motor Control

Permanent magnet synchronous motors necessitate sinusoidal waveforms for operation, leading to increased control complexity. These motors require a dedicated control system such as a variable frequency drive or a servo drive (Figure 9.12).

Figure 9.12. *Permanent magnet synchronous motor control.*

Source: https://www.linquip.com/blog/permanent-magnet-synchronous-motors/.

There are numerous control techniques used in control systems, with the optimal choice depending on the specific task facing the electric drive. For simple tasks, trapezoidal control with Hall sensors is typically used, while for more demanding tasks that require maximum performance, field-oriented control is usually preferred.

Advantages

The advantages of permanent magnet synchronous motor include;

- Dynamic performance in both high and low-speed operation
- High power density
- Low rotor inertia makes it easy to control
- No torque ripple when the motor is commutated
- High and smooth torque
- High efficiency at high speeds
- Resistant to wear and tear
- Available in small sizes at different packages
- Easy maintenance and installation
- Capable of maintaining full torque at low speeds
- High reliability
- Efficient dissipation of heat
- Less noisy

Disadvantages

The disadvantages of permanent magnet synchronous motors are:

- High initial cost
- Very expensive
- Difficult to start up because it is not a self-starting motor

Applications

The permanent magnet synchronous motors applications are:

- Air conditioners
- Refrigerators
- AC compressors
- Direct-drive washing machines
- Automotive electrical power steering
- Machine tools
- Control of traction
- Data storage units
- Servo drives
- Electric vehicle drivetrain
- Large power systems to improve leading, and lagging power factor
- Industrial applications like robotics, aerospace, and many more.

9.3.2. Switched Reluctance Motor (SRM)

The switched reluctance motor (SRM), also known as the variable-reluctance motor, is becoming increasingly popular in industrial sectors such as wind energy systems and electric vehicles. This is due to its straightforward and durable design, high-speed operation capabilities, ability to withstand high temperatures, and fault tolerance features. While SRM motors were previously used in clocks and phonograph turntables, they are now being more widely used in various applications such as appliances, industrial equipment, and commercial and vehicular uses. This low-cost electric motor operates using reluctance torque, with power being supplied to the stator instead of the rotor. This simplifies the mechanical design but makes the electrical design more complex, requiring a switching system to distribute power to the different windings.

Manufacturers of switched reluctance motors claim to offer superior performance, reliability, efficiency, and affordability compared to standard induction or other adjustable speed motors. These motors operate by switching currents in the stator windings in response to changes in the magnetic circuit formed by the rotor and stator. While the stator of a switched reluctance motor contains windings similar to a brushless DC motor, the rotor is simply made of steel shaped into salient poles, with no windings or magnets. This sets it apart from induction motors, which have rotor windings or permanent magnets. Since switched reluctance motors lack rotor bars, there is no torque-producing current flow in the rotor. They can be a viable alternative to induction motors in situations where operating conditions do not favor the latter. The most common method of powering a switched reluctance motor is using an asymmetric bridge converter, which allows for a lower switching frequency compared to AC motors. This converter corresponds to the motor phases, activating each phase by turning on the power switches on either side. Once the current reaches a certain level, the switch turns off, and the energy stored in the winding maintains the current direction, known as back EMF. This back EMF is redirected through the diodes to the capacitor for reuse, ultimately improving efficiency.

9.3.3. Stepper Motor

A stepper motor is an electric motor that rotates its shaft in fixed increments, or steps, due to its unique internal structure. This allows for precise control over the motor's angular position without the need for a sensor. The stepper motor's step-wise movement makes it versatile and suitable for a variety of applications.

9.3.3.1. Stepper Motor Working Principles

Like all electric motors, stepper motors consist of a stationary part known as the stator and a moving part known as the rotor. The stator contains teeth with wired coils, while the rotor can be either a permanent magnet or a variable reluctance iron core. We will explore the various rotor structures in more detail later on. In Figure 9.13, a drawing is presented that illustrates a cross-section of the motor, with the rotor being a variable-reluctance iron core.

The basic working principle of the stepper motor is the following: By energizing one or more of the stator phases, a magnetic field is generated by the current flowing in the coil and the rotor aligns with this field. By supplying different phases in sequence, the rotor can be rotated by a specific amount to reach the desired final position. Figure

9.14 shows a representation of the working principle. At the beginning, coil A is energized and the rotor is aligned with the magnetic field it produces. When coil B is energized, the rotor rotates clockwise by 60° to align with the new magnetic field. The same happens when coil C is energized. In the pictures, the colors of the stator teeth indicate the direction of the magnetic field generated by the stator winding.

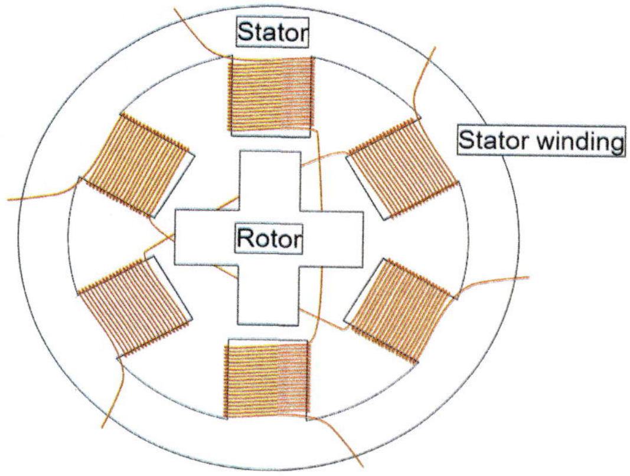

Figure 9.13. *Cross-section of a stepper motor.*

Source: https://www.monolithicpower.com/learning/resources/stepper-motors-basics-types-uses.

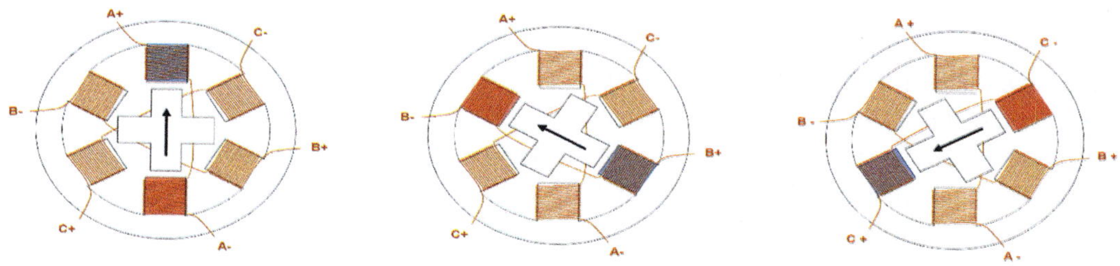

Figure 9.14. *Stepper motor steps.*

https://www.monolithicpower.com/learning/resources/stepper-motors-basics-types-uses.

9.3.3.2. Stepper Motor Types and Construction

The resolution, speed, and torque of a stepper motor are impacted by its construction details, which also determine its controllability. Stepper motors vary in internal structure, with different rotor and stator configurations.

- **Rotor:** For a stepper motor, there are basically three types of rotors:
  - **Permanent Magnet Rotor:** The rotor, a permanent magnet, aligns with the magnetic field generated by the stator circuit, ensuring both good torque and detent torque. This allows the motor to resist a change in position, even if only to a small degree, when a coil is energized. However, this solution has drawbacks including lower speed and resolution compared to

other types of motors. Figure 9.15 depicts a section of a permanent magnet stepper motor.

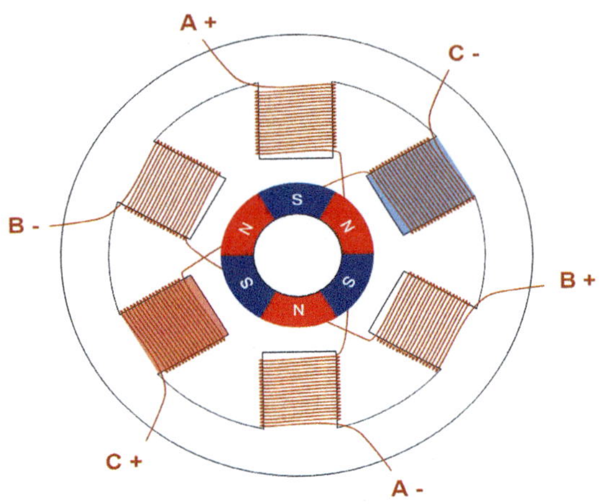

Figure 9.15. *Permanent magnet stepper motor.*

Source: https://www.monolithicpower.com/learning/resources/stepper-motors-basics-types-uses.

- **Variable Reluctance Rotor:** The rotor, constructed from an iron core, is designed with a specific shape that enables alignment with the magnetic field, as illustrated in Figures 9.13 and 9.14. Although this design facilitates achieving higher speeds and resolutions, it typically results in lower torque and lacks detent torque.
- **Hybrid Rotor:** This type of rotor is a combination of permanent magnet and variable reluctance designs, with a specific construction that includes two caps with alternating teeth magnetized axially. This design offers the benefits of both types of rotors, including high resolution, speed, and torque. However, the increased performance of this rotor comes with a higher cost due to its more complex construction. Figure 9.15 provides a simplified example of this motor's structure, showing how the N-magnetized tooth of the rotor aligns with the S-magnetized tooth of the stator when coil A is energized. Real motors have a more intricate structure with a greater number of teeth than illustrated, but they operate on the same principle as the stepper motor. The increased number of teeth enables the motor to achieve a small step size, potentially as low as 0.9° (Figure 9.16).

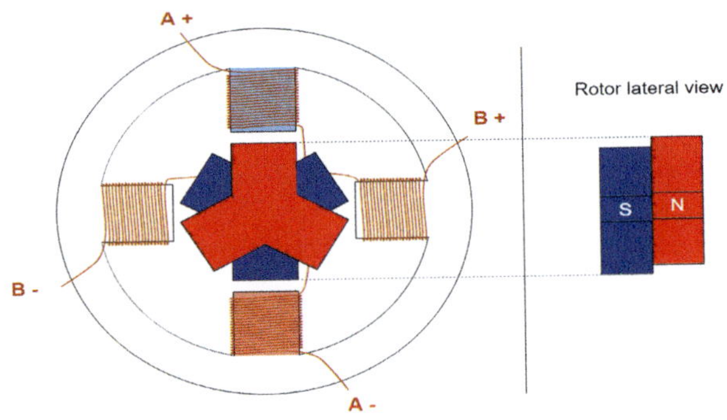

Figure 9.16. *Hybrid stepper motor.*

Source: https://www.monolithicpower.com/learning/resources/stepper-motors-basics-types-uses.

- **Stator:** The stator is responsible for generating the magnetic field that aligns the rotor. Key features of the stator circuit include the number of phases, pole pairs, and wire configuration. The number of phases refers to the independent coils, and the pole pairs indicate the number of teeth occupied by each phase. Two-phase stepper motors are most popular, with three-phase and five-phase motors being less common. See Figure 9.17 and Figure 9.18 for visual examples.

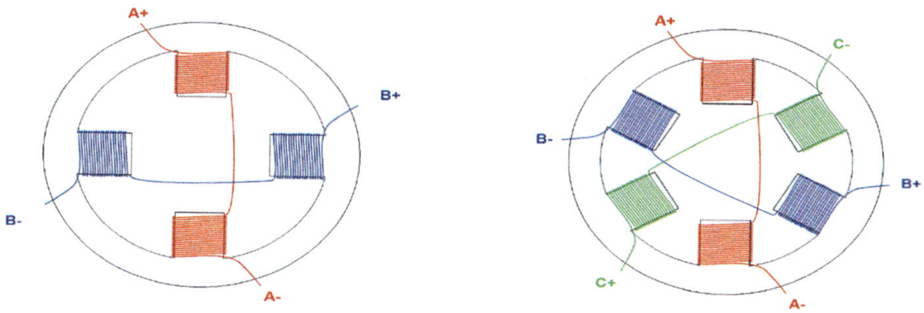

Figure 9.17. *Two-phase stator winding (left), three-phase stator winding (right).*

Source: https://www.monolithicpower.com/learning/resources/stepper-motors-basics-types-uses.

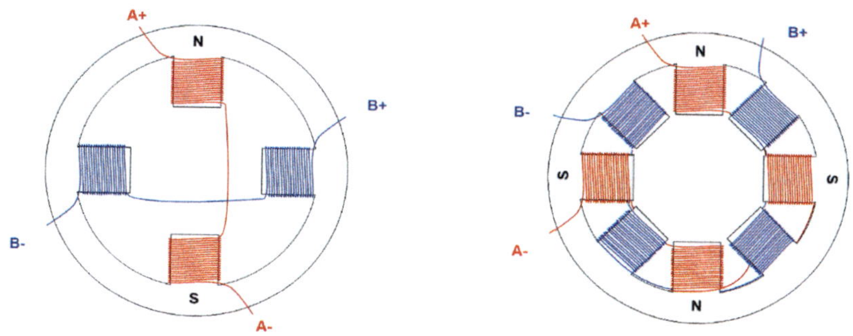

Figure 9.18. *Two-phase, single-pole pair stator (left) and two-phase, dipole pair stator (right). The letters show the magnetic field generated when positive voltage is applied between A+ and A–.*

Source: https://www.monolithicpower.com/learning/resources/stepper-motors-basics-types-uses.

9.3.3.3. Stepper Motor Control

As previously observed, the motor coils must be energized in a precise order to create the magnetic field needed for the rotor to align. Various devices are utilized to provide the required voltage to the coils, ensuring the motor operates correctly.

Starting from the devices that are closer to the motor we have:

- A transistor bridge is the device physically controlling the electrical connection of the motor coils. Transistors can be seen as electrically controlled interrupters, which, when closed allow the connection of a coil to the electrical supply and thus the flow of current in the coil. One transistor bridge is needed for each motor phase.
- A pre-driver is a device that controls the activation of the transistors, providing the required voltage and current, it is in turn controlled by an MCU.
- An MCU is a microcontroller unit, which is usually programmed by the motor

user and generates specific signals for the pre-driver to obtain the desired motor behavior.

Figure 9.19 shows a simple representation of a stepper motor control scheme. The pre-driver and the transistor bridge may be contained in a single device, called a driver.

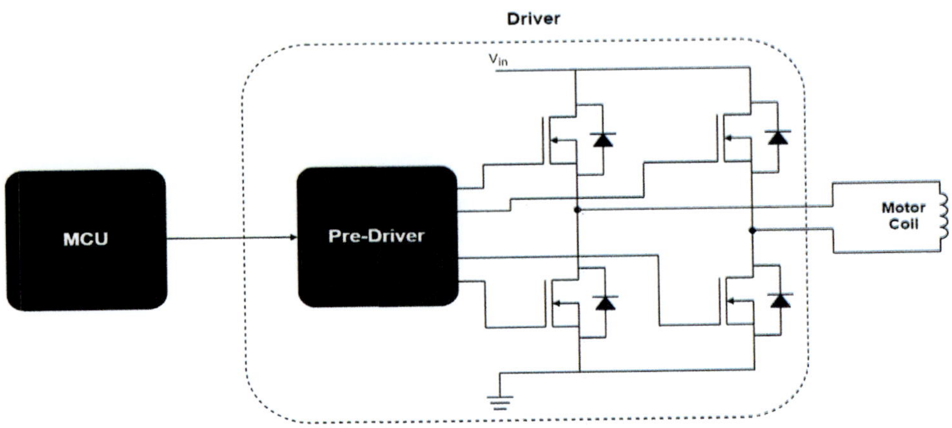

Figure 9.19. *Motor control basic scheme.*

Source: https://www.monolithicpower.com/learning/resources/stepper-motors-basics-types-uses.

9.3.3.4. Stepper Motor Driver Types

Various stepper motor drivers are available for purchase, each tailored to specific applications and offering unique features. The most important characteristics include the input interface. The most common options are:

- **Step/Direction:** By sending a pulse on the Step pin, the driver changes its output such that the motor will perform a step, the direction of which is determined by the level on the Direction pin.
- **Phase/Enable:** For each stator winding phase, Phase determines the current direction and triggers Enable if the phase is energized.
- **PWM:** Directly controls the gate signals of the low-side and high-side FETs.

Another important feature of a stepper motor driver is if it is only able to control the voltage across the winding, or also the current flowing through it:

- With voltage control, the driver only regulates the voltage across the winding. The torque developed and the speed with which the steps are executed only depend on motor and load characteristics.
- Current control drivers are more advanced, as they regulate the current flowing through the active coil in order to have better control over the torque produced, and thus the dynamic behavior of the whole system.

9.3.3.5. Unipolar/Bipolar Motors

The layout of the stator coils in the motor plays a crucial role in controlling the direction of the current, which ultimately determines the magnetic field's direction. Properly energizing and controlling the current direction is essential for achieving rotor motion. See Figure 9.20 for a visual representation. In stepper motors, the issue of controlling the current direction is solved with two different approaches.

Figure 9.20. *Direction of the magnetic field based on the direction of the coil current.*

Source: https://www.monolithicpower.com/learning/resources/stepper-motors-basics-types-uses.

In unipolar stepper motors, one lead is connected to the central point of the coil. This setup allows for easily controlling the current's direction with a simple circuit and components. The central lead (A_M) is connected to the input voltage V_{IN}. When MOSFET 1 is active, the current flows from A_M to A+. When MOSFET 2 is active, the current flows from A_M to A–, creating a magnetic field in the opposite direction. This simpler driving circuit only requires two semiconductors, but using only half of the coil's copper at a time means the magnetic field intensity is halved compared to using all the copper. Additionally, constructing these motors can be more challenging as more leads are needed for motor inputs.

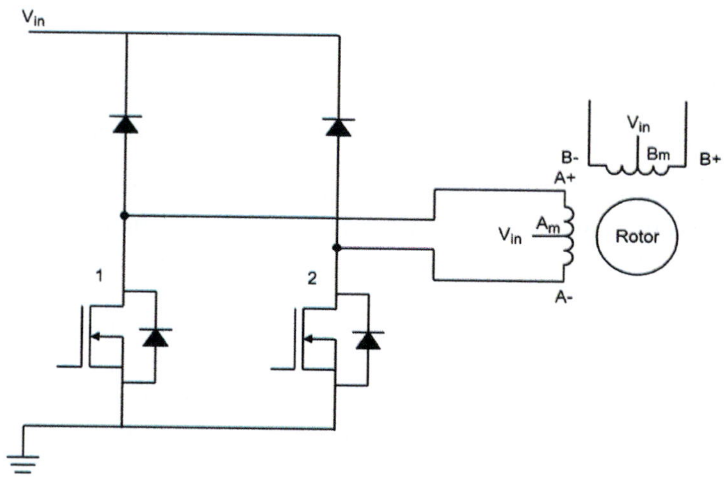

Figure 9.21. *Unipolar stepper motor driving circuit.*

Source: https://www.monolithicpower.com/learning/resources/stepper-motors-basics-types-uses.

In bipolar stepper motors, each coil has only two leads available, and to control the direction it is necessary to use an H-bridge (see Figure 9.22). As shown in Figure 9.21, if MOSFETs 1 and 4 are active, the current flows from A+ to A–, while if MOSFETs 2 and 3 are active, current flows from A– to A+, generating a magnetic field in the opposite direction. This solution requires a more complex driving circuit, but allows the motor to achieve the maximum torque for the amount of copper that is used.

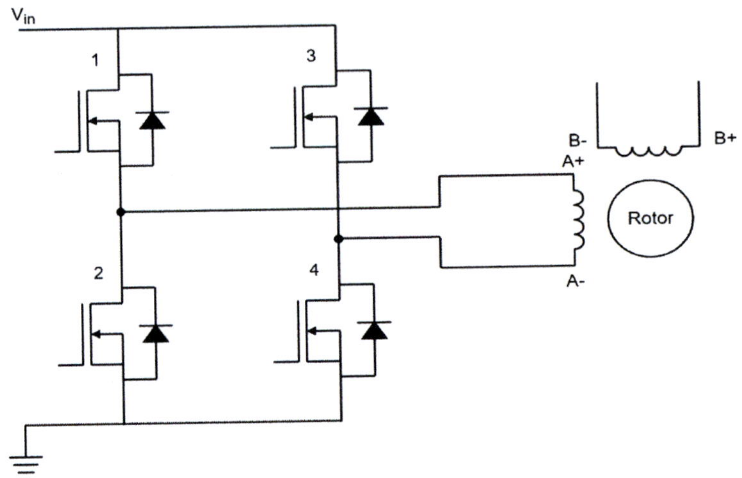

Figure 9.22. *Bipolar stepper motor driving circuit.*

Source: https://www.monolithicpower.com/learning/resources/stepper-motors-basics-types-uses.

With technology progress, the advantages of unipolar are becoming less relevant, and bipolar steppers are currently the most popular.

9.3.3.6. Stepper Motor Driving Techniques

There are four different driving techniques for a stepper motor:

- In wave mode, only one phase at a time is energized (see Figure 9.23). For simplicity, we will say that the current is flowing in a positive direction if it is going from the + lead to the – lead of a phase (e.g., from A+ to A–); otherwise, the direction is negative. Starting from the left, the current is flowing only in phase A in the positive direction and the rotor, represented by a magnet, is aligned with the magnetic field generated by it. In the next step, it flows only in phase B in the positive direction, and the rotor spins 90° clockwise to align with the magnetic field generated by phase B. Later, phase A is energized again, but the current flows in the negative direction, and the rotor spins again by 90°. In the last step, the current flows negatively in phase B and the rotor spins again by 90°.

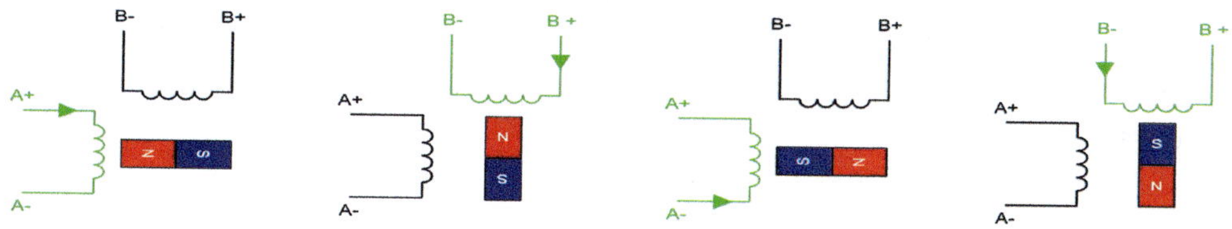

Figure 9.23. *Wave mode steps.*

Source: https://www.monolithicpower.com/learning/resources/stepper-motors-basics-types-uses.

- In full-step mode, two phases are consistently powered simultaneously. Figure 9.24 illustrates the various steps involved in this driving mode. These steps resemble those of the wave mode, with the main distinction being that in this mode, the motor can generate a higher torque due to increased current flow and a stronger magnetic field.

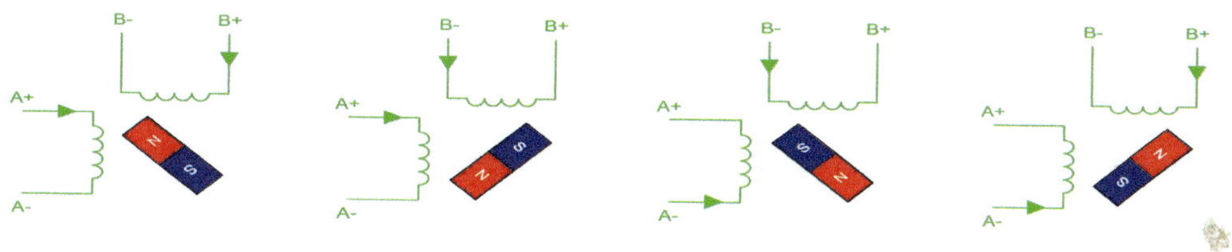

Figure 9.24. *Full-step mode steps.*

Source: https://www.monolithicpower.com/learning/resources/stepper-motors-basics-types-uses.

- Half-step mode is a blend of wave and full-step modes as shown in Figure 9.24. By utilizing this combination, the step size can be halved, resulting in a 45° step instead of 90°. However, the drawback of this mode is that the torque generated by the motor is not consistent. It is stronger when both phases are energized and weaker when only one phase is energized.

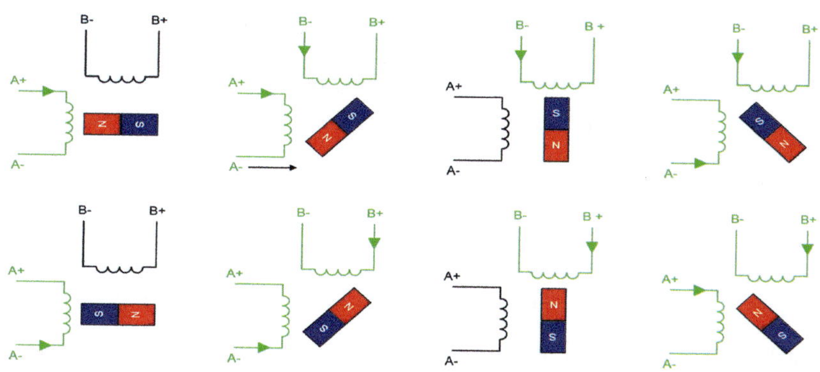

Figure 9.25. *Half-step mode steps.*

Source: https://www.monolithicpower.com/learning/resources/stepper-motors-basics-types-uses.

- Microstepping is an improvement upon half-step mode, as it allows for even smaller step sizes and consistent torque output by controlling current intensity in each phase. Implementing this mode requires a more advanced motor driver than previous solutions. Figure 9.26 illustrates how microstepping functions, with the maximum current flowing in a phase denoted as I_{MAX}. In the first figure shown, $I_A = I_{MAX}$ and $I_B = 0$ when starting from the left side. In the following stage, the currents are adjusted in order to achieve $I_A = 0.92 \times I_{MAX}$ and $I_B = 0.38 \times I_{MAX}$, resulting in a magnetic field that is shifted by 22.5° clockwise from the previous position. This process is then repeated with varying current values to reach the 45°, 67.5°, and 90° positions. By doing so, the step size is effectively halved compared to the half-step mode, with the potential to further refine positioning accuracy. However, this increased precision comes at a cost – utilizing microstepping requires a more intricate control system for the motor, and results in reduced torque output with each step. The torque is directly proportional to the sine of the angle between the stator and rotor magnetic fields, meaning that as step size decreases, torque diminishes accordingly.

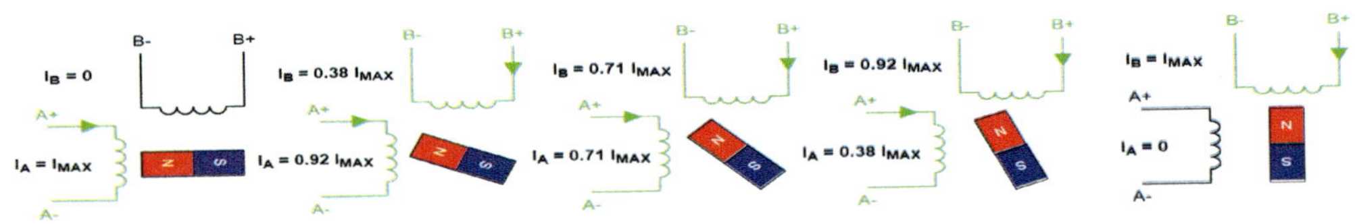

Figure 9.26. *Microstepping.*

Source: https://www.monolithicpower.com/learning/resources/stepper-motors-basics-types-uses.

9.3.3.7. Stepper Motors Advantages and Disadvantages

Having gained an understanding of how stepper motors function, let us now examine their advantages and disadvantages in comparison to other types of motors.

Advantages:

- Due to their internal structure, stepper motors do not require a sensor to detect the motor position. Since the motor moves by performing "steps," by simply counting these steps, you can obtain the motor position at a given time.
- In addition, stepper motor control is pretty simple. The motor does need a driver, but does not need complex calculations or tuning to work properly. In general, the control effort is lower compared to other motors. With microstepping, you can reach high position accuracy, up to approximately 0.007°.
- Stepper motors offer good torque at low speeds, are great for holding position, and also tend to have a long lifespan.

Disadvantages:

- They can miss a step if the load torque is too high. This negatively impacts the control, since there is no way to know the real position of the motor. Using microstepping makes stepper motors even more likely to experience this issue.
- These motors always drain maximum current even when still, which makes efficiency worse and can cause overheating.
- Stepper motors have low torque and become pretty noisy at high speeds.
- Finally, stepper motors have low power density and a low torque-to-inertia ratio.

9.3.4. PMBLDC Motor

Brushless DC electric motors, also referred to as electronically commutated motors (ECMs or EC motors), prioritize primary efficiency. This is because the rotor, which carries the magnets, requires no power and eliminates the need for connections, a commutator, and brushes. Instead, BLDC motors utilize control circuitry in place of traditional components. To determine the rotor's position at specific times, BLDC motors are equipped with either rotary encoders, Hall sensors, or controllers.

9.3.4.1. Construction of Brushless DC Motor

This motor utilizes permanent magnets on the rotor and current-carrying conductors on the stator. Electrical commutation is employed to transform electrical energy into mechanical energy. The primary distinction in design between brushed and brushless motors is the elimination of a mechanical commutator in favor of an electric switch circuit. A brushless motor, or BLDC Motor, operates as a synchronous motor with the stator and rotor magnetic fields rotating at the same frequency. Unlike brushed motors, brushless motors do not contain current-carrying commutators. Instead, the magnetic field inside a brushless motor is controlled by an amplifier triggered by a commutating device such as an optical encoder.

The layout of a DC brushless motor can vary depending on whether it is in "Out runner" style or "Inrunner" style.

- **Outrunner:** The field magnet is a drum rotor which rotates around the stator. This style is preferred for applications that require high torque and where high rpm isn't a requirement.
- **In Runner:** The stator is a fixed drum in which the field magnet rotates. This motor is known for producing less torque than the out runner style, but is capable of spinning at very high rpm.

9.3.4.2. Working Principle of Brushless DC Motor

The BLDC motor operates based on a similar principle to that of a Brushed DC motor. According to the Lorentz force law, when a current-carrying conductor is placed in a magnetic field, it experiences a force. This results in a reaction force, causing the magnet to experience an equal and opposite force. In a BLDC motor, the current-carrying conductor remains stationary while the permanent magnet moves. When the stator coils are supplied with power, they become electromagnets and create a consistent field in the air gap. Even though the power source is DC, the motor generates an AC voltage waveform with a trapezoidal shape through switching. The interaction between the electromagnet stator and permanent magnet rotor causes the rotor to rotate. By switching the windings between High and Low signals, the corresponding windings are energized as North and South poles. This alignment between the North and South poles of the permanent magnet rotor and the stator poles causes the motor to rotate.

Advantages of Brushless DC Motor

- Less overall maintenance due to absence of brushes
- Reduced size with far superior thermal characteristics
- Higher speed range and lower electric noise generation.
- It has no mechanical commutator and associated problems
- High efficiency and high output power to size ratio due to the use of permanent magnet rotor
- High speed of operation even in loaded and unloaded conditions due to the absence of brushes that limits the speed
- Smaller motor geometry and lighter in weight than both brushed type DC and induction AC motors.
- Long life as no inspection and maintenance is required for commutator system
- Higher dynamic response due to low inertia and carrying windings in the stator
- Less electromagnetic interference
- Low noise due to absence of brushes

Limitations of Brushless DC Motor

- These motors are costly
- Electronic controller required control this motor is expensive
- Requires complex drive circuitry
- Need of additional sensors

Applications of Brushless DC Motor

Brushless DC motors (BLDC) use for a wide variety of application requirements such as varying loads, constant loads and positioning applications in the fields of industrial control, automotive, aviation, automation systems, health care equipments etc.

- Computer hard drives and DVD/CD players
- Electric vehicles, hybrid vehicles, and electric bicycles
- Industrial robots, CNC machine tools, and simple belt driven systems
- Washing machines, compressors and dryers
- Fans, pumps and blowers.

CASE STUDY

TROUBLESHOOTING INDUCTION MOTOR SHAFT ISSUES

Troubleshooting induction motor shaft issues involves identifying symptoms and addressing various potential problems. Misalignment typically causes vibrations, noise, and excessive wear on couplings and bearings, which can be resolved by realigning the motor and load. A bent shaft often results in unusual vibrations and uneven bearing wear, necessitating either replacement or professional straightening. Shaft imbalance, indicated by excessive vibrations and noise, can be corrected by balancing the rotor and shaft assembly. Shaft wear, which leads to increased clearance and potential bearing failure, requires measuring and replacing the shaft if it exceeds acceptable limits, while also investigating the root cause of the wear. Bearing issues, characterized by noise, heat, and increased vibrations, can be resolved by inspecting for wear or damage and ensuring proper lubrication. Loose or worn couplings cause noise, vibrations, and slippage, and should be inspected, tightened, or replaced as needed.

Situation

One of our valued clients recently purchased a new F-Class Induction Motor of our company to power their machinery. However, to their surprise, the motor's shaft broke after running for a short period of time.

Concerned about the issue, the operator promptly brought the motor to our service center. As the motor was still under warranty, we immediately replaced the rotor and conducted standard inspections before returning the motor to the operator.

Unfortunately, just one day later, the operator returned with the motor once again experiencing a broken shaft.

Recognizing the recurring problem, our support team decided to send an engineer to the client's factory to investigate the root cause behind the repeated shaft failures.

Our F-Class induction motor is renowned for its high quality and reliability. In the event of any issues with our motors or pumps, we strongly recommend contacting us through our toll-free number to report the problem.

This allows us to dispatch a trained professional who can conduct a thorough inspection and identify the underlying cause of the problems.

1. Customers can register a complaint by calling our toll-free number, after which they will receive a text message with information about the complaint as well as the contact details of our service engineer.

2. Our Engineer usually calls up to fix an appointment as per the electricity schedule of the area – as there are frequent power disruptions in some places.

Observation

Upon reaching the site, our service engineer immediately commenced the inspection to validate the reported problem and identify its root cause.

He first assessed the motor's foundation to ensure stability.

- The foundation of an F-Class induction motor should be solid and stable to ensure proper installation and operation. It should be constructed using a concrete base, designed to withstand the motor's weight and vibrations. Adequate reinforcement and leveling should be implemented to maintain the motor's alignment and prevent any movement or misalignment during operation.
- When attaching a pulley and belt-drive to an F-Class induction motor, it is important to ensure proper alignment and tension. The pulley should be securely mounted on the motor shaft with accurate alignment to minimize vibration and prevent belt slippage. The belt-drive system should be properly tensioned, maintaining the manufacturer's recommended tension levels, to ensure efficient power transmission and avoid excessive wear on the belt and pulley surfaces.
- Our engineer then went on to thoroughly examine the He then assessed the motor's speed (using a tachometer) and compared its value to the speed recommended by the equipment manufacturer.
- This comparison allows for a verification of whether the motor is operating within the prescribed speed range as advised by the manufacturer of the machinery.

Findings

While inspecting the foundation for the motor, our engineer found that its surface was uneven and unbalanced.

The motor had inadequate support and the materials used for base support were of sub-standard quality.

The engineer also observed that both the pulley and the belt drive exhibited signs of significant wear.

While checking the speed (RPM) of the machine, he observed that it exceeded the RPM recommended by the manufacturer.

Solution

Our engineer's visit proved instrumental in identifying multiple factors contributing to the motor's failure and damage.

Our engineer recommended the following remedial measures to the operator to avoid any future issues.

- Firstly, he emphasized the importance of constructing a sturdy and well-leveled foundation for the motor.
- Secondly, he advised the incorporation of heavy C Section Channels as a robust base plate or support for the machine.
- Next, he recommended them to replace the pulley and the belt as they had both worn out.
- Finally, he suggested them to buy a new pulley of appropriate size to ensure the motor operates within the manufacturer's recommended RPM range.

The operator diligently adhered to our engineer's recommendations and implemented all necessary changes. Upon installation of the motor, it was observed that the motor operated smoothly, devoid of any vibrations or undesirable noise.

Remarks

1. The F-Class induction motor is a very powerful machine which plays a significant role in a groundnut processing factory. It is typically utilized to power various machinery and equipment involved in groundnut processing operations. This motor provides the necessary power and torque to drive machines such as groundnut cleaning and sorting equipment, oil extraction machines, and other auxiliary systems within the factory.
2. The recommended power supply for a three-phase F-Class induction motor is 280 to 415 Volts.
3. Ensure that the belt and pulley are properly fitted and that the tension is appropriate before starting the motor.
4. Make sure your motor operates at the load as declared by the manufacturer. Please contact our customer support team to learn more.
5. Regular lubrication of the motor is crucial to ensure its optimal performance.
6. Regular inspection and servicing of the motor can help detect many issues beforehand.

For smooth and uninterrupted working of your motor, always follow the instructions and standard operating procedure issued by the manufacturer in the user manual.

CLASS ACTIVITY

THREE PHASE AC CIRCUITS

1. Suppose you need to design a three-phase electric heater to dissipate 15 kW of heat when powered by 480 VAC. Your options are to build a delta-connected heater array or a wye-connected heater array:

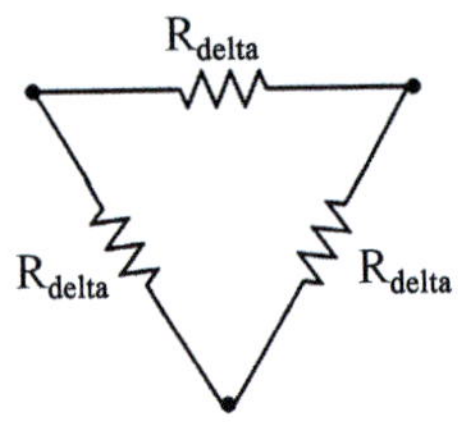

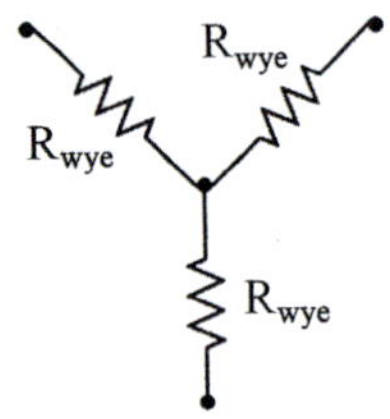

Calculate the proper resistance value for each array, to achieve the desired heat output:

R_{delta}=

R_{wye}=

2. A three-phase electric motor operating at a line voltage of 4160 VAC (RMS) draws 27.5 A of current (RMS) through each of its lines. Calculate the amount of apparent power consumed by this motor in both kVA electrical and HP mechanical units.

SUMMARY

- The induction motors especially three-phase induction motors are by and large used AC motors to convey mechanical power in modern applications. 80% of the motor is a three-phase induction motor among all motors used in undertakings. Consequently, the induction motor is the fundamental motor among any excess types of motor.
- When the rotor is stationary the frequency of rotor current is the same as the supply frequency. But when the rotor starts revolving, then the frequency depends upon the relative speed or on slip speed.
- Stator is made up of a number of stampings which are slotted to receive the windings. The stator carries a 3- phase winding and is fed from a 3-phase supply. It is wound for a definite number of poles, the exact number of poles being determined by the requirements of speed.
- The resistance of a squirrel-cage rotor is fixed & small as compared to its reactance which is very large especially at the start because at stand till, the frequency of the motor currents equals the supply frequency.
- The starting torque of such a motor is increased by increasing its power factor by adding external resistance in the rotor circuit from the star connected rheostat, the rheostat resistance being progressively cut out as the motor gains speed.
- The slip rings are mounted on the shaft. One end of each phase winding, after connecting the winding in star or delta is connected to the slip ring. Thus there are three slip rings mounted on the shaft with brushes resting on them.
- An induction motor can be used as an induction generator, or it can be unrolled to form a linear induction motor which can directly generate linear motion. The generating mode for induction motors is complicated by the need to excite the rotor, which begins with only residual magnetization.
- Polyphase motors have rotor bars shaped to give different speed-torque characteristics. The current distribution within the rotor bars varies depending on the frequency of the induced current.
- The method of changing the direction of rotation of an induction motor depends on whether it is a three-phase or single-phase machine. A three-phase motor can be reversed by swapping any two of its phase connections.
- Many useful motor relationships between time, current, voltage, speed, power factor, and torque can be obtained from analysis of the Steinmetz equivalent circuit, a mathematical model used to describe how an induction motor's electrical input is transformed into useful mechanical energy output.

REVIEW QUESTIONS

1. Explain the role of the changing magnetic field in generating electromotive force (EMF) in an induction motor.
2. What is difference between synchronous motor and induction motor?
3. Describe the advantages of using 3-phase power systems over single-phase power systems.
4. What is part of 3-phase induction motor?
5. How are the windings arranged in the stator of a 3-phase induction motor?
6. What is the function of the rotor bars and end rings in a squirrel cage rotor?
7. Explain the concept of slip and its significance in the operation of a 3-phase induction motor.
8. Describe how alternating current (AC) in the single coil produces a magnetic field.

REFERENCES

1. Alger, P. L., & Arnold, R. E. (1976). The history of induction motors in America. *Proceedings of the IEEE, 64*(9), 1380–1383.
2. Behrend, B. A. (1901). *The Induction Motor: A Short Treatise on Its Theory and Design, With Numerous Experimental Data and Diagrams*. McGraw Publishing Company / Electrical World and Engineer.
3. Boy de la Tour, H. (1906). *The Induction Motor: Its Theory and Design, Set Forth by A Practical Method of Calculation* (C. O. Mailloux, Trans.). McGraw Pub. Co.
4. Cook, D. M. (2003). *The Theory of The Electromagnetic Field* (p. 158). Courier Dover.
5. Ferraris, G. (1885). Electromagnetic rotation with an alternating current. *Electrician, 36*, 360–375.
6. Fogiel, M. (2002). *Basic Electricity* (p. 76). Research & Education Association.
7. Keljik, J. (2009). The three-phase, squirrel-cage induction motor. In *Electricity 4: AC/DC Motors, Controls, and Maintenance* (9th ed., pp. 112–115). Delmar, Cengage Learning.
8. Kennelly, A. E. (1893). Impedance. *Transactions of the American Institute of Electrical Engineers, X*, 172–232.
9. Loeb, L. B. (2007). *Fundamentals of Electricity and Magnetism* (Reprint of Wiley 1947 3rd ed.).
10. Tesla, N. (1888). A new system for alternating current motors and transformers. *AIEE Transactions, 5*, 308–324.
11. Thompson, S. P. (1895). *Polyphase Electric Currents and Alternate-Current Motors* (1st ed.). E. & F.N. Spon.

CHAPTER

10 Alternator and Synchronous Motor

LEARNING OBJECTIVES

After studying this chapter, you will be able to:

- Understand the basics of alternator
- Define alternator voltage regulation
- Describe alternator on load
- Discuss about synchronous motor
- Describe armature winding

KEY TERMS FROM THIS CHAPTER

Alexanderson alternator
Armature resistance
Electrical energy
Hysteresis motors
Mechanical energy
Permanent magnet generator (PMG)
Rotating field
Terminal voltage
Amplidyne
Combustion engines
Electrical generator
Lap winding
Permanent magnet alternator
Phasor diagram
Synchronous motors
Voltage regulator (VR)

INTRODUCTION

An alternator is a type of electrical generator that transforms mechanical energy into electrical energy in the form of alternating current. In most cases, alternators utilize a rotating magnetic field with a stationary armature for cost-effectiveness and simplicity. However, there are instances where a linear alternator or a rotating armature with a stationary magnetic field is employed. While technically any AC electrical generator can be classified as an alternator, the term is commonly used to describe small rotating machines powered by automotive and other internal combustion engines. A magneto is an alternator that utilizes a permanent magnet for its magnetic field. Alternators in power stations powered by steam turbines are referred to as turbo-alternators. The majority of the world's electric power is generated by large 50 or 60 Hz three-phase alternators in power plants, which is then distributed through electric power grids.

The synchronous motor is a popular AC rotating electrical machine that is commonly used to convert AC power from the grid into mechanical power. Featuring a salient pole synchronous rotor, it automatically synchronizes with the frequency of the power supply and operates at the synchronous speed.

Synchronous motors are used in environments that demand accurate speed control, such as oil pipelines, rolling mills, machine tools, production lines, and conveyors where fluctuations in speed could impact product quality. These motors operate using electromagnetic induction, with a 3-phase design that includes a field system producing a rotating magnetic field, along with a wound rotor or damper bars to create torque.

10.1. ALTERNATOR

An alternators is an electrical generator that converts mechanical energy to electrical energy in the form of alternating current. For reasons of cost and simplicity, most alternators use a rotating magnetic field with a stationary armature. Occasionally, a linear alternator or a rotating armature with a stationary magnetic field is used. In principle, any AC electrical generator can be called an alternator, but usually the term refers to small rotating machines driven by automotive and other internal combustion engines.

Simple forms of alternating current generating systems have been around since the discovery of magnetic induction in the 1830s. Initially, rotating generators naturally produced alternating current, but because it was not widely used, it was typically converted into direct current using a commutator in the generator. Pioneers like Michael Faraday and Hippolyte Pixii played a role in developing early machines, including Faraday's rotating rectangle with heteropolar operation. Lord Kelvin and Sebastian Ferranti also contributed to the development of early alternators with frequencies ranging from 100 to 300 Hz.

In the late 1870s, the first large-scale electrical systems were introduced with central generation stations to power Arc lamps. These lamps were used to illuminate entire streets, factory yards, and the interiors of large warehouses. Some lamps, such as the Yablochkov arc lamps introduced in 1878, operated more efficiently on alternating current. The development of these early AC generating systems also brought about the first use of the term "alternator."

In the early systems, engineers had to rely on their skills to regulate the proper amount of voltage supplied by generating stations. However, in 1883, the Ganz Works introduced the constant voltage generator, which could maintain a specified output voltage regardless of the load. The introduction of transformers in the mid-1880s popularized the use of alternating current, requiring the use of alternators to generate it. By 1891, polyphase alternators were introduced to supply currents with multiple phases. As technology advanced, alternators were designed to accommodate a range of alternating current frequencies for various applications such as arc lighting, incandescent lighting, and electric motors. Specialized radio frequency alternators like the Alexanderson alternator were developed for longwave radio transmission during World War 1, but were eventually replaced by vacuum tube transmitters in high power wireless telegraphy stations (Figure 10.1).

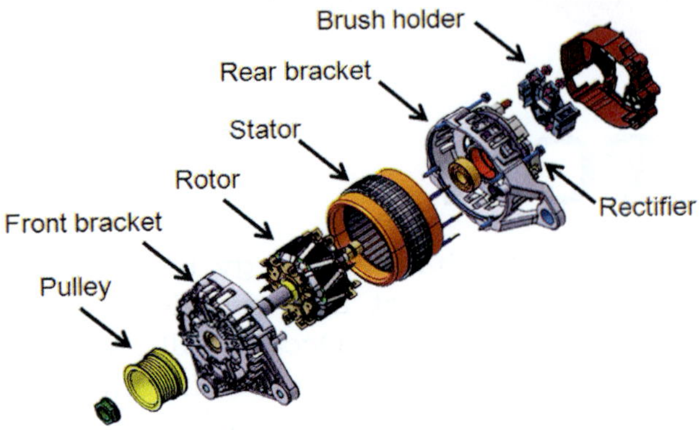

Figure 10.1. *Exploded view of an alternator.*

Source: https://www.researchgate.net/figure/Exploded-view-of-an-alternator-12_fig1_317158608.

10.1.1. Principle of Operation

When a conductor moves in relation to a magnetic field, an electromotive force (EMF) is produced (according to Faraday's Law). This EMF changes its direction when the conductor passes under magnetic poles of different polarity. Usually, a rotating magnet, known as the rotor, spins within a fixed group of conductors, called the stator, which are wound around an iron core. The magnetic field cuts through the conductors, creating an induced EMF as the mechanical energy causes the rotor to rotate.

The AC voltage in the stator windings is created by the rotating magnetic field. A synchronous generator becomes an alternator because the currents in the stator windings change in sync with the rotor's position.

The magnetic field of the rotor can be created by either permanent magnets or a field coil electromagnet. In automotive alternators, a rotor winding is used to adjust the voltage output by changing the current in the rotor field winding. Permanent magnet machines do not have the loss of magnetizing current in the rotor, but they are limited in size due to the expense of the magnet material. As the permanent magnet field remains constant, the terminal voltage of the generator changes in direct correlation with its speed. Brushless AC generators are typically larger than those found in automotive settings (Gandzha, Sogrin, & Kiessh, 2015).

The automatic voltage control device adjusts the field current to maintain a steady output voltage. If demand causes a drop in voltage from the armature coils, the voltage regulator (VR) increases current to the rotating field coils. This boosts the magnetic field around the coils, generating a higher voltage in the armature coils and restoring the output voltage to its initial level.

Alternators in central power stations are responsible for managing field current to regulate reactive power and stabilize the power system during temporary faults. Typically, they contain three sets of stator windings that are physically offset to produce a three-phase current, each phase shifted by one-third of a period relative to the others.

10.1.2. Classifications

There are different ways to classify alternators, including by their method of excitation, number of phases, rotation type, cooling method, and intended application.

- **By Excitation:** There are two primary methods for generating the magnetic field in alternators. One way is by utilizing permanent magnets that produce their own constant magnetic field, while the other method involves the use of field coils. Alternators that utilize permanent magnets are known as magnetos.

In some alternators, wound field coils create an electromagnet for generating a rotating magnetic field. An alternator that utilizes permanent magnets to produce alternating current is known as a permanent magnet alternator (PMA). A permanent magnet generator (PMG) can generate either alternating current or direct current with the presence of a commutator.

- **Direct-Connected Direct-Current (DC) Generator:** This excitation method involves a DC generator mounted on the same shaft as the alternator. The DC generator produces a small amount of power to stimulate the field coils of the alternator, allowing it to generate electricity. Another version of this system involves an alternator that initially uses DC power from a battery for excitation upon startup, and then becomes self-excited.
- **Transformation and Rectification:** This technique relies on the residual magnetism in the iron core to produce a slight magnetic field, resulting in a small voltage being created. This voltage is then utilized to stimulate the field coils, enabling the alternator to produce a more powerful voltage during its initial buildup phase. Once the AC voltage has been established, the field is then powered by rectified voltage from the alternator.
- **Brushless Alternators:** A brushless alternator consists of two alternators connected end-to-end on a single shaft. Prior to 1966, alternators utilized brushes with a rotating field. The development of semiconductor technology made it possible to create brushless alternators. While smaller brushless alternators may appear as a single unit, the main alternator and exciter are clearly visible in larger versions. The exciter contains stationary field coils and a rotating armature, while the main alternator uses a rotating field and stationary armature configuration. A rotating rectifier assembly, known as a bridge rectifier, is mounted on the rotor, eliminating the need for brushes and slip rings. This design reduces the number of wearing parts in the alternator, with the main alternator featuring a rotating field and stationary armature for power generation.

Adjusting the current in the stationary exciter field coils changes the 3-phase output from the exciter. This output is converted to DC by a rotating rectifier assembly, located on the rotor. The DC then powers the rotating field of the main alternator, affecting the overall alternator output. In essence, a small DC exciter current indirectly governs the main alternator's output.

- **By Number of Phases:** Alternators can also be classified based on the number of phases in their output voltage. This can be either single phase or polyphase. Among polyphase alternators, three-phase alternators are the most prevalent, though they can also have two phases, six phases, or even more.
- **By Rotating Part:** The alternator can have either a revolving armature or a revolving magnetic field as its rotating component. In the revolving armature type, the armature is wound on the rotor and moves through a stationary magnetic field. This type is not commonly used.

The revolving field type, on the other hand, has a magnetic field on the rotor that rotates through a stationary armature winding. This design has the advantage of requiring less power in the rotor circuit compared to the armature circuit, resulting in smaller and less expensive slip ring connections. For direct-current rotors, only two contacts are needed, whereas rotors with multiple phases and sections would require more connections. The stationary armature can be wound for various medium voltage levels, but slip ring connections for voltages over a few thousand volts are costly and inconvenient to manufacture.

- **Cooling Methods:** Several alternators are cooled by surrounding air, pushed through the enclosure by a fan attached to the same shaft that powers the alternator. In certain vehicles like transit buses, a significant load on the electrical system may necessitate an oil-cooled alternator. Water-cooling is also utilized in marine applications. High-end automobiles may opt for water-cooled alternators to accommodate demanding electrical systems.

10.1.3. Specific Applications

- **Electric Generators:** Many power generation stations utilize synchronous machines as their generators. To connect these generators to the utility grid, synchronization conditions must be fulfilled.
- **Automotive Alternators:** Modern internal combustion engine automobiles utilize alternators to both charge the battery and supply power to the electrical system while the engine is operating.

Before the 1960s, cars were equipped with DC dynamo generators featuring commutators. However, with the introduction of cost-effective silicon-diode rectifiers, alternators became the preferred choice.

- **Diesel-Electric Locomotive Alternators:** In newer diesel-electric trains, the prime mover powers an alternator that generates electricity for the traction motors, whether they are alternating current (AC) or direct current (DC).

The traction alternator typically includes built-in silicon diode rectifiers to supply the traction motors with a maximum of 1,200 volts DC.

The original diesel electric locomotives, as well as those that are still in use today, were equipped with DC generators for easier speed control of the DC traction motors. In many cases, these locomotives were built with two generators, one for producing excitation current for the main generator.

The generator can also provide head-end power (HEP) or power for electric train heating, if desired. To enable the HEP option, the engine must maintain a steady speed of 900 revolutions per minute (r/min) for a 480-volt, 60 Hz application, even when the locomotive is stationary.

- **Marine Alternators:** Marine alternators used in yachts are specifically designed for the salt-water environment, similar to automotive alternators but with necessary modifications. These alternators are built to be explosion proof in order to prevent brush sparking from igniting any explosive gas mixtures in the engine room.

They can be either 12 or 24 volt, depending on the yacht's electrical system. Larger marine diesels may have multiple alternators to meet the high electrical demand of a modern yacht. In systems with a single alternator, power can be divided between the engine starting battery and the domestic battery using a split-charge diode or a voltage-sensitive relay. Because large house battery banks are expensive, Marine alternators typically utilize external regulators. Multistep regulators are used to manage the field current in order to optimize charging efficiency and prolong battery lifespan. These regulators can be customized for various types of batteries. Additionally, two temperature sensors can be installed – one for the battery to adjust charging voltage and another to protect the alternator from overheating.

10.2. ALTERNATOR VOLTAGE REGULATION

Voltage regulators are used to automatically control the terminal voltage of the generator. It should be mentioned that AC machines have higher internal reactance as compared to DC machines. Magnetic amplifying components are often used in regulators to achieve this voltage control.

The purpose of automatic regulator equipment is to:

- Control the system voltage within prescribed limits.
- Regulate the division of reactive power shared between machines running in parallel.
- Control the field circuit closely to keep the machine in synchronism with the system when operating at near unity or leading power factor.
- Boost the excitation under system fault conditions to keep the machine in synchronization with the system.

The output voltage can only be maintained approximately constant, and this regulation is determined by two key measurements:

- Load regulation, which signifies the change in output voltage for a given change in load current.
- Line regulation, also known as input regulation, refers to how much the output voltage fluctuates when there are changes in the input voltage. It is typically expressed as a ratio of the output change to the input change.

The output voltage of an AC generator is controlled by regulating the strength of the rotating DC field. Increasing the field strength increases the AC output voltage, whereas decreasing the DC field strength decreases the output voltage. As the load on the generator increases, it is necessary to increase the field strength in order to keep the output voltage constant. (Guo Min, Jin Qingren, Yao Zhiyang, & Chen Weidong, 2020) (Figure 10.2).

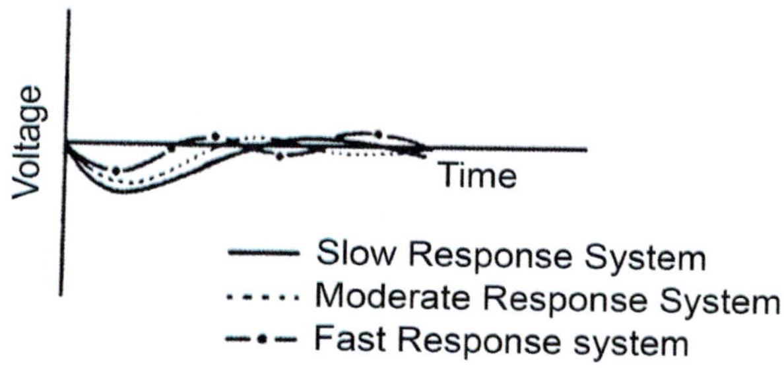

Figure 10.2. *Restoration of voltage as a function of response time.*

Source: https://ecampusontario.pressbooks.pub/electrotechnology/chapter/part-4-alternator-voltage-regulation/.

Voltage regulation methods:

i. Direct-acting rheostat regulator
ii. Electromechanical
iii. Electronic
iv. Amplidyne
v. Power amplifier

i. **Direct-Acting Rheostat Regulator:** Direct-acting rheostat regulators rely on adjusting variable resistances in the main exciter's field circuit.

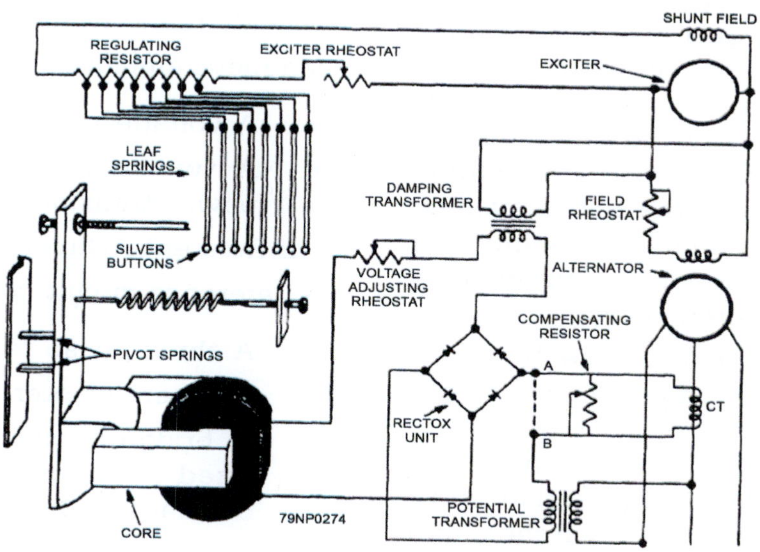

Figure 10.3. *Silverstat voltage regulator on an AC generator.*

Source: https://electronicstechnician.tpub.com/14344/Figure-8–5-A-Silverstat-Voltage-Regulator-On-An-Ac-Generator-334.html.

The Silverstat voltage regulator uses silver-buttoned leaf springs to control voltage, operating as a direct-acting rheostatic type of regulator. The leaf springs close or open contacts in a series to adjust the voltage (Figure 10.3).

ii. Electromechanical Voltage Regulators

Disadvantages

- The response time of the regulator and the equipment necessary to change the voltage variation signal into a mechanical force and act on the main exciter field is too great.
- The dead band is not acceptable for steady-state conditions when generators are running near the stability limit.
- As machines get larger in output, the increasing power in the excitation circuits makes it difficult for the regulator contacts to handle

iii. Electronic Regulators: The initial electronic regulators were designed with a bridge circuit containing both fixed and variable arms. This bridge was powered by a reference voltage, with the error voltage determined by the two remaining bridge terminals, showing the variance between the current value and the desired setting.

The error voltage from the bridge is input into an electronic amplifier, which can be constructed using a grid of thyratrons to regulate the flow of field current. Thyratrons are specialized gas-filled tubes that serve as high-energy electrical switches and controlled rectifiers.

In a standard power amplifier setup, a DC generator is designed to use 1 watt in the field circuit, while expecting an output load of 100 watts. This output can be sent to the field of a second generator, creating another 100:1 ratio, resulting in a total power amplification of 10,000:1.

Nevertheless, a major downside to this configuration is the extensive time needed to implement the change in output. The time delay in a two-stage amplification process is around 1 second, which is deemed too sluggish.

iv. Amplidyne: The DC generator is uniquely designed to achieve both higher amplification than a two-stage amplifier and a response speed ten to twenty times faster.

The Amplidyne is a unique motor generator that serves as a power amplifier. It has a separately excited field that enables fine adjustments in excitation power, known as the input or control signal. These small alterations in the input signal lead to significant changes in the output power. By utilizing this technique, power amplification ratios of up to 250,000:1 can be attained.

v. Magnetic Amplifier Regulator

Magnetic amplifier regulators consist of:

- A voltage reference circuit.
- A two-stage magnetic amplifier, power transformer, and rectifier.

vi. Reference Circuit

A reference circuit consists of:

- A three-phase rectifier
- A Potentiometer
- A bridge circuit made up of two fixed resistors and two glow-discharge tubes

A VR tube, also known as a glow-discharge tube, is used to regulate voltage in a circuit by creating a constant voltage despite changes in line voltage or load. When a specific voltage is applied to the electrodes of the tube, the gas becomes ionized, producing a glow discharge near the cathode electrode. This ionization turns the VR tube into a negative resistance device. As the current passing through the tube increases, the level of ionization also increases, leading to a decrease in resistance. This allows the VR tube to conduct sufficient current to maintain the desired voltage across its terminals.

Advantages

- Minimum of moving parts
- No electronic components
- Regulator is built entirely of static components
- Time response is very rapid, and there is no dead band.
- This regulator operates by amplifying the difference between the generator terminal voltage and a fixed reference voltage using magnetic amplifiers in series. The amplified voltage is then used to directly adjust the main exciter field current.

10.3. ALTERNATOR ON LOAD

The terminal voltage of the alternator will fluctuate when the load changes, primarily due to three factors: armature resistance IR_a causing voltage drop, armature leakage reactance IX_L causing voltage drop, and voltage drop due to armature reaction.

10.3.1. Effect of Load on Alternator

- **Voltage Drop Due to Armature Resistance:** The armature winding resistance per phase results in a voltage drop of IRa per phase.

  The voltage drop caused by the armature resistance is synchronized with the armature current I. In reality, this voltage drop is so minuscule that it can be disregarded.
- **Voltage Drop Due to Armature Leakage Reactance:** As current passes through the armature conductors, the magnetic flux will begin to move through the armature core. Some of the flux will follow alternate paths and will not pass through the air gap, leading to what is known as leakage flux (Figure 10.4).

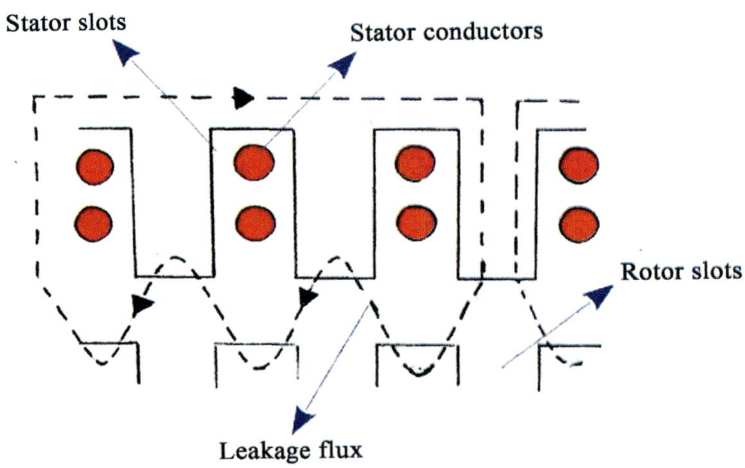

Figure 10.4. *Voltage drop due to armature leakage reactance.*

Source: https://www.electrically4u.com/alternator-on-load/#google_vignette.

In this scenario, the amount of leaked energy is determined by the current passing through the conductor and its alignment with the terminal voltage. This leakage energy creates an electromotive force (emf) due to self-inductance, known as reactance emf. This reactance emf causes the armature current I to lag by 90°.

Therefore, the armature winding is characterized by a leakage reactance XL, resulting in a voltage drop of IXL. The generated emf must surpass this voltage drop caused by the leakage reactance in order to produce its output.

$E = V + I(R_a + jX_L)$ (1)

This is illustrated in Figure 10.5.

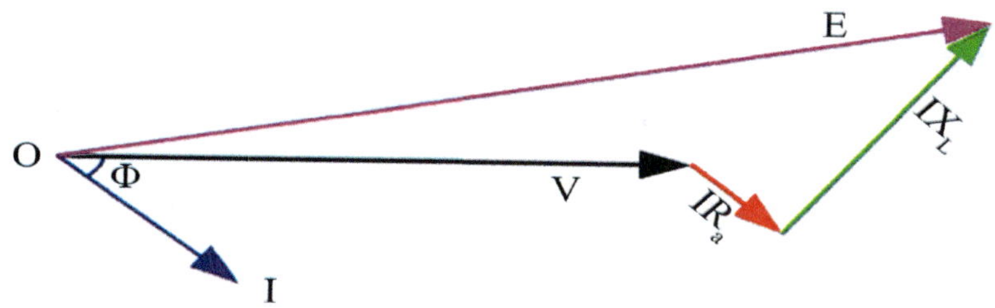

Figure 10.5. *Phasor diagram of voltage drop due to armature leakage reactance.*

Source: https://www.electrically4u.com/alternator-on-load/#voltage-drop-due-to-armature-resistance.

The above phasor diagram is constructed as below;

- The voltage phasor is taken as the reference phasor.
- The armature current lags behind the voltage by an angle Φ. Hence the current phasor is drawn at an angle Φ from the voltage phasor.
- The phasor for armature resistance drop is drawn parallel to the current phasor from the extremity of Voltage phasor V.
- Leakage reactance drop is drawn perpendicular to the current phasor from the extremity of IR_a phasor.
- Join 0 and the extremity of IX_L phasor to get E_b.
- **Voltage Drop Due to Armature Reaction:** Armature reaction is the influence of the armature flux on the main field flux. This phenomenon can also be observed in DC generators.

When comparing to a DC generator, the power factor of a load in an alternator significantly impacts the armature reaction. When discussing power factor under loading conditions, three cases are considered.

- Unity power factor load.
- Zero power factor lagging load
- Zero power factor leading load.

The alternator's armature reaction results in various effects including cross-magnetizing, demagnetizing, and magnetizing effects, all of which induce distortion in the main field flux and impact the generated electromotive force. The voltage drop caused by armature reaction can be approximated by considering the presence of a fictitious reactance X_a, known as armature reactance. This voltage drop is expressed as IX_a.

The leakage reactance X_L and armature reaction reactance X_a together called synchronous reactance X_S.

$$X_S = X_L + X_a \quad (2)$$

Therefore, the total voltage drop in an alternator when under load is the combined result of voltage drops caused by armature resistance, armature leakage reactance, and armature reaction reactance.

$$V = I R_a + j I X_L + j I X_a = I (R_a + j X_L + j X_a) = I (R_a + j (X_L + X_a)) \quad (3)$$

$$V = I (R_a + j X_S) = I Z_S \quad (4)$$

Where; Z_S is known as the synchronous impedance of an alternator.

The above discussions indicate that changes in load result in alterations to the terminal voltage of the alternator, which can be attributed to the synchronous impedance of the alternator.

Now let us look at the phasor diagrams of alternator for different load conditions.

10.3.2. Phasor Diagrams of Alternator on Load

In order to create the phasor diagrams, it is important to understand the terms used in the diagrams below.

E_0 is the no-load voltage. It is the maximum voltage induced in the armature without giving any load.

E is the load voltage. It is the induced voltage after overcoming the armature reaction. E is vectorially less than the no-load voltage.

I is the armature current per phase

V is the terminal voltage. It is vectorially less than E by IZ and also vectorially less than E_0 by IZ_S.

Φ is the cosine angle between terminal voltage and current.

The impedances are given by:

$$Z = R_a + jX_L = \sqrt{(R_a^2 + X_L^2)} \quad (5)$$

$$Z_S = R_a + jX_s = \sqrt{(R_a^2 + X_S^2)} \quad (6)$$

where X_L is the leakage reactance, X_a is the armature reaction reactance and X_S is the synchronous reactance and Z_S is the synchronous impedance.

10.3.2.1. Unity Power Factor Load

The phasor diagram of an alternator for unity power factor load is shown Figure 10.6.

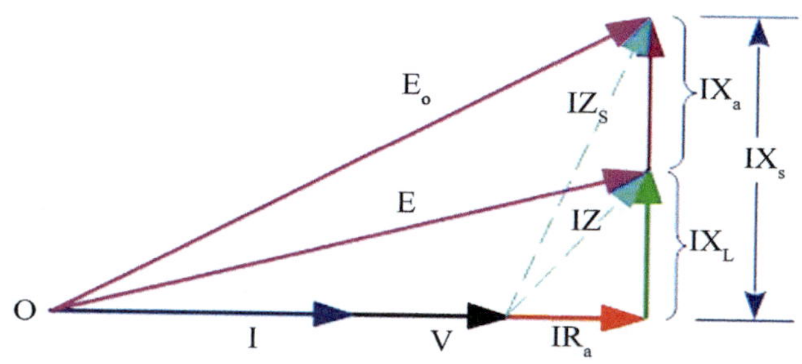

Figure 10.6. *Phasor diagram of an alternator for unity power factor load.*

Source: https://www.electrically4u.com/alternator-on-load/#phasor-diagrams-of-alternator-on-load.

- How the Phasor Diagram Is Drawn? Follow the Below Given Procedure
    1. Voltage phasor V is taken as the reference phasor.
    2. For unity power factor load, V and I phasor are in phase. So, the Current phasor I is drawn on the voltage phasor V.
    3. The phasor of Armature resistance drop IRa is drawn parallel to the current phasor from the extremity of V phasor.
    4. The armature leakage resistance drop IXL is drawn perpendicular to the current phasor, from the extremity of IRa phasor.
    5. Join V phasor and IXL phasor to get IZ phasor (shown as a dotted line).
    6. Join O and the extremity of IZ to get E (shown as a pink color line)
    7. Draw the armature reaction reactance drop phasor IXS perpendicular to the current phasor from the extremity of IXL phasor.
    8. Join V phasor and IXS phasor to get IZS phasor (shown as a dotted line).
    9. Join O and the extremity of IZS to get E0(shown as a pink color line)

10.3.2.2. Lagging Power Factor Load

The phasor diagram of an alternator for lagging power factor load is shown Figure 10.7.

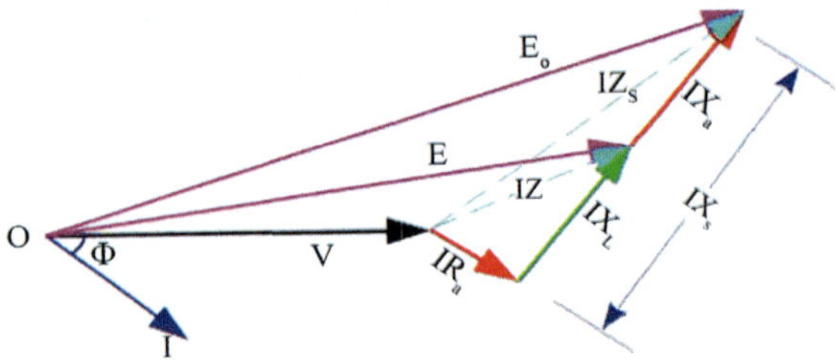

Figure 10.7. *Phasor diagrams of alternator on load.*

Source: https://www.electrically4u.com/alternator-on-load/#phasor-diagrams-of-alternator-on-load.

The phasor diagram above is created using the same method detailed for unity power factor.

The sole difference is that the current is now lagging behind the voltage by an angle Φ. Therefore, draw the current phasor with an angle Φ in relation to the voltage phasor V.

10.3.2.3. Leading Power Factor Load

The phasor diagram of an alternator for leading power factor load is shown Figure 10.8.

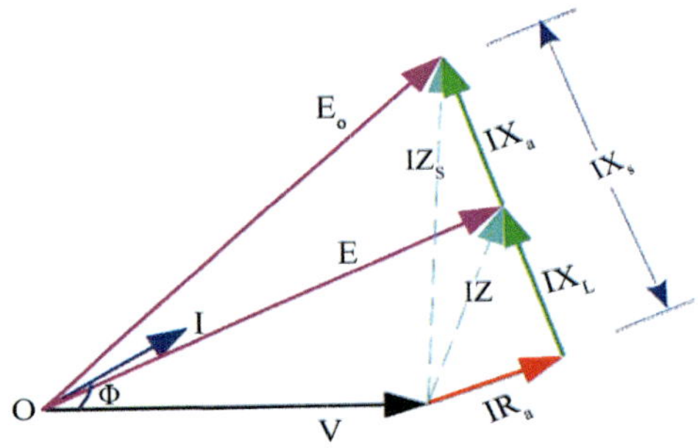

Figure 10.8. *Phasor diagram of an alternator for leading power factor load.*

Source: https://www.electrically4u.com/alternator-on-load/#phasor-diagrams-of-alternator-on-load.

When dealing with the leading power factor load, the phasor diagram is drawn in a manner similar to that of unity power factor. The main distinction is that in this case, the current leads the voltage by an angle Φ. Therefore, the current phasor is depicted at an angle Φ in relation to the voltage phasor V.

10.4. SYNCHRONOUS MOTOR

A synchronous electric motor is an AC motor where the shaft rotates in sync with the supply current frequency. Electromagnets are used in the stator to create a magnetic field that aligns with the current oscillations. The rotor, equipped with permanent or electromagnets, turns in unison with the stator field, creating two synchronized rotating magnet fields. If the motor is powered by independently excited multiphase AC electromagnets on both the rotor and stator, it is referred to as doubly fed (Figure 10.9).

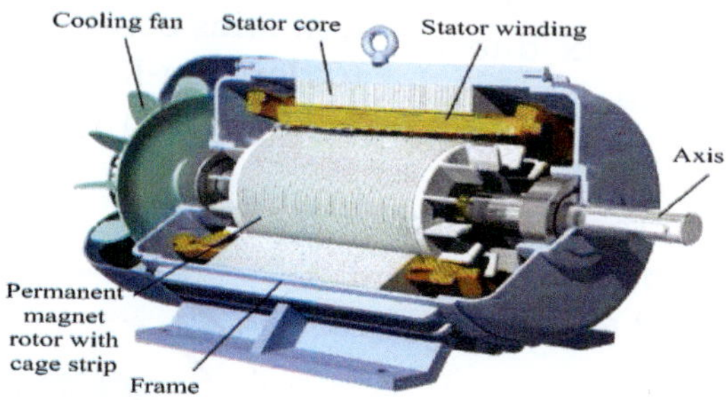

Figure 10.9. *Structure of synchronous motor.*

Source: https://www.researchgate.net/figure/Structure-of-synchronous-motor_fig3_368944283.

Synchronous and induction motors are the two most commonly utilized AC motors. Synchronous motors operate at a speed synchronized with the line frequency as they do not rely on induction for the rotor's magnetic field. On the other hand, induction motors require slip, where the rotor must rotate slightly slower than the AC frequency to induce current in the rotor. (Cai, Hao, Wang, et al., 2021).

Synchronous clocks, appliances timers, tape recorders, and precision servomechanisms require small synchronous motors which must operate at a precise speed. The accuracy of these motors depends on the power line frequency, which is controlled in large interconnected grid systems.

Synchronous motors come in a variety of sizes, from self-excited fractional to industrial sizes. In the lower horsepower range, they are often used to maintain a consistent speed. These motors are commonly found in analog electric clocks, timers, and similar devices.

In industrial applications, synchronous motors are a highly efficient way to convert AC energy into work. Larger motors typically have an electrical efficiency of over 95%, and they can operate at leading or unity power factor to help correct power factor issues.

10.4.1. Types of Synchronous Motor

Synchronous motors fall under the category of synchronous machines that also includes synchronous generators. When the field poles are advanced ahead of the air-gap flux by the prime mover, it results in generator action. Conversely, when the field poles are lagging behind the air-gap flux due to the shaft load's retarding torque, it causes motor action.

There are two main categories of synchronous motors based on how the rotor is magnetized: non-excited and direct-current excited.

10.4.1.1. Non-Excited Motors

In motors without excited components, the rotor is constructed from steel and revolves in synchronization with the stator's rotating magnetic field, resulting in a nearly consistent magnetic field within it. The external stator field causes the rotor to become magnetized, creating the necessary magnetic poles for its rotation. The rotor is typically composed of high-retentivity steel, such as cobalt steel. These are manufactured in permanent magnet, reluctance and hysteresis designs:

10.4.1.1.1. Permanent-Magnet

A motor that uses permanent magnets in the rotor to create a constant magnetic field is known as a permanent-magnet synchronous motor (PMSM). The stator is equipped with windings that are connected to an AC electricity supply in order to produce a rotating magnetic field, similar to an asynchronous motor. When operating at synchronous speed, the rotor poles align with the rotating magnetic field. PMSMs are comparable to brushless DC motors, with neodymium magnets being the most commonly used. However, fluctuations in neodymium magnet prices have led to research in ferrite magnets. Machines using ferrite magnets require a magnetic circuit that can concentrate the magnetic flux, often resulting in the use of spoke type rotors. While machines with ferrite magnets have lower power and torque densities compared to neodymium machines. (Orlowska-Kowalska, Wolkiewicz, Pietrzak, et al., 2022) (Figure 10.10).

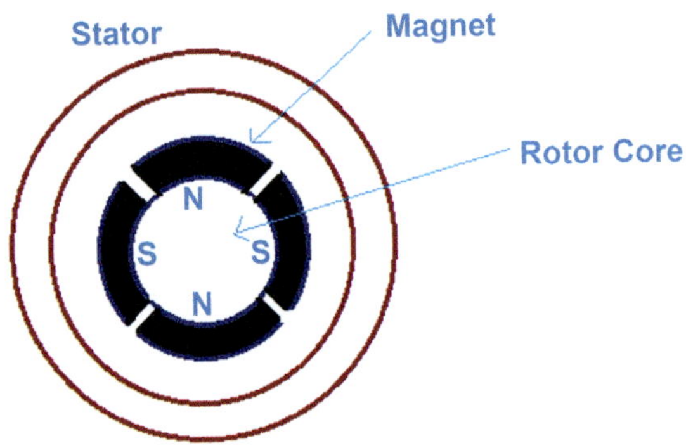

Figure 10.10. *Permanent magnet synchronous motor.*

Source: https://electricalbaba.com/permanent-magnet-synchronous-motor-pmsm-construction-working-principle/.

Permanent magnet synchronous motors (PMSMs) have been utilized as motor options for gearless elevators since the year 2000. While many PMSMs need a variable-frequency drive for initial activation, some models incorporate a squirrel cage in the rotor for self-starting capabilities, known as line-start or self-starting models. These

PMSMs are commonly chosen as more efficient replacements for induction motors due to their lack of slip, but it is essential to ensure that the synchronous speed is achieved and the system can handle torque ripple during startup. The control of PMSMs is typically achieved through direct torque control and field oriented control methods.

10.4.1.1.2. Reluctance

Reluctance motors have a rotor made of solid cast steel with salient toothed poles that project out. Typically, there are fewer rotor poles than stator poles in order to reduce torque variations and prevent all poles from aligning at once, which would result in no torque being generated. The air gap in the magnetic circuit and thus the reluctance is smallest when the rotor poles align with the rotating magnetic field of the stator, and increases as the angle between them grows. This difference in alignment creates torque that causes the rotor to align itself with the nearest stator pole. Once the rotor reaches synchronous speed, it becomes "locked" to the rotating stator field, preventing the motor from starting. To provide torque below synchronous speed, the rotor poles are usually equipped with squirrel-cage windings. The motor initially operates as an induction motor until it nears synchronous speed, at which point the rotor is pulled in and locks to the stator field.

Reluctance motors can be rated from a few watts to 22 kW. Smaller reluctance motors are suitable for instrumentation applications due to their low torque. Medium torque, multi-horsepower motors typically utilize squirrel cage construction with toothed rotors. When paired with an adjustable frequency power supply, all motors in a drive system can run at the same speed since the motor operating speed depends on the power supply frequency.

10.4.1.1.3. Hysteresis

Hysteresis motors feature a sturdy, smooth, cylindrical rotor made from high-coercivity magnetically "hard" cobalt steel. This material has a broad hysteresis loop, requiring a strong magnetic field to reverse magnetization once it is magnetized in a specific direction. As the stator field rotates, each section of the rotor experiences a changing magnetic field, causing the magnetization phase to lag behind the applied field. This lag results in the rotor's magnetic field axis trailing behind the stator field's axis by a constant angle δ, generating torque as the rotor attempts to align with the stator field. When the rotor is operating at a speed below synchronous speed, each particle within the rotor encounters a magnetic field that oscillates at the slip frequency, guiding it through a hysteresis loop. This results in the rotor field falling behind and generating torque. The rotor is constructed with a 2-pole low reluctance bar structure. As the rotor nears synchronous speed and the slip decreases to zero, it becomes magnetized and aligns itself with the stator field, leading to the rotor being synchronized with the rotating stator field.

One of the main benefits of the hysteresis motor is its ability to maintain a constant torque from the moment it starts up until it reaches synchronous speed, thanks to the fact that the lag angle δ is not dependent on speed. This allows the motor to be self-starting without requiring an induction winding, although some designs include a squirrel-cage conductive winding structure in the rotor to provide additional torque during start-up.

Hysteresis motors are produced in sub-fractional horsepower capacities, mainly for servomotors and timing motors. While more costly than reluctance motors, hysteresis motors are preferred for applications where consistent and precise speed control is essential.

10.4.1.2. Externally excited motors

These motors are typically manufactured in larger sizes, usually greater than 1 horsepower or 1 kilowatt, and require direct current (DC) for rotor excitation. The most common method of supplying DC to these motors is through the use of slip rings.

A brushless AC induction and rectifier arrangement can also be used.

The power may be supplied from a separate source or from a generator directly connected to the motor shaft.

10.4.2. Control Techniques

A control system, such as a VFD or servo drive, is necessary for the operation of a permanent magnet synchronous motor or reluctance motor. Various control methods for synchronous machines are available and can be selected based on the specific construction and application of the electric motor.

10.4.2.1. Scalar Control

Scalar control of an AC electric motor allows for variable speed operation by adjusting the supply voltage or current ("magnitude") and frequency without considering the motor's magnetic field orientation. It is typically open-loop, using equations for steady-state operation and limited current sensing. While scalar control is still common in low-performance motors due to its simplicity and cost-effectiveness, higher-performance motors now often utilize vector control for improved handling of transient processes. The industry expects vector control to eventually become the standard for all motors. The variants of the scalar control include open-loop control and closed-loop control.

- **Open-Loop:** In an open-loop system, the voltage V is typically controlled proportionally to the frequency f (known as V/f control, V/Hz control, Constant Volts/Hertz, or CVH). The V/f method has the advantage of maintaining a constant magnetic flux within the stator, ensuring consistent motor performance at all speeds. A voltage boost is often utilized at lower frequencies to offset the resistance of the coils.

  An open-loop V/f control system is effective for applications where the load torque remains constant and there are slow changes in rotational speed. These controllers are commonly referred to as general purpose AC drives.
- **Closed-Loop:** When sensors are used in a closed-loop configuration to improve transitional response time, a common method involves employing a rotational speed sensor, also known as closed-loop V/Hz control. The speed error is sent through a proportional-integral controller to generate the accumulated slip difference, which is then combined with the direct speed sensor reading to create a frequency control signal.

In a torque-control variant (TC), not to be confused with direct torque control (DTC), the motor torque remains constant in the steady state, but this method requires a current sensor. In TC, the control signals for frequency and flux (voltage or current, depending on the drive type) are separated. The flux control is determined by the flux estimate, and the frequency control is derived from the torque estimate and speed sensor data. Although this approach offers improved performance, it also introduces additional complexity and potential stability challenges.

10.4.2.2. Vector Control

Vector control, also known as field-oriented control (FOC), is a method of controlling a variable-frequency drive (VFD) for a three-phase AC or brushless DC electric motor. In this method, the stator currents of the motor are identified as two orthogonal components that can be represented as a vector. One component represents the magnetic flux of the motor, while the other represents the torque. The control system of the drive calculates the current component references based on the flux and torque references provided by the speed control. Proportional-integral (PI) controllers are typically used to ensure that the measured current components match their reference values, with pulse-width modulation determining the transistor switching based on the output of the PI current controllers.

Field-oriented control (FOC) is utilized for regulating AC synchronous and induction motors. Initially designed for demanding motor tasks which necessitate seamless operation across the entire speed spectrum, delivering maximum torque at zero speed, and boasting superior dynamic performance like rapid acceleration and deceleration. Nevertheless, FOC is gaining popularity in less demanding applications due to its ability to reduce motor size, cost, and power consumption. As microprocessors continue to boost in computational power, it is anticipated that FOC will eventually supplant single-variable scalar control methods such as volts-per-Hertz (V/f control) in a widespread manner.

10.4.3. Synchronous Speed

The speed at which an AC motor operates synchronously is influenced by both the frequency of the power source and the number of poles in the motor. To calculate the motor's revolutions per minute (RPM), you can multiply the frequency by 60 and then divide by the number of pole pairs.

The synchronous speed of a synchronous motor is given:

in RPM, by:

$$N_s = 60\frac{f}{P} = 120\frac{f}{p} \tag{7}$$

and in rad·s^{-1}, by:

$$\omega_s = 2\pi\frac{f}{P} = 4\pi\frac{f}{p} \tag{8}$$

Where; f is the frequency of the AC supply current in Hz; p is the number of magnetic poles; and P is the number of pole pairs (rarely, planes of commutation), $P = p/2$.

Examples

The synchronous motor, which has 4 poles (2 pole-pairs), is running on a single-phase AC supply with a frequency of 50 Hz.

The number of pole-pairs is 2, so the synchronous speed is:

$$N_s = 60\frac{\text{rpm}}{\text{Hz}} \times \frac{50\text{ Hz}}{2} = 1500\,\text{rpm} \tag{9}$$

A three-phase, 12-pole (6-pole-pair) synchronous motor is operating at an AC supply frequency of 60 Hz. The number of pole-pairs is 6, so the synchronous speed is:

$$N_s = 60\frac{\text{rpm}}{\text{Hz}} \times \frac{60\text{ Hz}}{6} = 600\,\text{rpm} \tag{10}$$

The number of magnetic poles, represented as p, in a motor is equivalent to the number of coil groups per phase. To calculate the number of coil groups per phase in a 3-phase motor, simply divide the total number of coils by 3. This process can become complicated when the coils span multiple slots in the stator core. For example, if you count 12 coil groups in a 3-phase motor, it will have 4 magnetic poles. In a 12-pole 3-phase machine, there will be a total of 36 coils. It is important to note that the number of magnetic poles in the rotor is always equal to the number of magnetic poles in the stator.

10.4.4. Construction

Electric motors consist of two main components: the stator and the rotor. The construction of synchronous and induction motor stators is quite similar. In the case of a synchronous motor, the construction is similar to that of a synchronous alternator. The stator frame typically includes a wrapper plate, along with circumferential ribs and keybars. Frame mounts and footings are necessary to support the weight of the machine. The synchronous stator winding usually consists of a 3-phase winding, supplied with a 3-phase power supply, while the rotor requires a DC power supply.

DC excited motors need brushes and slip rings to connect to the excitation source. The field winding can also be excited by a brushless exciter. Cylindrical rotors, also called non-salient pole rotors, are utilized for motors with up to six poles.

A salient pole rotor is often used in machines that require a large number of poles. In most synchronous motor designs, the armature remains stationary while the field winding rotates. This construction offers advantages over DC motors, which use a rotating armature.

10.4.5. Operation

Electric motors function by utilizing the interaction between the magnetic fields of the stator and rotor. In synchronous motors, the stator carries three-phase currents, creating a rotating magnetic flux that generates a rotating magnetic field. The rotor aligns with this magnetic field and rotates accordingly, leading to synchronization. While it is possible for a single-phase or two-phase stator to be utilized, the direction of rotation may not be defined and the motor could start in either direction unless specific measures are taken during startup.

10.4.5.1. Amortisseur Winding

The amortisseur winding, also known as the damper winding, is a squirrel-cage winding located on the rotor of a standard synchronous electric machine. Its purpose is to reduce transient oscillations and assist with the start-up process.

Because the design of a damper winding is similar to that of an asynchronous motor, it allows for direct-on-line starting and can also be utilized for motor operation in asynchronous mode.

The damper winding was first created by Maurice Leblanc in France and Benjamin G. Lamme in the US to address the issue of hunting oscillations that occurred when early generators were powered by steam engines with pulsating torque. In modern designs, generators are now driven by turbines, making hunting less of a concern. However, motors, such as those driving piston compressors, still encounter pulsating torque.

Building the damper windings is a complicated process that relies heavily on practical experience. A standard damper winding includes short-circuit bars that, in machines with cylindrical rotors, are located in the same slots as the field windings. In machines with salient pole rotors, the bars are situated in specific slots on the pole shoes. The quadrature axis area of salient pole machines does not have any bars. The bars are connected to rings or plates that encircle the rotor.

10.4.6. Starting Methods

Once a certain size threshold is reached, synchronous motors are unable to start on their own. This is because of the rotor's inertia, which makes it unable to immediately match the rotation of the stator's magnetic field. Because synchronous motors do not generate any average torque when at a standstill, they require an additional mechanism to help them reach synchronous speed.

Big motors running on commercial power have a squirrel-cage induction winding that offers enough torque for acceleration and helps dampen motor speed oscillations. As the rotor approaches synchronous speed, the field winding becomes energized and the motor synchronizes. In some cases, very large motor systems might have a "pony" motor that speeds up the synchronous machine when it's not loaded. Motors that are electronically controlled can be accelerated from zero speed by adjusting the frequency of the stator current.

Compact synchronous motors are frequently utilized in plug-in electric clocks or timers that rely on the power line frequency to operate the gear mechanism at the precise pace. These small synchronous motors can initiate without aid as long as the rotor's moment of inertia and its mechanical load are minimal enough. The motor progresses from slip speed to synchronous speed in an accelerating half cycle of the reluctance torque. Unlike shaded-pole types, single-phase synchronous motors in electric wall clocks have the ability to rotate in both directions freely.

Costs are a crucial factor for beginners. Rotor excitation can help address this issue. Furthermore, starting techniques for large synchronous machines involve repeatedly reversing the polarity of the rotor poles during startup.

10.5. ARMATURE

The armature in electrical engineering refers to the winding or windings in an electric machine that carry alternating current. These windings are capable of conducting AC even in DC machines, thanks to the commutator action or electronic commutation found in brushless DC motors. Depending on the type of electric machine, the armature can be found on either the rotor (the rotating part) or the stator (the stationary part).

The armature windings engage with the magnetic field in the air-gap, which can be created by either permanent magnets or electromagnets produced by a conducting coil.

The armature must conduct current, making it either a conductor or a conductive coil positioned perpendicular to both the magnetic field and the direction of movement, torque in a rotating machine, or force in a linear machine. The armature serves two main purposes. The first is to conduct current through the field, producing shaft torque in a rotating machine or force in a linear machine. The second purpose is to generate an electromotive force (EMF).

The relative motion between the armature and the field in a machine creates an electromotive force (EMF) in the armature. When the machine is functioning as a motor, this EMF resists the armature current, allowing the armature to convert electrical power into mechanical power in the form of torque, which is then transmitted through the shaft. Conversely, when the machine is operating as a generator, the armature EMF drives the armature current, converting the movement of the shaft into electrical power. In an induction generator, power is generated from the stator.

10.5.1. Armature Winding

An armature winding is typically made up of conductors that are coated with a single layer of cotton, a double layer of cotton, cotton fiberglass, or enamel. The coils of the armature winding are usually secured together with cotton tape before being dipped in varnish and dried. The conductors in the armature winding are housed and protected within the armature slots and properly connected. These windings generate a helpful electromotive force (e.m.f) across the brushes.

In a DC machine, the armature winding is typically wound using two techniques known as Lap Winding and Wave Winding.

1. **Lap Winding:** In this winding configuration, the conductors are connected so that their parallel poles and paths are identical. The end of each armature coil can be connected to the adjacent section of the commutator. The number of brushes in this winding is the same as the number of parallel paths, and these brushes are evenly distributed between positive and negative polarity windings. Lap winding is commonly used in high-current, low voltage machines. Lap windings are classified into three types which include the following (Figure 10.11):

i. Simplex Type Lap Winding
ii. Duplex Type Lap Winding
iii. Triplex Type TLap Winding

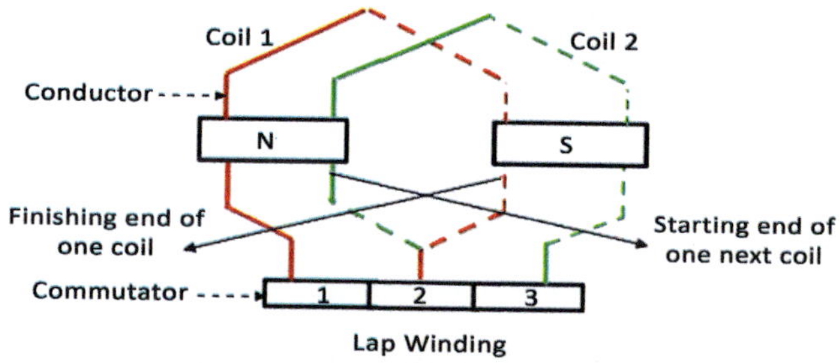

Figure 10.11. *Lap winding.*

Source: https://circuitglobe.com/lap-and-wave-winding.html.

i. **Simplex Type Lap Winding:** In this type of winding, one end of the primary coil is connected to the commutator section, while the other end of the secondary coil is placed in the same pole position. The number of parallel lanes is equal to the number of poles in the windings.

ii. **Duplex Type Lap Winding:** In this winding configuration, there are twice as many parallel paths between poles as there are poles. Lap winding is commonly used in high current applications. This type of winding is achieved by placing two identical windings on the same armature, connecting the even-numbered commutator bars to the primary winding and the odd-numbered bars to the secondary winding.

iii. **Triplex Type Lap Winding:** This winding arrangement involves winding around one-third of the commutator bars. The triplex lap winding has multiple paths, making it suitable for high current applications. However, a disadvantage is that it requires a larger number of conductors, increasing the overall cost of the winding.

2. **Wave Winding:** This type of wave winding utilizes only two parallel paths for the positive and negative brushes. The end of the first armature coil is connected to the beginning of the second armature coil's commutator segment at a distance. Conductors in this winding are connected to the two parallel paths of the machine poles. The number of parallel ports matches the number of brushes. This type of winding is best suited for high-voltage, low-current machines (Figure 10.12).

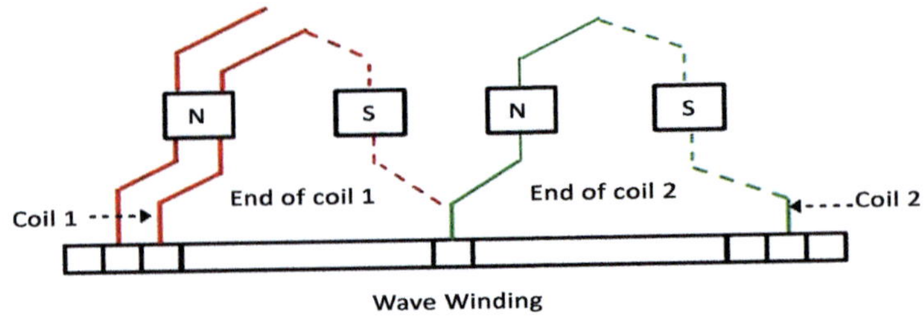

Figure 10.12. *Wave winding.*

Source: https://circuitglobe.com/lap-and-wave-winding.html.

Once the armature winding completes one full rotation, it shifts into a slot to the left of its starting point, resulting in a retrogressive winding. Conversely, if the armature winding shifts one slot to the right, it is known as a progressive winding.

Assuming there are two winding layers and that the AB conductor is located in the higher semi of the slot either on the right or left side. Let YF and YB represent the front and back pitches, which are approximately equal to the winding pole pitch. The average winding pitch can be calculated using the following equation.

10.5.2. Armature Reaction in an Alternator or Synchronous Generator

Faraday's law is the principle that governs the operation of all rotating electrical machines. All electrical machines consist of a magnetic field and a coil, known as the armature, which must have relative motion between them. In an alternator, electricity is supplied to the poles to create a magnetic field, and power is then collected from the armature. As the field and armature move relative to each other, the armature conductors cut through the magnetic flux, resulting in a changing flux linkage. According to Faraday's law of electromagnetic induction, this action induces an electromotive force (emf) in the armature. Therefore, when the load is connected to the armature terminals, a current flows through the armature coil. This current creates a reverse effect on the main field flux of the alternator or synchronous generator, known as armature reaction.

In an alternator, just like in all other synchronous machines, the impact of armature reaction is determined by the power factor, which is the phase difference between the terminal voltage and the armature current.

When a generator supplies a lagging load, it is providing magnetic energy to the load through reactive power. This results in a reduction of reactive power in the generator due to the excitation of the synchronous machine. As a result, the armature reaction in the generator becomes demagnetizing. Conversely, when the generator supplies a leading load, it receives lagging reactive power from the load, causing the armature reaction to have a magnetizing effect. In the case of a purely resistive load, the armature reaction is only cross magnetizing.

The phase difference between these two quantities, i.e., Armature current and voltage may vary from – 90° to + 90°

If this angle is θ, then;

$$-90^{\circ} \geq \theta \geq +90^{\circ}$$

To understand actual effect of this angle on armature reaction of alternator, we will consider three standard cases:

When θ = 0

ii. When θ = 90°

When θ = – 90°

10.5.2.1. Armature Reaction of Alternator at Unity Power Factor

When operating at unity power factor, the angle between the armature current (I) and the induced emf (E) is zero, indicating that they are in phase with each other. It is important to note that the emf induced in the armature is a result of the changing main field flux that is linked with the armature conductor.

As field is excited by DC, the main field flux is constant in respect to field magnets, but it would be alternating in respect of armature as there is a relative motion between field and armature in alternator. If main field flux of the alternator in respect of armature can be represented as:

$$\phi_f = \phi_{fm} \sin \omega t \qquad (11)$$

Then induced emf E across the armature is proportional to, dφf/dt

Now

$$\frac{d\phi_f}{dt} = -\omega\phi_{fm} \cos\omega t \qquad (12)$$

Now, armature flux φ_a is proportional to armature current I. Hence, armature flux φ_a is in phase with armature current I.

When the power factor is at unity, I and E are in phase. Therefore, at unity power factor, the armature flux, φ_a, is in phase with the induced EMF E. This means that the armature flux is in phase with the induced EMF E and the field flux is in quadrature with E. As a result, the armature flux, φ_a, is in quadrature with the main field flux, φ_f.

Since the two fluxes are perpendicular to each other, the armature reaction of the alternator at unity power factor is solely of the distorting or cross-magnetizing type.

As the armature flux pushes the main field flux perpendicularly, distribution of main field flux under a pole face does not remain uniformly distributed. The flux density under the trailing pole tips increases somewhat while under the leading pole tips it decreases.

10.5.2.2. Armature Reaction of Alternator at Lagging Zero Power Factor

When the power factor lags at zero, the current in the armature lags 90° behind the induced emf. The emf in the armature coil is induced by the main field flux, with a 90° lead over the main field flux.

We get, the field flux

$$\phi_f = \phi_{fm} \sin \omega t \qquad (13)$$

Therefore, induced emf $E \propto -\frac{d\phi_f}{dt}$

Hence, at ωt = 0, E is maximum and φ_f is zero.

At ωt = 90°, E is zero and φ_f has maximum value.

At ωt = 180°, E is maximum and φ_f zero.

At $\omega t = 270°$, E is zero and φ_f has negative maximum value.

Here, φ_f got maximum value 90° before E. Hence φ_f leads E by 90°.

Now, armature current I is proportional to armature flux φ_a, and I lags E by 90°. Hence, φ_a lags E by 90°.

So, it can be concluded that, field flux φ_f leads E by 90°.

As a result, the armature flux and field flux are in direct opposition to each other. This means that the armature reaction of an alternator at a lagging zero power factor is purely demagnetizing in nature. In other words, the armature flux directly reduces the strength of the main field flux.

10.5.2.3. Armature Reaction of Alternator at Leading Power Factor

In the optimal power factor scenario, the armature current I leads the induced emf E by 90°. Additionally, it has been demonstrated that the field flux φ_f leads the induced emf E by 90° as well.

Once more, the armature flux φ_a is directly proportional to the armature current I. Therefore, φ_a is synchronized with I. This means that the armature flux φ_a leads E by 90°, just as I leads E by 90°.

In this scenario, both the armature flux and field flux in an alternator cause an induced emf E that is offset by 90°. This suggests that the field flux and armature flux are aligned in the same direction. Therefore, the resultant flux is simply the sum of the field flux and armature flux. Ultimately, it can be said that the armature reaction of an alternator operating at a purely leading electrical power factor is primarily of a magnetizing nature.

10.5.3. Advantages of Stationary Armature–Rotating Field Alternator

Practically, most of the synchronous generators or alternators have stationary armature and rotating field.

A stationary armature-rotating field alternator has several advantages over a rotating armature type alternator, as given below:

- When an alternator uses a stationary armature, the output current can be accessed directly from the fixed terminals on the armature without the need for slip rings, brushes, and other components.
- Because the armature windings of the rotating field alternator remain fixed in place, they are not affected by any vibrations or centrifugal forces.
- It is simple to insulate a stationary armature for the high voltage that the alternator is intended for, which can reach up to 33 kV.
- When the stationary armature is used, the armature windings can be braced better mechanically against the high electromagnetic forces due to large short circuit currents.
- Direct current is supplied to the rotating field, needing only two slip rings for power. Typically, the field voltage ranges from 100 to 500 volts, making insulation of the low voltage slip rings from the shaft a simple task.
- The weight of the armature windings is greater than the windings of the rotating field poles. Therefore, the rotating field type alternator has smaller size than a rotating armature type alternator.

- a stationary armature may be cooled easily because the size of the armature can be increased to provide more cooling ducts.
- Since the rotating field is comparatively light, it can be constructed for high speed rotation.
- The forced cooling with gas or liquids can be easily provided in stationary armature type alternators.
- The cost of the rotating field type alternator is low as compared to the rotating armature type alternator.

CASE STUDY

ANALYSIS OF SYNCHRONOUS MACHINE EXCITATION SYSTEMS: COMPARATIVE STUDY OVERVIEW

Excitation system is widely used to provide direct current to the synchronous machine field winding. It also helps to control the field voltage, field current and reactive power flow of the system. It enhances the stability during the start-up of the synchronous machines. Furthermore, in a power system, the protection functions of the excitation system enables to improve the rated capacity limits of the synchronous machines.

The main components of an excitation system are automatic voltage regulator (AVR), exciter, measuring elements, power system stabilizer (PSS) and protection unit. The excitation systems can be divided into different categories.

Types of the Excitation Systems

Synchronous machine excitation systems can be classified into three major groups based on the power supply used as source of excitation. These are DC excitation, AC excitation and static excitation systems

1. **DC Excitation System:** This excitation system utilizes DC generators as source of excitation power. It also provides current to the rotor of synchronous machine through slip rings. The exciter can be driven by a motor or shaft of a generator. It can be either self-excited or separately excited. When it is separately excited, the exciter field current is supplied by a pilot exciter comprising of a permanent magnet generator.

   DC excitation systems represent early systems, spanning the years from the 1920s to the 1960s. They lost favor in the mid1960s and were superseded by ac excitation systems.

2. **AC Excitation Systems:** These excitation systems utilize alternators as source of generator excitation power. Usually, the exciter is kept on the same shaft as the turbine generator. The AC output of the exciter is rectified by either controlled or diode rectifiers to produce the direct current which is needed for the generator field. The rectifiers can be stationary or rotating.

AC excitation systems can take many forms depending on the rectifier arrangement, method of exciter output control, and source of excitation for the exciter. Currently, stationary and rotating AC rectifier systems are widely used in AC excitation systems. In stationary rectifiers, the DC output is fed to the field winding of the generator through the slip rings. On the other hand, in rotating rectifiers there is no need of slip rings and brushes. The DC supply is directly fed to the generator field as the armature of the exciter and rectifiers rotate with the generator field. Such systems are known as brush-less systems and were developed to avoid the problems with brushes when extremely high field currents are applied to large generators.

3 **Static (ST) Excitation Systems:** All the components of these systems are either static or stationary. Such systems directly provide synchronous generator field winding with excitation current by means of slip rings. Rectifiers in ST systems gain the power from generator through auxiliary windings or a step-down transformer. In such systems, generator is a source if power which means that the generator is self-excited. As the generator cannot produce any voltage without excitation voltage, the generator must have auxiliary power source to provide field current and energize the generator. Station batteries are usually used as additional power sources and the process is known as field flashing.

From the excitation power gain point of view the excitation systems can be further divided into independent and dependent excitation systems. The independent exciter is not connected to the grid. Thus, its excitation parameters have no direct relationship with grid parameter. The dependent exciter utilizes either part of generator power or it is connected to the grid. Its excitation parameters are dependent on grip parameters.

Control and Protection of Excitation System

A modern excitation control system includes a number of control, limiting, and protective functions. Any given excitation system can include only some or all of these functions, depending on the requirements of the specific application and the type of exciter.

The main components of excitation system are voltage regulators, excitation system stabilizers, power system stabilizers, voltage sensing and load compensator, under-excitation limiters, over-excitation limiters, and volts-per-hertz limiters. AC voltage regulator is used to maintain generator stator voltage. It also helps to control the generator excitation voltage. DC voltage regulator is used to hold generator excitation voltage on constant level and it is manually controlled. The regulator is mainly used during tests, start-ups and to cover the AC regulator outages.

Review

Excitation systems stabilizing circuits are used to improve the dynamic performance of the excitation system. The DC and AC excitation systems have elements with significant time constants. Feedback compensation system can used to minimize the phase shift caused by elements. This contributes to generator stable operation conditions such as prior to the synchronization or after load rejection. The power system stabilizer (PSS) uses special stabilizing signals in order to control the excitation system and to improve the dynamics of the power system. The basic input signals of PSS are shaft speed, frequency and power. The stabilizer damps rotor oscillations through excitation control.

In order to get oscillation damping stabilizers, we have to produce appropriate electric torque component.

Load compensation system can be used to control voltage variation at a point which is either external or internal to the generator. The compensator's impedance is required to be adjustable in order to simulate electrical distance between the generator terminals and the point at which the voltage to be controlled. Voltage regulation at the point that is external to the generator is commonly used to provide proper sharing of the reactive power between generators which are connected together. On the other hand, voltage regulation at the point that is internal to the generator is used to compensate the voltage drop on the step up transformers. Under excitation limiter (UEL) is used to prevent generator excitation from dropping to the limit at which generator stability is lost. The limiter input signal is either generator voltage and current, or active and reactive power. The limits are determined by the input signals exceeding the reference level. The function of over excitation limiter (OEL) is to prevent generator from overheating due to long term excitation. The OEL senses over-current and after some time delay, it reduces the excitation to pre-determined value. There are two types of time delay: Fixed time and inverse time delay. Fixed time limiter operates when excitation current exceeds reference value during preset time. Inverse time limiter operates with the delay that matches field thermal condition. Over-flux limiter protects generators and step-up transformers from excessive magnetic flux due to low frequency or over-voltage. Excessive magnetic flux can cause overheating of generator or transformer which leads to permanent damage. Generator protection is applied when Volt-per-hertz regulator exceeds pre-determined value during specified time.

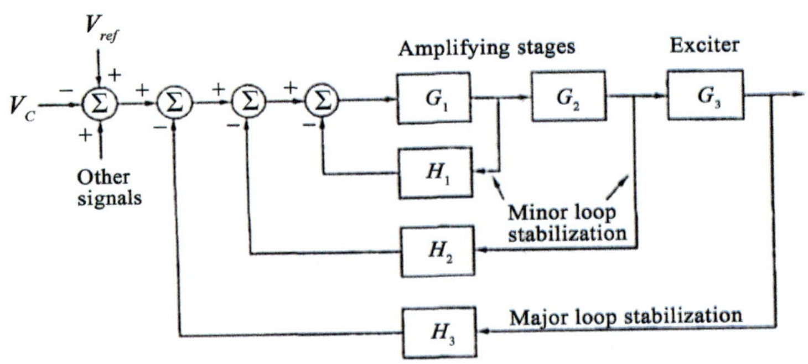

Figure 10.13. *Structure of a detailed excitation system model.*

Source: https://publications.waset.org/10007291/analysis-of-synchronous-machine-excitation-systems-comparative-study.

Model of the Excitation Systems

The mathematical models of excitation systems are used to estimate the parameters of control and protection circuit configurations. It also helps to coordinate the whole system stability. Figure 10.13 depicts the general structure of excitation system model. This model has the advantage of retaining a direct relationship between model parameters and physical parameters. In this study, a model reduction technique is used to simplify and obtain practical model for the system under consideration.

The parameters of the simplified model are selected such that its gain and phase characteristics matches that of the detailed model over the frequency angle from 0 to 3 Hz. All significant non-linearities that affect the system stability are also taken into consideration. However, the direct correspondence between the model parameters and the actual system parameters is generally lost in the simplified model.

IEEE has standardized 12 model structures for representing wide variety of excitation systems. In this work, two models, AC4A and DC1A, are considered for analysis.

A. **Amplifier Model:** The excitation system amplifier can be a magnetic amplifier, rotating amplifier, or electronic amplifier. The amplifier is represented by its gain K_A and a time constant τ_A. Its transfer function is given by:

$$\frac{K_A}{T_A s} = \frac{V_R(s)}{V_e(s)} \tag{1}$$

Typical values of KA lies in the range of 10 to 400. The time constant of the amplifier is very small, in the range of 0.02 to 0.1 second, and is often neglected.

B. **Generator Model:** The emf generated from synchronous machine is a function of its magnetization curve and terminal voltage. It is also dependent on the generator load. In the simplified model, the transfer function that relates the generator terminal voltage and its field voltage can be represented by a gain K_G and time constant τ_G. Its transfer function is given by:

$$\frac{K_G}{T_G s} = \frac{V_T(s)}{V_F(s)} \tag{2}$$

These constants are load-dependent; K_G can vary from 0.7 to 1, and τ_G lies between 1.0 and 2.0 seconds from full-load to no-load.

C. **Sensor Model:** The voltage is sensed by a potential transformer and it is rectified using a bridge rectifier. The sensor is modeled by a simple first order transfer function as given below:

$$\frac{K_R}{\tau_R s} = \frac{V_S(s)}{V_T(s)} \tag{3}$$

where K_R denotes the gain of the sensor and τ_R is its time constant. The value of τ_R is very small in the range of 0.01 to 0.06 second. The block diagram of automatic voltage regulator (AVR), which uses the aforementioned component models.

The AVR system of a generator has the following parameters The DC1A exciter model represents field controlled DC commutator exciters, with continuously acting voltage regulators. The exciter can be separately excited or self- excited, the latter type being more common. When self-excited, the excitation gain (KE) is selected so that initially voltage regulation (VR) is equal to zero. Then it tracks the voltage regulator by periodically trimming the shunt field rheostat to set point. Sample data for DC1A exciter model of self-excited DC exciter are $K_A = 187$, $\tau_A = 1.15$, $\tau_E = 1.15$, AEX = 0.014, BEX = 1.55, $K_F = 0.058$, $\tau_F = 0.62$, $\tau_B = 0.06$, $\tau_C = 0.173$, $\tau_R = 0.05$, $VR_{MAX} = 5$ V, $V_{RM\ IN} = -1.7$ V.

The AC4A exciter model represents an alternator supplied controlled rectifier excitation system. It has a high initial response excitation system utilizing bridge rectifier. Excitation system stabilization is usually provided in the form of a series lag-lead network (transient gain reduction).

The time constant associated with the regulator and firing of the controlled rectifier is represented by τA. The overall gain is represented by KA. The effects of rectifier regulation on exciter output limits are accounted for in the model by constant KC. Sample data for AC4A exciter model and regulator are given as follows. $K_A = 200.0$, $\tau_A = 0.04$, $\tau_C = 1.0$, $\tau_B = 12.0$, VRMAX = 5.64, $V_{RM\ IN} = 4.53$, $K_C = 0$, $VI_{MAX} = 1.0$, VIMIN = −10.

Conclusion

Comparison of two IEEE standard model structures representing synchronous machines, generators in particular is presented. Excitation system model for synchronous machines is very important to deliver field current and to give protection functions. But the most crucial question here is which excitation model structure and exciter is better, reliable and stable. Based on Simulink simulation results of synchronous machine excitation systems, DC1A IEEE model structure representing synchronous machine excitation system is better, reliable and stable than AC4A.

CLASS ACTIVITY

Using Three-Phase Synchronous Motor/Generator, Power Supply, AC Ammeter, connect the circuit shown in Figure. Note that the three stator windings are wye- connected to the fixed 208 V 3-phase output of the power supply, terminals 1, 2, and 3.

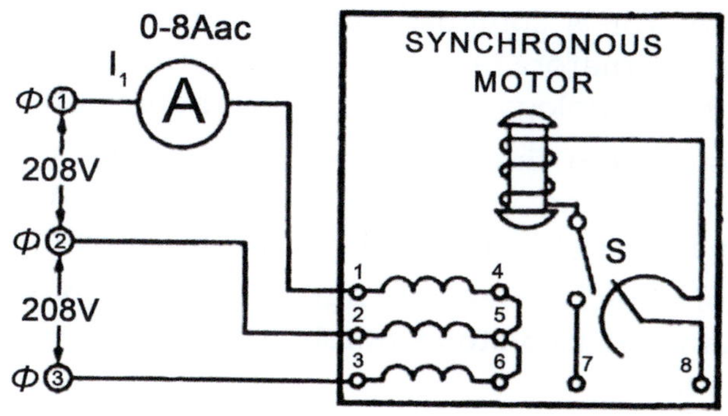

SUMMARY

- An alternators is an electrical generator that converts mechanical energy to electrical energy in the form of alternating current.
- The rotating magnetic field induces an AC voltage in the stator windings. Since the currents in the stator windings vary in step with the position of the rotor, an alternator is a synchronous generator.
- A device that uses permanent magnets to produce alternating current is called a permanent magnet alternator (PMA). A permanent magnet generator (PMG) may produce either alternating current, or direct current if it has a commutator.
- A brushless alternator is composed of two alternators built end-to-end on one shaft. Until 1966, alternators used brushes with rotating field. With advancement in semiconductor technology, brushless alternators are possible.
- Voltage regulators serve the purpose of automatically controlling the terminal voltage of the generating machine. It's important to note that AC machines inherently possess a higher internal reactance in comparison to DC machines.
- DC generator of special design that combines in one machine a greater amplification than the two-stage amplifier together with ten or twenty times the speed of response.
- Armature reaction is the effect of armature flux on the main field flux. The effect of the armature reaction can be seen in the DC generator as well.
- The synchronous motor is an AC rotating electrical machine widely used to convert AC power drawn from the grid into mechanical power.
- Synchronous motors use electromagnets as the stator of the motor which create a magnetic field that rotates in time with the oscillations of the current.
- A permanent magnet synchronous motor and reluctance motor requires a control system for operating (VFD or servo drive).
- Electric motors generate power due to the interaction of the magnetic fields of the stator and the rotor. In synchronous motors, the stator carries three phase currents and produces three phase rotating magnetic flux (and therefore a rotating magnetic field).
- In electrical engineering, the armature is the winding (or set of windings) of an electric machine which carries alternating current. The armature windings conduct AC even on DC machines, due to the commutator action (which periodically reverses current direction) or due to electronic commutation, as in brushless DC motors.

REVIEW QUESTIONS?

1. What is the basic principle of operation of alternators?
2. Describe voltage regulation methods.
3. Discuss the effect of load on alternator.

4. What is the basic structure of a synchronous motor? Discuss.
5. Differentiate between non-excited motors and externally excited motors.
6. How does a synchronous motor operate? Explain.
7. Why synchronous motor is not self-starting?
8. Discuss the starting methods of synchronous motors.
9. Deal with the armature reaction in an alternator or synchronous generator
10. Why damper windings are used in synchronous motor? Or how synchronous motor is started as squirrel cage motor?

REFERENCES

1. Boulter, E. (2004). Historical development of rotor and stator winding insulation materials and systems. *IEEE Electrical Insulation Magazine, 20*(3), 25–39.
2. Cai, B., Hao, K., Wang, Z., Yang, C., Kong, X., Liu, Z., Ji, R., & Liu, Y. (2021). Data-driven early fault diagnostic methodology of permanent magnet synchronous motor. *Expert Systems with Applications, 177*, 115000. https://doi.org/10.1016/j.eswa.2021.115000.
3. Emadi, A., Rajashekara, K., Williamson, S. S., & Lukic, S. M. (2005). Topological overview of hybrid electric and fuel cell vehicular power system architectures and configurations. *IEEE Transactions on Vehicular Technology, 54*(3), 763–770.
4. Gandzha, S. A., & Kiessh, I. E. (2015). Application brushless machines with combined excitation for small and medium power windmills. *Procedia Engineering, 129*, 191–194.
5. Gandzha, S. A., Sogrin, A. I., & Kiessh, I. E. (2015). The comparative analysis of permanent magnet electric machines with integer and fractional number of slots per pole and phase. *Procedia Engineering, 129*, 408–414.
6. Guo, M., Jin, Q., Yao, Z., & Chen, W. (2020). Analysis on the reason of low voltage problem and the effectiveness of voltage regulation in a distribution area. *IOP Conference Series: Earth and Environmental Science, 440*(3), 032128.
7. Kalsi, S., Weeber, K., Takesue, H., & Lewis, C. (2004). Development status of rotating machines employing superconducting field windings. *Proceedings of the IEEE, 92*(10), 1688–1704.
8. Kassakian, J. G., Wolf, H. C., Miller, J. M., & Hurton, C. J. (1996). Automotive electrical systems circa 2005. *IEEE Spectrum, 33*(8), 22–27.
9. Orlowska-Kowalska, T., Wolkiewicz, M., Pietrzak, P., Skowron, M., Ewert, P., Tarchala, G., Krzysztofiak, M., & Kowalski, C. T. (2022). Fault diagnosis and fault-tolerant control of PMSM drives–state of the art and future challenges. *IEEE Access, 10*, 59979-60024. https://doi.org/10.1109/ACCESS.2022.3180153.
10. Schmuelling, S., Kreischer, C., & Gołebiowski, M. (2015). Comparison of different methods for excitation of synchronous machines. *KOMEL, 107*, 89–93.

11. Tapia, J. A., Leonardi, F., & Lipo, T. A. (2003). Consequent pole permanent magnet machine with field weakening capability. *IEEE Transactions on Industry Applications, 39*(6), 1704–1709.

12. Wang, W., Tian, W., Wang, Z., Hua, W., & Cheng, M. (2020). A fault diagnosis method for current sensors of primary permanent-magnet linear motor drives. *IEEE Transactions on Power Electronics, 36*, 2334–2345. https://doi.org/10.1109/TPEL.2020.3011125.

13. Zhang, P., & Williamson, S. S. (2008). Recent status and future prospects of integrated starter-generator based hybrid electric vehicles. *IEEE Vehicle Power and Propulsion Conference*. https://doi.org/10.1109/VPPC.2008.4677707.

Index

D

E

F

Printed and bound by CPI Group (UK) Ltd, Croydon, CR0 4YY

18/09/2026

14965063-0002